ENGINEERING DESIGN

A Survival Guide to

Senior Capstone

ENGINEERING DESIGN

A Survival Guide to

Senior Capstone

1ST EDITION

Cory J. Mettler, M.S.E.E.

 Springer

Cory J. Mettler, M.S.E.E.
Montana State University
Bozeman, MT, USA

ISBN 978-3-031-23311-1   ISBN 978-3-031-23309-8   (eBook)
https://doi.org/10.1007/978-3-031-23309-8

This Springer imprint is published by the registered company Springer Nature Switzerland AG
The registered company address is: Gewerbestrasse 11, 6330 Cham, Switzerland

Support for this Textbook

Senior capstone was one of the most industry-applicable classes I experienced in college. Besides working together with colleagues to design a cutting-edge product for a company, the material in this course taught useful life-long skills on project management. Greatest of all, I learned to communicate both verbally and on paper. Communication is the most important and sought-after skill in prospective employees today.

– Andrew Hora, P.E. (Pro-Tech Power Sales Engineer)

Cory Mettler understands the transition from academic learning to practical, real-world application. He steps through the coursework and lays it out in a way that's relevant, so you actually absorb the knowledge, skills, and tools. Of all the courses I took, Senior Design has been by far the most applicable to my professional career post-graduation.

– Brad Penney, P.E. (Vantage Point Project Manager)

The soft skills you are about to learn in Senior Design have translated to my job more than I ever imagined. Whether writing reports, taking notes, or interacting with clients, communicating properly is crucial to the success of all my projects. As a student, I didn't expect to write much but now, as an engineer, I find myself writing in some form every day. Notetaking does not seem important when working on a single project since that project is your sole focus, but when managing 10+ projects, notes and project documentation are not only helpful, but essential to keep all details straight. I use the skills I developed in Cory Mettler's Senior Design curriculum in these efforts on a daily basis.

– Grant Metzger (DGR Design Engineer)

During the first five years of my career, I can absolutely look back on Senior Design and think of various scenarios that have helped me succeed thus far. Jumping into a Project Management position was definitely easier [after experiencing the exact same material as taught in this textbook because] I was able to manage schedules, budgets, and owner needs and requirements. I am more confident as a Project Manager because of the training I had during Senior Design and I'm thankful for the complexity of the project I was given because I would not be in the same position I am now without it.

– Katie Dunphy (Project Manager)

The methods outlined in this book describe a proven framework that will help guide design teams towards a rewarding and successful capstone experience. The process will help teams create a plan to ensure success, show them how to fully explore the design space to find the optimal solution, and stay organized through the project. I have witnessed several multidisciplinary teams use these methods with high levels of success.

– Dr. Todd Letcher (Associate Professor, Mechanical Engineering)

I have seen students thrive and grow under Cory Mettler's Project Management based approach to Senior Design. Under these methods students of all skill-levels were able to succeed on more complicated projects than I would have believed possible based on their prior academic performance. Employers report that these students are better able to "hit the ground running" and succeed as practicing engineers. These Senior Design projects are regularly discussed during the interview process and have even been a deciding factor when career-positions are being offered.

– Dr. Robert (Bob) Fourney (Associate Professor, Electrical Engineering)

Foreword

There are many qualities that help make a good engineer. Yes, technical skills are important, but learning soft skills through the experiences and elements outlined from this textbook elevates students into great engineers! All of our development areas and engineering projects are part of a larger system which must be managed cohesively. The need for communication skills across various disciplines and teams is critical to successful product deployments and well-architected systems. A skill that is often underappreciated is listening. How do we truly hear our customers, understand the issues they are experiencing, and connect with the ideas they have that can improve efficiency and make their lives better without listening? With careful listening, we have an opportunity to solve great challenges! Another concept we value greatly is respectful debate. Respectful debate within development teams encourages ideas and fosters team involvement. Carefully listening to our customers and respectful debate among our team members are key areas helping us to grow and succeed. The concepts presented in this textbook support the development of these soft skills in a way that traditional classes do not.

The skills taught from this textbook are proven to help engineers develop talents to be successful in the workplace and become leaders. We have partnered with Cory Mettler for many years sponsoring Senior Design projects. We quickly realized that skills learned during a well-structured Senior Design class mirror the necessary qualities we require our design engineers and developers to possess while supporting our team dynamics. Students who have the opportunity to learn with this structure and team environment attain critical communication skills and project management experiences. We have hired many design and software engineers from the Senior Design teams we have sponsored. These team members are doing wonderfully! They truly are doing great things, solving great challenges, and making a positive impact on the world!

Engineering Manager & University Relations, Raven   Shane Swedlund
Sioux Falls, SD, USA

Acknowledgments

This textbook would not have been possible without the support of my parents. I have enjoyed discussing management styles with my father and learning from his experience for many years. I was excited and honored to collaborate with him on this project. My mother has encouraged higher education in our family all my life and saw her children, her husband, and herself gain advanced degrees. She contributed an invaluable point of view while developing this material.

There is no one "right" way to do project management or complete the design life cycle. The material taught in this textbook is comprised of best practices from many people who have influenced me throughout my career. Most notably, I appreciate the mentorship of Dennis and Ali. The project management content in this textbook was further influenced by numerous Senior Design project sponsors and engineering decision makers. Shane, Dale, Erica, and Jared, your support has been invaluable to me, as well as to countless students.

The curriculum taught in this textbook was directly influenced by collaboration with my colleague, Wataru. Thank you for your support and patience on this project! I have learned much from you.

The list of students who deserve recognition for providing educational material to this textbook is far too long to include here. Please know that if I used your name or project in this textbook, you have stood out in my experience as special, unique, and highly valued. Thank you so much for lending your experiences to enrich the education of the cohorts that follow behind you.

Contents

List of Figures

List of Tables

List of Examples

List of Common Pit Falls

List of Industry Point of Views

PART I: INTRODUCTION

> *"Scientists investigate that which already is;*
> *Engineers create that which has never been."*
> — Albert Einstein, physicist

> *"Art without Engineering is dreaming;*
> *Engineering without Art is calculating."*
> — Steven K. Roberts, author

> *"[Engineering] is a great profession. There is the fascination of watching a figment of*
> *the imagination emerge through the aid of science to a plan on paper. Then it moves to*
> *realization in stone or metal. Then it elevates the standard of living…That is the*
> *engineer's high privilege."*
> — Herbert Hoover, 31st President of the United States

You are on the precipice of an important evolution in your professional development; you have been a student, but in the near future you will be an engineer. Have you ever stopped to contemplate the storied profession you are about join? Consider for a moment some of the great figures of ancient history – likely most names on your list fall into one of two categories: Conquerors/Politicians or Philosophers/Engineers. Why is this? It is because these two groups of people have an amazing ability to change the world – this textbook will focus on skills used by the second group.

A short list of influential engineers and their contributions include the following:

- Archimedes and his invention of levels, pulleys, and screws to move heavy objects
- Galileo's functional improvements to early telescopes and related contributions to astronomy
- Alexander Graham Bell's invention of a telephone and the advent of the communication age
- Thomas Edison and Nikola Tesla's contributions to the understanding of electricity and development of the modern power grid
- William Shockley's invention of the transistor and the modernization of the computer.

These historical engineers started with a figment of imagination, created a new reality, and elevated all of society. Today's engineers are still making equally impactful contributions. Engineers at *Intuitive Surgical* created the da Vinci robot which, when approved in 2000, allowed surgeons to begin to perform medical procedures remotely. *First Solar* employs engineers who are making important advancements in the development and application of renewable energy technologies while leaving the smallest possible carbon footprint. *SpaceX* engineers are redefining what was thought to be possible in space exploration. Where will YOU apply your skills and what will you contribute? The possibilities are endless, and exciting.

Albert Einstein claimed that the difference between scientists and engineers is that engineers create what has never been created before. That sounds inspiring, but daunting. How does one go about accomplishing this in a practical sense? Creating something new will require a delicate balance between imaginative art and rigorous implementation.

Have you ever considered why the word "art" is used so often in engineering? Today's lexicon often pits those two skills against each other: creative vs. methodical, right brain vs. left brain, mathematical vs. loquacious. Yet, engineers regularly talk about "state of the art" to describe a new technology, "prior art" to refer to something that was already designed, and, in a patent application, "exemplary embodiments of the art" describe typical use-cases of a new design.

Impactful innovation requires artistic skills such as creative thinking, ingenuity, and imagination to see what can be. However, implementation of that new idea requires methodology, structure, and adherence to standards. As you can see, there is a dichotomy of skills sets that must be masterfully balanced to become an effective engineer.

Unfortunately, most of your education up to this point has not sufficiently prepared you to handle this dichotomy; in fact, until now you probably did not even realize that the dichotomy existed. Most material covered in 200- and 300-level engineering courses are related to the technical fundamentals of your particular discipline. Like in most professions, the fundamentals are necessary but inadequate to base an entire career upon. The Senior Design experience is intended to bridge the gap between the fundamental knowledge you developed in previous courses and the skill sets required to be an effective engineer and help you balance this dichotomy.

Consider some of the ways a Senior Design project will be different than previous projects you have experienced.

- A heat transfer homework-problem has a known set of inputs and parameters. A design project has set of inputs which, depending on the design methodology, may be more or less important; you will have to decide which to use and which to discard.
- A mesh-current circuit analysis homework-problem has a relatively strict process which can be applied to solve the problem. An engineering project has no such process or guidelines; priorities and methods will need to be intelligently selected to produce an optimal solution.
- A calculus test-problem has a well-known solution which the teacher can directly compare to a student's response. An engineering problem has many possible solutions and the engineer must convince the reviewer that the provided response sufficiently solves the design problem.
- A robotics-team challenge has a specified goal, and when the team can demonstrate that the goal is met, the challenge is considered successful. An engineering design project has real-world budget, resource, and profit concerns. The solution must not only show that it meets the goal, but convincing evidence must also be provided to prove that the solution is the lowest-cost, most-robust solution with the highest profit margin.

Senior Design is an exciting course – you will finally have the opportunity to develop something real, something that will make a difference, something that is completely yours. However, this course is also challenging because you will be assessed on aspects of your work you likely have never been critiqued on before. This is necessary because the inherent challenge of creating something that "has never been" is the fact that examples, templates, or guidelines might not exist. Thus, you will have to take on a more proactive role in determining and explaining your solutions. Since this is not something most courses force you to do, it can be somewhat overwhelming, and it may, at times, feel like you are just trying to survive the course. But do not worry.

This textbook was written as a survival guide for your experience!

The order of material in this textbook has been laid out to follow the standard engineering Design Life-Cycle. The textbook will present new concepts in the order that you will need to use them as you progress through your project. Four types of additional support are included throughout the text in order to support your understanding of new concepts.

- EXAMPLES of how previous students successfully navigated particular concepts are regularly provided throughout the text. These examples all come from real students who completed the exact curriculum you are about to start.
- AVOIDING COMMON PITFALLS are provided by the author, who at the time of writing this textbook, has supported over 300 student design teams. Many of the students on those teams experienced similar problems. The AVOIDING COMMON PITFALLS feature explains where these students have stumbled and how they might have been more successful.
- AN INDUSTRY POINT OF VIEW is provided throughout the textbook when appropriate. This year you, the student, will be consistently asked to put yourself in the position of an entry-level engineer. This is probably not too much of a stretch since you are less than 1 year away from actually being one. However, to do your job well, you often have to understand concepts from a reviewer's point of view as well. These reviewers are typically senior engineers, managers, and CEOs – the Decision Makers. These INDUS-TRY POINT OF VIEW excerpts are stories as told by engineering managers who have experienced working with both effective and ineffective engineers. Hopefully, after reading these you will be able to put your own work into the context of how a Decision Maker will view that work and make more informed decisions accordingly.
- CASE STUDIES of three projects will be presented at the end of most chapters. The textbook will follow these projects, for better or worse, throughout the entire engineering Design Life-Cycle. Although these teams were exceptionally successful, they were not perfect. The CASE STUDIES will present their work exactly as the student teams presented it. In each case, we will review why the teams were successful and what they could have done better.

The chapters of Part I of this textbook will introduce you to Senior Design and the Engineering Design Life-Cycle, and introduce you to the four CASE STUDY teams.

Chapter 1: **Introduction to Senior Design**

Welcome to Senior Design! This experience will be dramatically different than your other engineering courses. It will be both frustrating and rewarding, humbling and confidence-inspiring, and exhausting and refreshing – in a word, it will be real. This experience will require you to *apply* much of what you already know toward solving a real engineering problem – this is what the last many years have been developing you for. Senior Design will also expose you to many engineering issues which are *not* taught in technical courses Project Managers leverage during the Design Life-Cycle.

There is rarely such a thing as a "right" answer in engineering design. You will not find a solution set in the back of this textbook. So how will you know whether you have solved the problem or not? Well, my friend, this is what engineering is all about – when you truly understand the answer to that question, you will have become an engineer. Spoiler alert: You will know you have the answer by using quantitative analysis to make design decisions and subjecting those choices to rigorous testing to validate them in such a way that all Project Stakeholders are convinced that you have solved the problem. The previous sentence is easy to understand, but putting that statement into context is difficult until you have personally experienced the entire Design Life-Cycle. One purpose of Senior Design is that you have that complete experience before entering your career.

The design process puts a lot of responsibility on you – the engineer – which makes engineering design both exciting and a bit scary. The exciting part is that this will be a chance for you to be the ultimate decision-maker, a chance to design and build something no one has ever done, and any success will be yours and yours alone. But therein lies the rub, while the success is entirely yours, so too are any problems or shortcomings – which can be intimidating. Don't worry, though! This textbook was developed to guide you through the process in an attainable, manageable, and enjoyable way. It will provide you with what you need to know and when you need to know it. It is full of commentary from industry members to help you understand the big picture and help you apply the presented guidelines. It even has many examples of both successful and unsuccessful projects to help you avoid common pitfalls.

As you go through this process, keep in mind the following:

* You are extremely intelligent, or you would not have made it this far in an engineering program.
* This process was created specifically to challenge you in such a way that you experience a real engineering project in the "safe" academic space where there are still lots of people to support you – USE THEM.

In this chapter you will learn

---

* Why it is critically important to develop "soft" skills along with technical skills.
* What Project Management is and how it will apply to your Senior Design project.
* What the engineering Project Design Life-Cycle is.

C.J. Mettler, *Engineering Design*, https://doi.org/10.1007/978-3-031-23309-8_1

1.1 Developing "Soft" Engineering Skills

The Senior Design experience will require you to develop and use a set of skills which most engineering courses neglect. Unfortunately, this often gives students the very incorrect impression that these skills are not important or, at least, less important than the technical skills you have developed. This could not be further from the truth! In fact, it is *this* additional set of skills which will define your quality as an engineering employee far beyond your ability to analyze a circuit, program a microcontroller, or convolve two signals.

That is correct! The skills you develop in this course will be more widely applicable, more impactful, and more important than *any* other engineering skills you have developed thus far. So, what are these skills?

Critical Soft Skills for All Engineers

- Meticulous documentation methods
- Effective communication, both written and verbal
- Conducting efficient meetings
- Insightful requirement gathering
- Productive interpersonal skills

You might ask yourself, why are these soft skills so important in a technical career. Perhaps you were under the impression that students who were not good in English and Speech class were safe if they went into engineering, were they not? Well, consider for a moment the two scenarios presented in Example 1.1.

EXAMPLE 1.1 Importance of Soft Skills

Scenario 1

Nile was a C-level student in many of his sophomore and junior courses; he often struggled with technically challenging problems. However, he was extremely motivated to do well on his senior project. But, unfortunately, in the end, his project did not work.

Nile excelled in his soft skills. His final presentation included a thorough discussion about his design choices, backed up with references. He carefully explained his testing procedures, demonstrating them when appropriate, and analyzed a meticulously prepared dataset.

Experts in his audience understood the design approach, concurred with the analysis, and produced a number of suggestions on how to solve the problem. Although Nile was out of time to implement those suggestions, he carefully wrote about them in his final report and provided detailed guidance on how someone might integrate those suggestions into his design. The next year, a second team used that report to develop a successful solution to the design challenge.

EXAMPLE 1.1, continued

Scenario 2

Tyler was consistently an A-level student in his courses; he excelled in nearly every academic challenge up to Senior Design. He was a very motivated student.

Unfortunately, Tyler's soft skills were weak; he was disorganized, struggled to communicate, and forgot about deadlines. Surprisingly, at the end of his project, he did have a working prototype. Unfortunately, he had lost much of his documentation regarding how the design was built and so did not include it in the final report.

When the sponsor tried to produce a second version of the product, the printed circuit board (PCB) had to be manually traced to determine what the circuit was and then reproduce it. The code was not commented and had to be reversed engineered. After the second prototype was produced, it was determined that the first device had been so finely tuned that the system was incapable of working with more than one device at a time, rendering the project useless.

Analysis: Who produced the more valuable results?

Although Nile was ultimately unsuccessful, he produced something valuable at the end of his project. His sponsors were able to leverage his work and produce a successful result on the second attempt. Conversely, although Tyler ultimately met his project objectives, his sponsors were forced to repeat all his work, essentially doubling the cost of the project. If someone has to repeat the original work to end up with the same result, then the original work was completely unnecessary.

Conclusion: Nile's work, in the long run, was much more valuable than Tyler's. "The long run" is not an aspect which most technical engineering courses consider very deeply, but it will be of paramount interest to your sponsor.

Avoiding Common Pit Falls 1.1 Keep an Open Mind

It is common for some students to think that the soft skills presented in this textbook are a "waste of time." They say things like the following:

- "I have never had to do this before."
- "My instructor is being *really* picky."

Maybe you have heard similar comments from students in other courses.

- The ninth grade Algebra student: "I have never had to use letters in my mathematical equations before, I shouldn't have too now!"
- The sophomore Circuits Analysis student: "There are so many more steps to a Superposition problem, I think I will just keep doing this with Mesh Analysis."

Avoiding Common Pit Falls 1.1, continued

Here are some suggestions to avoid these types of thoughts from limiting your learning opportunities:

- Remember, you are trying to become an engineer – the entire focus of your career will be learning things you "have not done before."
- Trust that the material presented in this textbook is the culmination of experience from many engineering managers.
- Try rephrasing your doubts into questions; rather than think "This is a waste of time," ask "How can I apply this to a real project."

Industry Point of View 1.1 Soft Skills Are Important

Often, an employee with soft skills is more valuable than one with technical skills. When management has to make hard choices, the employees who work well with the team will be rewarded over employees who only focus on the technology.

Joe was a very gifted firmware engineer who worked for me. Given any problem, he developed an excellent solution. I knew I could count on him. However, his monthly status reports were poorly written with little useful information. Management could never tell whether his project was on schedule or not. He never accepted the importance of communication skills, and so he did not put any energy into improving this aspect of his skillset. While he was an extremely capable engineer, he typically received an average or poor performance rating in his annual review.

Trent was another engineer on the same team. He was your run-of-the-mill firmware engineer. He was competent but typically required more support than Joe. However, his status reports were well written and informative; I never had to waste time asking for additional information from him. He was also the spokesperson for the ten-person team. Whenever a presentation was made to management, Trent was the guy. He was articulate and could speak to management as well as the engineering team. Management did not have to second guess or interpret his message. Trent always received an above-average rating in his annual review.

In 2012, the company had to lay off several firmware engineers. Management decided that either Joe or Trent would be on that list. A few managers argued that Trent was less skilled than Joe, so he had to be the one laid off. But the overwhelming majority of the decision-makers appreciated Trent's ability to communicate and get the job done. In the end, it was the director of Firmware Engineering that had to make the call. He knew Trent from all the presentations he had given over the years. The only time he heard Joe present, the presentation was so poorly done that it left everyone confused and frustrated. The director decided that while Joe was the better engineer, Trent was more valuable employee.

Honestly, I was relieved. I liked Joe and he really did amazing work, but overall, he made my job much more complicated. My team could solve *any* problem in those days as long as they worked together. We did not need the "rock star"; we needed someone who could keep everyone working together. Trent's skills were significantly more valuable to the team.

> **Industry Point of View 1.1, continued**
>
> Of course, management wants to hire the best engineers with excellent communication skills, but those people are rare – nobody is perfect. A well-functioning team will always be more successful than any individual, no matter how good they are. So, when a choice has to be made, competent engineers with effective soft skills will almost always be preferred over a genius who lacks the ability to communicate with their team.

1.2 A Case for Project Management

You may have already guessed that in order to accomplish this large task you are about to undertake, you will need to manage your project effectively. Many people *think* they understand Project Management without really understanding what that means. Just because someone can change the oil in their car, it does not mean they are an auto mechanic. Many people successfully manage aspects of small engineering projects without rigorous Project Management but that will not work as the scopes become larger or the resources more difficult to track.

In order for you to be successful, your Senior Design project was intentionally scoped to be a relatively *small* project. Therefore, sometimes it might *feel* like you do not need to apply rigorous Project Management or certain Design Life-Cycle strategies to complete the design. However, this is similar to the Introduction-to-Physics student learning to calculate forces of a "box" sliding down a triangular "hill." That problem is intentionally constrained to be relatively simple so that the student can develop the desired skills. When the student faces a similar problem of calculating forces enacted on a skier's knees while bouncing over moguls, they will have the necessary fundamental skillset to apply to a much more complicated problem. As the problems become more complicated, the skills become more necessary.

While the goal of this textbook is not to make you a Project Manager, it will give you some insight into that career field and introduce you to a few of the tools Project Managers use on a regular basis. We will focus on specific aspects that will directly apply to your senior project and which are universally applicable in all engineering careers.

Definition of Project Management

Before defining Project Management, we should first define what is meant by the term "project." According to the Project Management Institute (PMI), a certification body for Project Managers, a project can be defined as the following:

> A temporary endeavor undertaken to create a unique product, service, or result [1]

It is interesting, and important, to note that a project is *temporary* and *unique*. By "temporary," we are limiting the term "project" to efforts that have a defined beginning and end in time. This is important because it inherently means the scope of the project and the number of resources that can be applied toward the project are also limited.

By "unique," we are limiting our "projects" to efforts that are not routine/repetitive ones but rather efforts that produce a *singular* goal. Figuring out how to mow your 100-acre lawn is not a project, although it might be hard work, because once you have finished mowing, it will not be long before you have to start over again. Even though managing the company contracted to mow 100 1-acre lawns is not managing a project, there would be a lot to manage considering the different machines, travel, and employees that would all need to work together to successfully accomplish the task. A small landscaping design effort would be a good example of a project with a singular goal. That effort would require a vision of the final outcome and the management of resources, budget, and time – and, most importantly, once finished, the final product could be assessed against the original goal.

Example 1.2 will provide a few examples of projects versus engineering endeavors which may sound like project but which do not fit the formal definition of a "project."

EXAMPLE 1.2 Examples of Projects

Which of these are projects? For those which are not projects, can the wording of the goal be modified to make them projects?

- Developing a sensor for the Department of Defense
- Managing a microelectronic division for a small R&D company
- Developing course materials for Senior Design I
- Assessing curriculum content for an engineering college

Analysis:

Developing a sensor for a sponsor has a clear end goal – the delivery of a functional sensors – therefore it is a project.

Managing a division of a company is not a project – a company is an ever-evolving entity and the goal would always be to make it better. This statement could be modified to "Establishing a microelectronics division..." It now has a singular, time restricted goal and is therefore a project.

Developing course material is a project – once the material has been produced, it can be delivered to students.

Assessing curriculum content is not a project; it is an endeavor the college is required to undertake every year. Modifying this statement to "Assessing curriculum content for the upcoming ABET accreditation visit" would make this a project with an end goal.

Another way of defining a "project" is as follows:

An effort that is bounded by scope, time, and cost [2]

This concept is known as the *triple point constraint theory*, Fig. 1.1, which says every project operates within the boundaries of scope, time, and cost and that these three legs must be balanced to maintain the quality of the result. For example, if your sponsor wants more features (scope leg increases), your project will require either more time, more cost, or reduced quality. If your team requires longer-than-expected to complete the design of a subsystem (time leg increases), you will need to balance that by reducing the scope or increasing the cost if you still wish to finish on the original schedule.

In recent years, authors have started to point out short comings to the triple point constraint theory and adapt it in a variety of ways [3]. Primarily, it should be noted that most projects which are not completed on time (time leg extended) also result in a project which is over budget (cost leg extended). While this is true, it is often the result of ineffective Project Management. Therefore, one definition of Project Management is as follows:

An effort to balance the Triple Point Constraint such that a project is completed effectively and efficiently

Figure 1.1 Diagram of the triple point constraint

A more formal definition of Project Management according to PMI is as follows:

The application of knowledge, skills, tools, and techniques toward project activities required to meet the project requirements [1]

All too often in the real world, the amount of time required to complete the project is underestimated, and not enough resources are allocated to the project. It is also common for upper-level management to add scope partway through the project. Quality cannot possibly survive in this environment. The Project Manger's responsibility is to balance the triple point constraint until the desired quality can be achieved. This might mean increasing the project timeline when resources are scarce or possibly reducing the scope when time is limited.

Importance of Project Management

A Standish Group survey in 1999, Fig. 1.2, studied the results of system-level projects when using traditional Project Management techniques. They found that even when projects *are* managed well, they rarely finish on time. Can you imagine the results if these projects were attempted *without* Project Management?

More recent studies continue to show the same results even though PM techniques continue to evolve. Many software projects are managed with a style of PM called Agile Project Management. Results improve, Fig. 1.3, but not drastically so [4]. The term "challenged" on this figure refers to completed projects which finished late, similar to those specifically defined in the previous figure.

It is also interesting to note *why projects fail.* A study conducted in 2019, Fig. 1.4, on Information Technology projects shows that the top four reasons project fail can be directly affected by effective Project Management techniques: (1) requirement definitions, (2) scope definitions, (3) risk management, and (4) communications problems.

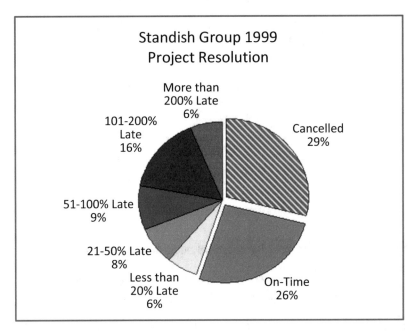

Figure 1.2 *Results of traditional Project Management*

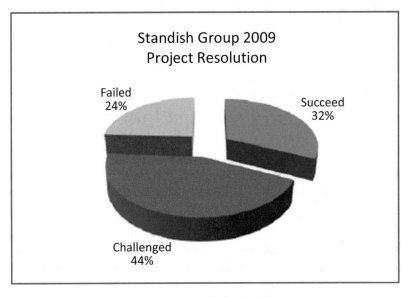

Figure 1.3 *Results of agile Project Management*

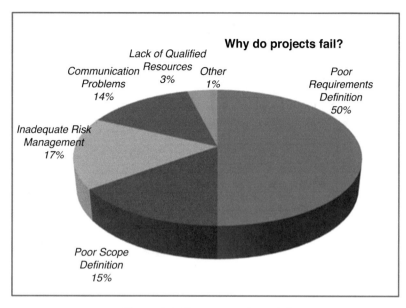

Figure 1.4 Reasons projects fail

Interestingly, the four primary issues that cause projects to fail are directly related to three of the Nine Knowledge Areas of PM, Fig. 1.5 (Requirement Definitions and Scope Definitions both are addressed by what is called Scope Management) [5]. By developing effective Project Management techniques, you will be able to increase the chances of success by directly affecting the most common reasons projects fail.

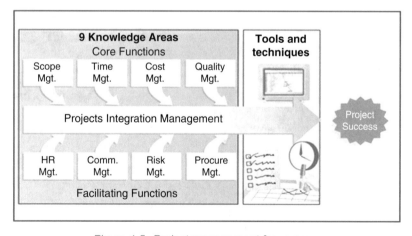

Figure 1.5 Project management focus areas

Industry Point of View 1.2 Project Management Is Important

Project Management is an important skill for all engineers to develop. It is extremely difficult to coordinate large projects without it.

In the late 1980s, I was the procurement engineer for a team responsible for designing the scanning head of the next generation of hard drives. PMI had only just begun to provide a Project Management certification, and the concept was not widely employed or understood in many companies. My team had yet to incorporate Project Management as we understand the discipline today.

There were many different parts of this team each responsible for different features of the design. This included the optical reader, the firmware to control the head and read data, and the mechanical arm that moved the reader, just to name a few.

Bill, a lead engineer on the team, was responsible for the conceptual design of the mechanical portion of the system. I was responsible for coordinating with venders to manufacture Bill's design.

Our management team provided us with a loose timeline and sufficient budget, but the scope of the design was poorly defined and there was not much effort to coordinate the multiple designs. This caused some serious problems!

After Bill produced the first version of the design, I began working with a vendor in London, England, to develop a price quote. But by the time I had a quote, Bill had changed the design and my process started again. Soon after that, one of the other designers made a system change which required Bill to modify the design yet again.

This continued for months. Without a clear scope, it was difficult for us to know what order to make decisions, who had priority, or even what the final result was supposed to look like. Each engineer was competent and did their best, but without a dedicated Project Management plan, the design process required a significant amount of trial and error, which is different than rigorous engineering design.

Eventually, the vendor could not wait for us any longer. They told us to make a final decision in the next month or we would lose our position in their processing line. Bill submitted a final design and the vendor got to work. This required the rest of the team to work around Bill's final design. Although the project was ultimately successful, this workaround caused numerous difficulties along the way.

Ironically, after a delay of 24 months and countless engineering hours, Bill's final design was not significantly different than the original in any way. Worse, our competition beat us to market, and we lost 100 s of thousands of dollars of revenue.

We all learned a lot from that experience. The company began to value Project Management, and I had the opportunity to gain Project Management skills of my own over the next many years. Ten years later I became the manager of a department with three distinct teams working on very different projects. To keep track of the different project's schedules, resource requirements, scope, etc., I had to rely on the Project Management skills that I had obtained. More importantly, I passed those skills on to each of the team leads. Each team was soon running more efficiently, and all levels of management knew exactly the status of each project. I and most everyone on each team were rewarded with promotions and/or awards.

Overview of Project Management Covered in this Textbook

It is important to understand that there is no "correct" way of performing Project Management. Project Management is a set of guidelines and skills which must be adapted for a wide variety of use-cases. What is necessary for IBM (a Fortune-50 company in 2020 [6]) may not work for a NASA-funded satellite calibration lab at South Dakota State University. What works for the Northern Plains Power Technologies (NPPT), a distributed generation consulting firm with seven engineers in 2010, would not necessarily be appropriate for the American Science and Technology (AST), an R&D Department of Defense contractor also with seven engineers in 2008. Oh, and none of the strategies ideal for those entities would directly relate to an academic Senior Design project. The material in this textbook was developed by merging the strategies used by IBM, NASA, NPPT, and AST to provide an understanding of *universal* guidelines used by Project Managers, while making a few concessions to the academic environment. The concepts presented in this textbook are only a small portion of the concepts a Project Manager must know. The concepts that made the final cut were chosen to fulfill the following priorities:

- The material selected will directly assist you in completing your senior project.
- The material selected is applicable to all entry-level engineers' careers.
- The material selected is a limited subset of Project Management that will conveniently fit within the time constraints of a typical Senior Design sequence.
- The material selected will introduce you to a rewarding, flexible, and lucrative career field and hopefully spark an interest so that you pursue the material further.

This method of managing Senior Design projects has been successfully utilized by over 300 Senior Design teams. It has been reviewed and supported by numerous industry Project Managers and hiring executives. And, most importantly, countless alumni have reported back that the lessons they learned from this material positively impacted the first few years of their careers. You can be assured that the material you learn here will be some of the most applicable to your professional career.

1.3 The Engineering Design Life-Cycle

The engineering Design Life-Cycle, sometimes called the Design Life-Cycle or other similar names, is a conceptual way of tracking the stages of project development. There are probably as many graphical representations of the topic as there are styles of Project Management. However, they all are attempting to convey the same set of ideas. All successful projects start, develop, implement, and complete – but what we call each of those stages can vary widely from source to source.

Common Representations of Design Life-Cycles

One very common graphical representation of the Design Life-Cycle, Fig. 1.6, attempts to show how the effort required to complete the project changes during different stages [7]. In this representation, we can see that the "1. Concept" stage requires little effort, but during the "2. Development" stage, the effort ramps up. During the "3. Execution" stage, effort reaches its maximum and again tails off during the "4. Transfer" stage. Stages 1–3 are particularly applicable to Senior Design.

Another common representation of the Design Life-Cycle called the V-model, Fig. 1.7, is used by the Department of Defense to define its expectations on system engineering [8]. Although the figures are very different and the terminology has been changed, the concepts are still the same. On the V-model, we see that the process always starts, in the upper left, with a *need* (the previous figure called this the *goal*), needs drive *requirements* (or *objectives*), and requirements drive the *design* (covered in detail throughout the *development phase*). The design will produce a *product* which will be *validated* (during the *execution phase* of the previous figure) and shown to *deliver capability* (during the *transfer* phase).

These two examples are very common representations, but neither perfectly apply to Senior Design. They are presented here in order to illustrate that once you understand the general concepts, all the representations essentially show the same thing.

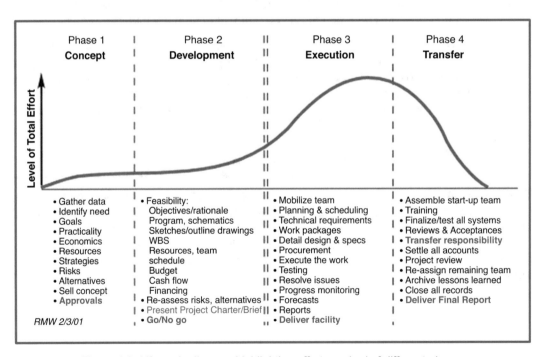

Figure 1.6 Life cycle diagram highlighting effort required of different phases

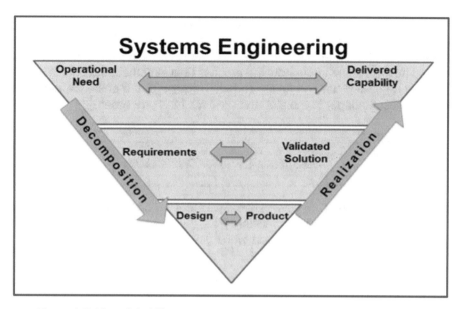

Figure 1.7 V-model of life cycle highlighting relationships of different phases

One major feature that both of these representations neglect is the cyclical nature of the design process. PMI uses a representation, Fig. 1.8, that neglects the effort shown in the first figure and some of the relationships presented in the second. However, it does highlight the cyclical nature of the process, and – more importantly – it uses terms that are probably the most universally recognized and, therefore, will be the basis of the terminology used through-out the rest of this textbook.

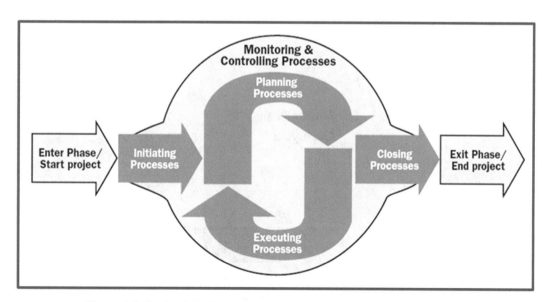

Figure 1.8 Design Life-Cycle highlighting the cyclical nature of the process

In PMI's version of the Design Life-Cycle, we see how a project must get started, process through an initiating phase, and then enter a "cycle." This cycle starts with some planning processes which are then executed. However, it is very likely that the planning steps will need to be revisited multiple times throughout the process until eventually the goal is achieved and the project can be closed out and finally ended. Throughout the entire life cycle, there is a need for the Project Manager to monitor and control all the processes.

Design Life-Cycle for Senior Design

The specific representation of the Design Life-Cycle that will be used for the remainder of this textbook, Fig. 1.9, was developed specifically for the academic environment to help you understand what will be expected of you while you progress through your senior project. This representation is heavily based on the PMI's version.

Here we can see the six major phases: STARTING, INITIATING, PLANNING, EXECUTING, CLOSING, and EXITING. The Monitoring and Controlling phases, shown on Fig. 1.8, will not be explicitly covered in this textbook. In Fig. 1.9, we can see the cyclical nature of the process; however rather than one large cycle, this figure has four smaller cycles – one after each of the primary checkpoints (Project Kickoff Meeting, Preliminary Design Review, PDR, Critical Design Review, CDR, and Final Design Review (FDR)). The following chapters will cover each of these phases in greater detail, but for now, let us look at some of the high-level concepts covered during each phase.

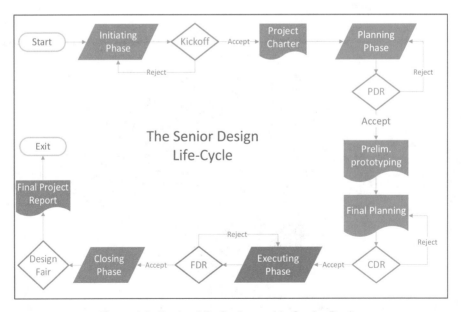

Figure 1.9 Design Life-Cycle used in Senior Design

The STARTING and INITIATING Phases

Maybe it is not surprising to find out that the most import part of the Design Life-Cycle is how the project gets underway; without a strong foundation, the project is likely doomed to fail. There are two parts to this getting a project off the ground, the STARTING phase and the INITIATING phase, Fig. 1.10.

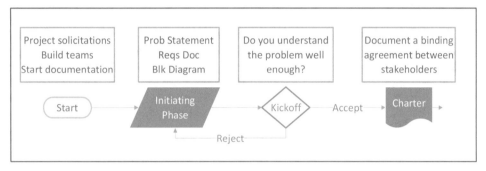

Figure 1.10   The STARTING and INITIATING phases of the Design Life-Cycle

The following are three main processes which occur during the STARTING phase.

- Collect Project Solicitations
- Build Design Teams
- Start Documentation

Collecting project solicitations was likely the job of your instructors before the semester started. Projects come from industry contacts, collegiate research units, students, and faculty. The projects can vary in scope, complexity, engineering disciplines required, etc., but all projects will require Project Management in order to successfully deliver the project to your stakeholder.

The first significant task which you are likely to be involved in is the team building process. Hopefully there is a wide variety of projects available to you and you are able to select a project which captures your interest. Chapter 3 will discuss a number of topics you will want to consider before selecting your project. After you are assigned a project team, there are a few processes that you can start immediately to strengthen your team and your ability to contribute to the team which will help increase your chances of success. 0 will aid you in those team building steps.

Project documentation is so important that proper implementation of it does not even wait until the project has officially started. 0 of this textbook will assist you in formatting and documenting your preliminary entries.

Once the project team has been set up and documentation formatting has been defined, we can enter the INITIATING phase and the real work can begin. During this phase, the project solicitation is formalized into a formal Problem Statement supported by a detailed Requirements Document – recall:

The number one reason project fail is poor requirement definitions!
Let us get this right!

After the team has worked through these processes, there are two major milestones that must be accomplished in order to move to the next phase:

- A meeting called the Project Kickoff Meeting
- A document called the Project Charter

During the Project Kickoff Meeting (PKM), the teams will explain their project to all the stakeholders in an attempt to convince the decision-makers the project is ready to move forward. Assuming everyone is in agreement, the team produces a written document called a Project Charter cementing the project in a legally binding agreement. However, if everyone is not in agreement, we will experience our first project "cycle." We will need to return to the initiating processes, address any of the disagreements from the PKM, and try again – while that is not an uncommon scenario in large, complicated projects, we should avoid that on our Senior Design projects by putting forth maximum effort to get things right early on.

Once the Project Charter has been accepted, we can move on to the next phases of the project where we get to be more creative and see our hard work pay off!

The PLANNING and EXECUTING Phases

The next stages are where the engineering fun is at!

During the PLANNING phase, Fig. 1.11, there are a number of important processes that are completed as follows:

- Alternative selections
- Initial system design
- Initial prototyping
- Final planning

During the first process, a few critical components are selected by carefully analyzing a number of potential alternatives and selecting the best choice. Then, a conceptual design will be established using proper engineering analysis and theory around these selections. A Preliminary Design Review (PDR) will be held so that the team has the opportunity to present their ideas to a panel of experts and receive feedback. Assuming the PDR is successful, a small amount of the project budget will be released to the team. This money can be used to validate the design and/or learn more about some unknown aspect of the design in order to improve the design. This is typically accomplished by building prototypes. Sometimes a single prototype of a critical portion of the design is produced. In some cases, the "prototype" may only involve purchasing a critical component to characterize. Other times, the prototype process may include developing certain capabilities which will be useful during the EXECUTING phase. The exact details of what *you* should accomplish during your initial prototyping process will be up to you to determine what is most useful for your planning efforts.

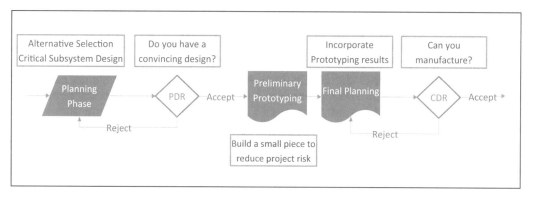

Figure 1.11 The PLANNING phase of the Design Life-Cycle

Once the team has learned enough from the Preliminary Prototyping process, the fabrication and verification plans can be completed during the final planning process. Any lessons learned during the Preliminary Prototyping process are incorporated into the system design. Also, details about how the product will be manufactured and validated or developed.

Finally, a Critical Design Review (CDR) is held to assess the final plans and prepare for the next phase. This is the team's last chance to propose changes to the Requirements Document.

Assuming your team's final plans are accepted at the CDR, the project is able to move into the EXECUTING phase, Fig. 1.12. The important processes during this phase include the following:

- Fabricating the system
- Validating the system

During this phase, the design is finally manufactured during the build system process. Each subsystem and feature are carefully verified during independent tests. As the subsystems are connected to one another, more testing is performed. Finally, when the system has been fully built, the system is validated in the validate system process. This is simply more testing, often repeating some of the previous tests, but at a complete system level. The EXECUTING phase officially ends when you successfully complete a Final Design Review.

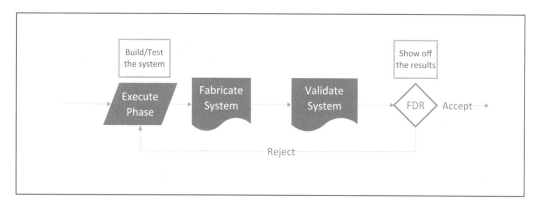

Figure 1.12 The EXECUTING phase of the Design Life-Cycle

The CLOSING and EXITING Phases

Much of the formal CLOSING and EXITING phases, Fig. 1.13, processes are not particularly applicable to Senior Design. These processes typically revolve around sponsor support and legacy documentation and are very important to an organization's long-term success. But considering that you probably will be focused on graduation, we will assume that you would rather get to the end quickly! So, we will only address a few topics during these phases. What this textbook will focus on is a public reveal, sometimes called a Design Fair or Design Expo, the Final Project Report (FPR), and the Project Post-Mortem.

At the end of most projects, there is often some sort of public event held for you to show off your projects to friends and family – the entire goal of all the hard Project Management work you have accomplished so far is to make *this* day a fun and relaxing event. In Senior Design, this is typically called the Design Fair.

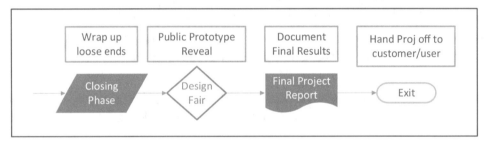

Figure 1.13 The CLOSING and EXITING phases of the Design Life-Cycle

Much of the CLOSING phase is dedicated to completing any necessary documentation required to wrap up the project or archiving the knowledge gained during the project. In Senior Design, this will be limited to what is called a Final Project Report (FPR). The FPR records *everything* you have experienced: all the design decisions, testing data, analysis, and even all the mistakes. Imagine what an arduous task that would be had you failed to consistently document all the little steps that occurred during a huge project! Fortunately, the Project Management process is designed to develop all the necessary documentations in small, incremental stages throughout the project so that writing the FPR is not a difficult task.

One of the most overlooked, yet most important, processes of the Design Life-Cycle is conducting the After-Action Retrospective meeting. This is a chance to reflect on what you have learned so that you are able to leverage those lessons on future projects. The ultimate purpose of conducting a project is rarely the project itself. Successful organizations only expend valuable resources on projects which support a larger purpose. That purpose may range from increasing stock values before the next shareholder's meeting to "let's go change the world." Forgetting the real purpose of conducting a project is easy (and normal) when you are so focused on the success of that particular project. But at the end of the project, it is critical for the health of the organization and the professional development of the individuals that a After-Action Retrospective meeting is conducted to advance that ultimate goal. In Senior Design, the ultimate purpose of conducting your project is to provide you a valuable learning experience. We will not let all the progress you make toward furthering your education go to waste at the end of the project.

And that is it! You are done, complete, outta here! All that is left is graduation, new jobs in entry-level positions, promotions, old jobs in senior-level positions, and retirement, but do not worry – after this experience, you will be ready for all that. Are you ready to get started?

1.4 Chapter Summary

In this chapter, you learned about the Design Life-Cycle and the importance of soft skills in the engineering career field. Here are the most important takeaways from the chapter:

- Soft skills, including communication and documentation, are important. Without these skills, technical achievements will not be understood or implemented by all the project stakeholders.
- Effective Project Management addresses the most common reasons why projects fail. By implementing proper Project Management techniques, you will dramatically increase the likelihood of project success.
- The Design Life-Cycle used for Senior Design includes the following phases:
  - STARTING phase
  - INITIATING phase
  - PLANNING phase
  - EXECUTING phase
  - CLOSING phase

One last thing to keep in mind before we move forward

The Senior Design course has a number of important learning objectives. It might surprise you to learn that none of these objectives include "completing an engineering project." The purpose of Senior Design is to teach you the engineering design process. This includes enhancing your communication, documentation, and management skills. The best way to learn these soft skills is to put them in practice within the context of a real project.

Occasionally, students will focus on completing their project without applying the appropriately Project Management techniques. While this may result in a complete project, this method will likely not result in a desirable grade. Fortunately, the entire point of Project Management is to increase the likelihood of project success while reducing project efforts.

Sometimes it will be difficult to see how a particular task reduces effort – it may seem like you would be able to advance the project design faster without "wasting" time on a certain task. This sentiment may actually be true – *in the short term.* But Project Management is a tried-and-true practice that reduces effort, iteration, and mistakes in the long run. You will not save time by rushing through a particular task only to find that you have to revisit the design decisions you made later on. Moreover, only a small portion of the grade is typically assigned to "did the product work," and a large portion of the grade is assigned to "how did you perform the project-related decisions."

So, keep in mind, the focus the focus of this course will be on learning the soft skills necessary to solve complex engineering problems in a team environment. Completing your project is important but only if you apply the skills appropriately.

Chapter 2: **Introduction of Case Study Teams**

Throughout this textbook, we will be following the development of three particularly interesting Senior Design projects that were conducted by students just like you between 2016 and 2018. These projects were completed using the exact Project Management methods presented in this textbook, meaning the lessons these teams learned during their Senior Design experience are still relevant to students reading this textbook today.

At the end of most upcoming chapters, the actual work these students produced will be presented. We will discuss both what these teams did well and where they could have improved. You are encouraged to learn from the examples these case studies provide. But, keep in mind, you are working on a unique project, for a different client, with new team members – what worked for these case study teams will not necessarily work exactly the same way for you. Use these examples as general suggestions of what your results should look like, but be creative and make your results your own!

It was a difficult choice to select these particular teams from the 350+ teams who have successfully used this material to complete Senior Design. The process started by focusing on projects which produced a successful product for an industry sponsor, in other words, teams who were successful in bridging the gap from academics to industry. The selection was further limited to projects which required a wide range of engineering skills; these teams were made up of electrical, mechanical, and software engineers and required both hardware and software developments.

Additionally, what set these particular teams apart was their understanding of how important Project Management skills were going to be in their careers and their willingness to accept and respond to feedback in order to improve this area of their education. Three of the engineers on these teams have completed Master of Science degrees in system design. Two of them are now certified Project Managers. Most have been given some sort of management responsibility of engineering projects.

These teams all produced successful (i.e., "A" level) results and delivered a working product to their client. It can be tempting to focus on their amazing results and miss the very realistic fact that they all struggled through this process. They all made mistakes; they experienced frustrations and even a few tears. They likely would have done things differently at the beginning of their projects had they known then what they eventually learned throughout their Senior Design experience. But that is okay; that is what being a student is all about. This experience ensured they did not make those mistakes when they began their careers. They worked the problem, learned the process, and produced excellent results, and all became especially successful engineers.

C.J. Mettler, *Engineering Design*, https://doi.org/10.1007/978-3-031-23309-8_2

Just like these exceptional teams, *you* are not expected to be perfect either. Many of the skills that are introduced here require years of practice to perfect. To be successful in Senior Design, you must recognize the importance of this material, do your best, and respond positively to the feedback and criticism you will undoubtedly receive. These case studies will review both what they did well and where they could have improved in an attempt to highlight this fact.

In this chapter you will

• Meet the case study teams

2.1 Augmented Reality Sandbox

The mission of the Kirby Science Discovery Center (KSDC) Children's Museum in Sioux Falls, SD, is to "encourage, amuse, instruct, and enhance the community by making art and science a part of people's lives." In 2016, the KSDC desired to update the museum with an augmented reality (AR) sandbox exhibit. The purpose of the exhibit was to educate children on basic concepts of topography, hydrology, and regional agriculture.

Augmented reality is similar to virtual reality (VR), except that at least a portion of the environment is physical, whereas a VR environment is entirely imagery. The AR sandbox is a literal sandbox, Fig. 2.1, left, for children to play with. However, as the children play, the AR portion of the system reads a three-dimensional scan of the sand and generates a topographic image layered on top of the sand, Fig. 2.1, right. This creates the illusion that children are able to make real-time changes to a topographic map.

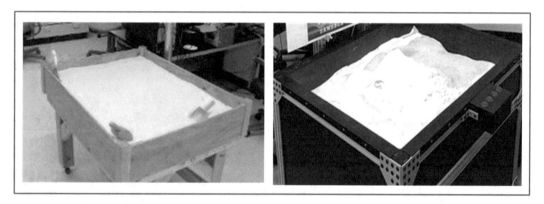

Figure 2.1 AR sandbox, (L) turned off during development, and (R) final product turned on

A feature was added to this particular AR sandbox so that after a child designed a landscape in the sand, virtual "rain" could be added to the environment. Children could then witness the hydrology of their "landscape" by watching the runoff paths of the water. The system also used predictive crop growth models developed at South Dakota State University to predict how much corn could be grown on the child's landscape. These calculations were performed in one-inch squares across the landscape and considered the elevation, provided rainfall, and predicted regional temperatures to predict yields. Results were displayed to the child using graphics of corn in varying states of health, Fig. 2.2.

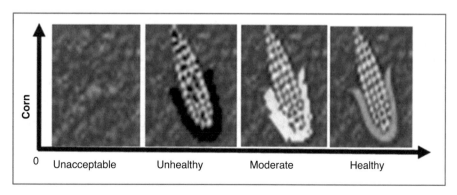

Figure 2.2 Images of corn indicating varying levels of health

An intuitive child could improve their overall yield by including irrigation paths in the sand, adjusting the slopes and elevations, or by modifying the amount of rain provided throughout the simulated growing season.

The student team, Fig. 2.3, successfully met all of their sponsor's requirements, and their product was installed at the KSDC in May 2017. The exhibit quickly became a crowd favorite and was heavily used by nearly 1,000,000 annual visitors over the next 4 years.

The team was comprised of three electrical engineers, James, Moe, and Logan, and two mechanical engineers, Trevor and Tucker. Overall, the team performed well and was quite successful. Their strengths included developing a strong working relationship with their sponsor and excellent presentation skills. They could have improved by including a software engineering teammate and by adhering to a more rigorous work schedule.

Figure 2.3 2016/2017 augmented reality sandbox student design team (from left to right: James, Moe, Trevor, Tucker, and Logan)

2.2 Smart Flow Rate Valve

Raven [9] supports farmers around the world in their efforts to grow more crop with less resources by producing precision agriculture products.

Utilization of GPS technologies has allowed precision agriculturists to develop so-called prescription maps of a particular field. These maps provide high-resolution details of the health of a crop, amount of fertilizer and pesticide requirements, and historical crop yield. Farmers are able to better manage their resources when they have this detailed information. For example, if one small area of a typically productive field produces a low yield, the famer could choose to apply extra fertilizer only in that specific area. Or, if a field is suffering from an insect infestation which originates from a swampy corner of the field, the farmer might choose to spray pesticide only in that corner rather than the whole field.

Traditional sprayers require a famer to manually adjust the spray volume. So, farmers might set the spray at one "average" volume and cover an entire field with that flow rate; this ensures that most of the fields either has insufficient or surplus chemicals applied. Even in the unlikely case where an entire field requires the exact same application across the entire field, if the farmer varies how fast they drive across the field, the chemical flow rate should also change. Raven sought to combine elements of the system into one compact package that could vary the flow rate for prescription map or speed change requirements. The system used feedback controls to carefully monitor and maintain the actual flow rate to match the desired flow rate.

In 2017, Raven had developed a proof of concept for this product using a number of discrete parts. The remaining challenge was to incorporate this concept into a self-contained product that was physically small enough to be installed into typical sprayer systems. A student design team, Fig. 2.4, was created which consisted of three electrical (Tyler, Keith, and Thein) and three mechanical (Gannon, Christian, and Mason) engineers. The electrical engineers optimized the component selection used in the initial proof-of-concept and developed the electronic control system. The mechanical engineers developed a single-unit housing to contain the necessary components and easily interface the product into a standard sprayer unit; they were also integral in optimizing the controllable flow valve component selection.

The Smart Valve team was comprised of six strong students with a wide range of technical strengths. As the project developed, each team member contributed substantially to the technical development of the project. However, initially the team struggled to operate as a cohesive unit; they consistently divided the project into two concepts – one responsible for the electrical and the other responsible for the mechanical aspects of the project. This unintentional division created communication issues among the team. Fortunately, as the team developed, this issue was resolved. By the end of the project, this team was a cohesive unit that delivered an excellent prototype to their Sponsor.

Figure 2.4 2017/2018 Smart Valve student design team (from left to right: Gannon, Christian, Mason, Tyler, Keith, and Thien)

2.3 Robotic ESD Testing Apparatus

High-end, elite hard drives are often called "enterprise" drives. Enterprise drives are not necessarily designed for your personal computer; they are designed for so-called high-end customers whose business depends upon data reliability. Such customers might include Master Card, Amazon, or State Farm Insurance. Hard drives purchased by these customers *cannot* fail, glitch, or in any way lose data. Therefore, each drive is put through an incredibly rigorous verification process before being sold.

One reason hard drives fail is caused by electrostatic discharge (ESD) "shocks." The most common cause of an ESD shock is when an operator touches the drive or storage rack and discharges the static electricity all humans naturally possess. As such, hard drive manufactures rigorously test their designs to prove that an ESD shock will not cause an enterprise drive to fail. The ESD verification test description for a single drive may include a matrix of test points which could require up to 500 individual shocks.

Prior to 2016, ESD verification was performed manually by a technician or intern. Unfortunately, errors and omissions can easily occur during such a long and complex testing procedure. Additionally, the manual documentation of 500 separate occurrences was an exceedingly tedious process.

During the 2016/2017 academic year, a large enterprise hard drive manufacturer sponsored a student team, Fig. 2.5, to develop a Robotic Electrostatic Discharge Testing Apparatus (RESDTA) to automate their ESD verification process. The apparatus was required to interface with the same ESD delivery "gun" that was used in the manual testing process. The apparatus was required to articulate the ESD gun to a variety of predefined test points and deliver the required ESD shock. A user interface was required to set up a specific test and collect the test results.

This project was sponsored by a local branch of a large international company. To avoid international intellectual-property complications, this company requested that their name not be used on their sponsored projects. All the information presented in this textbook has previously been publicly released and therefore is no longer subject to intellectual property law. However, this textbook will continue to respect the company's request of anonymity. The student team gave permission to use their personal work for this case study.

The 2016/2017 RESDTA team was comprised of two mechanical engineers (Kelsi and Easton), two electrical engineers (Leah and Kaitlin), and one software engineer (Patrick). Leah and Kelsi were students of the highest caliber; the technical ability these two students brought to the team far exceeded the expectations of a typical Senior Design student. However, their technical abilities alone are not what made this group so successful.

This team was chosen as a case study because they worked exceptionally well with each other. Each student demonstrated care and dedication to the project, the sponsor, and most notably to each other. Significant technical contributions were made by each student. Everyone on the team contributed to necessary Project Management tasks. And perhaps most importantly, Kaitlin pulled the team's contributions together as she developed into an exceptionally capable team leader. In fact, Kaitlin went into a career in System Engineering and continues to use the Project Management skills she developed during her senior experience.

Figure 2.5 2016/2017 Robotic ESD testing apparatus (from left to right: Patrick, Kaitlin, Kelsi, Easton, and Leah)

2.4 Chapter Summary

In this chapter, you were introduced to three teams which we will use as case studies throughout the rest of the textbook. Here are the most important takeaways from the chapter:

- These teams were comprised of students much like yourself.
- The primary reason these teams were successful was a willingness to learn and to implement the Project Management skills being taught in this textbook.
- These teams each made many mistakes along the way. They were not perfect, but they appreciated the feedback they received throughout the project and did their best to implement correction to avoid making the same mistake twice.

PART II: THE STARTING PHASE

"Houston, we have had a problem."
– Commander James (Jim) Lovell Jr

"Let's work the problem, people. Let's not make things worse by guessing."
– Gene Kranz, NASA Flight Director, April 1970

*"We had risen to probably one of the greatest challenges in human history, put a man on the moon within the decade. We'd created incredible technologies. But the most important factor was we'd created the **teams**, what I call the human factor."*
– Gene Kranz, NASA Flight Director, April 2000

On April 13, 1970, at 20:06 hours, oxygen tank #2 in the lunar service module of Apollo 13 exploded [10] and so began what many have coined "NASA's finest hour." Commonly referred to as a "successful failure," the Apollo 13 story has been an inspiration to Americans for over 50 years. Over the course of 87 hours, Gene Kranz's Mission Control team "worked the problem" in a manner which has become an exemplary case study of teamwork and crisis management.

The first moments after the explosion were filled with confusion as a stream of alarms and piles of data started to flood into Mission Control, but amazingly nobody wasted time panicking – they immediately got to work doing their jobs. The team began to self-organize; there was so much to do, there simply was not enough time for any individual to wait to be directed by management. Analysts poured over data to determine the extent of the problem, flight controllers began plotting possible routes to bring the crew home, and engineers designed solutions to reduce power consumption and filter the astronauts' breathable air to provide life support until an ultimate solution could be reached.

A group-of-individuals could never have solved so many complex problems in such a short time, but Apollo's Mission Control demonstrated conclusively that the right *team*, working together, is far superior than the sum of its parts. So, what was it that allowed these people to perform so admirably? The complete answer to that question is fascinating and worth further investigation, but this textbook will focus on three particular aspects as they relate to Senior Design: motivation of the individual, structure of the team, and the available resources.

Each individual on the Apollo 13 mission was a highly motivated and inspired person, but they were not authoritative experts in the way we think of "experts" today. The average age of Mission Control members was just 26 years, only a few years older than a typical Senior Design student. They were not "rocket scientists" and only few of them had advanced degrees. Most experienced engineers of the day did not believe landing on the moon was possible and therefore were difficult to recruit. So Kranz developed a team that balanced their lack of experience with an overt dedication to excellence [11].

Apollo 13 was not the first challenge Kranz oversaw; he was the flight director for the Apollo 1 disaster in February 1967 when a fire ignited in the cabin during a training exercise and killed the entire crew. In the aftermath of that experience, Kranz gave a speech coining a term that would become the "price of admission" for all those who worked in Mission Control thereafter. Kranz expected his team to be "Tough and Competent." Kranz further explains, "Tough because we… are forever accountable for what we do, or fail to do. Competent because we will never take anything for granted, we will never stop learning."

The culture Kranz developed for his team was uncompromising. Each individual team member was expected to master the basic science, applied technology, and advanced applications that supported their mission and that of their teammates. The next time something went wrong, Kranz's expectation of *competency* ensured that someone on the team had the ability to solve any problem.

Kranz also expected that each individual was *accountable.* What he meant was that everyone was expected to "step forward and take the lead [when the time came], and then return to their role as a member of a team [12]." Nobody was just along for the ride; everyone was expected to be fully competent and to make valuable contributions.

Although the Apollo 13 team had over 1000 hours of mission-specific training, nobody had specifically prepared for the oxygen tank explosion or the power failures that had actually occurred. Fortunately, Kranz had developed his teams well. He was able to "relinquish control of the situation and trust in his team of tough, competent, and accountable people…to develop the right solution" in an extremely high pressure, short timeframe [13] and bring the crew home. Many people have commented on how it seemed "everybody had a sense of what needed to be done and began moving in the right directions without being directed." That did not "just happen," it was "allowed to develop" because of the team structure already in place.

Highly motivated people in a well-structured team still require the appropriate resources to do their job well. Before the full extent of the damage caused by oxygen tank #2's explosion could be understood, each damaged component had to be assessed. Before a resolution could be developed, the resources available to the crew had to be inventoried. The quality of the engineering documentation produced during the design and development of Apollo 13 was a critical resource that allowed the team to analyze the issues and produce realistic solutions. This critical aspect of engineering is so often overlooked, but without the emphasis NASA placed on proper documentation, the Mission Control engineers had no chance of bringing the crew home.

The STARTING phase of the Design Life-Cycle is primarily about selecting a project, developing a team, and establishing a documentation structure. The following table lists the processes and related documents this textbook will focus on. The chapters in Part II of this textbook will guide you through these processes.

Processes of the Starting Phase	Presentation Structure
Generate or identify an engineering challenge to be solved.	Project Solicitation
Recruit an effective team.	Team Assignments
Set workload expectations.	Workload schedules
Define documentation procedures.	Project Notebook Guidelines

Chapter 3: **Choosing Your Design Project**

The first stage of the engineering Design Life-Cycle is aptly called the STARTING phase, and the first process in this phase is selecting a project on which to work. There is more variability on how this process is addressed than any other process discussed in this textbook. Often, detailed financial and risk analysis is performed before starting a project. Most of the formal STARTING phase processes are not appropriate materials for a typical Senior Design sequence; however, you should at least be aware of some of the considerations related to how projects are started.

Even in a Senior Design class, the way project development is handled can vary. Some projects are generated by student interest; these are often great projects but typically require that students work with the course instructor before the start of the semester to get the project accepted. Other projects come from university-sponsored competition teams such as the ASME "Human Powered Vehicle Challenge," the ASCE "Concrete Canoe Race," or IEEE "American Solar Challenge." Most projects are generated by either the faculty or industry partners; these projects have the distinct advantage of providing you the experience of working for a paying Sponsor with a vested interest in the project's outcome.

Regardless of the manner in which the potential projects are generated, a list of projects will be made available to you at the beginning of the semester, and you will have to quickly decide on the project you will dedicate many hours of time and effort over the next two semesters.

> This is an important decision, and it should not be taken lightly.

Your enjoyment and the sense of fulfillment you achieve over the next many months will depend upon this decision.

In this chapter you will learn:

- How different types of organizations start new projects.
- How to choose your design project.

C.J. Mettler, *Engineering Design*, https://doi.org/10.1007/978-3-031-23309-8_3

3.1 How Organizations Start Projects

The STARTING phase of the Design Life-Cycle can vary drastically between different organizations. Some generalized patterns can be observed, but they should not be considered guidelines of what is necessary or demanded.

Projects can be generated by many different methods. Some of these methods provide very clear and precise project definitions; others provide only a concept and require significant engineering work before design or development can commence. Some examples of how a project might originate follow.

An entrepreneur might identify an area of interest, identify many possible issues to address, choose one particular problem to solve, and hopefully design a profitable or impactful solution to that problem. Unfortunately, entrepreneurial efforts have an extremely high rate of failure. There are many reasons for this, but one important reason is a lack of understanding of how successful projects are managed throughout the Design Life-Cycle. To be successful, an entrepreneur should, at this stage, develop a mission statement or project definition that will guide future decisions. Without this clear statement of work, many start-ups get sidetracked and lose focus on the primary goal.

In a large corporation, a team of decision-makers typically will study markets for lucrative opportunities. Strategic plans are formulated, and directives are provided to engineering managers to carry out. Successful companies employ a hierarchy of technical leads, team leaders, project managers, operational managers, and executives to ensure the Design Life-Cycle is carried out effectively.

Consultants advertise their services and then respond to Sponsor requests. Statement of Work (SOW) agreements are typically formulated defining what the project is and who is responsible for each piece. Contractual agreements become very important documents in this method of conducting business.

Small Research and Development (R&D) companies might scour public sources for project solicitations and submit project proposals hoping to convince decision-makers to choose their company. An example of this would be government-sponsored Small Business Innovation and Research (SBIR) grants. These grants are funded by the Departments of Defense, Energy, and Health and other government entities.

Regardless of how a particular organization identifies potential project opportunities, the organization will eventually have to select a desired project and therefore reject other opportunities. This can be a critical decision to the health and longevity of the organization. Once the project has been selected, a team must be formed to address the problem.

> The STARTING phase ends once a viable project has been identified and a project team has been formed to solve the problem.

It might appear that this is a short phase which requires little effort. However, defining the "right" project can often be a long, arduous challenge. Knowing which projects to pursue and which to reject is a valuable skill to develop. There are many times when choosing to *not* pursue an opportunity is more valuable than undertaking the "wrong" project (of course, that does not apply to Senior Design assuming you would like to graduate someday soon).

Industry Point of View 3.1 Select Your Opportunities Carefully

I am a senior vice president for a company which develops digital displays like those you see in Times Square. A few years ago, we purchased a small competitor who produced an excellent graphics card we wanted to use. We wanted rights to their product, but more than that, we wanted their design capabilities. So, as part of the merger, we wanted to maintain a number of that company's executive leadership and senior engineers.

After the two companies merged, we tasked the new employees with ownership of our entire company's graphics capabilities. We never really performed any of the official processes of the STARTING phase of the Design Life-Cycle to decide this. Why would we? We believed in what they were doing enough to buy them out!

But, looking back, this was a mistake. There were definitely graphics-related design aspects which our company previously excelled at. Without careful consideration as to the strengths and weaknesses of both organizations, we missed important indicators which caused us trouble for the years to come. Our new employees continued to develop a technology which they had been developing prior to the merger. The technology worked but never aligned with our company's strategic goals. We ended up investing a lot of resources into a project which was never particularly useful or profitable.

Overall purchasing this company was a good decision. Maintaining their employees proved beneficial in the long run. However, not being selective about which projects we pursued was a pretty significant mistake. Carefully considering which projects you peruse, rather than just agreeing to the easiest or most obvious, is important for the success of the company. Overall, my company has been successful at selecting beneficial projects, but this was a case where we could have done it better.

3.2 Choosing Your Senior Design Project

In most Senior Design courses, the list of potential projects will have been generated from a variety of sources. You may be assigned a project or allowed to choose one. If you are allowed to choose your project, the selection will be an important choice that will impact your enjoyment and education over the next many months, so if you have a choice, choose wisely. Unfortunately, the wisdom required to choose wisely often comes with experience. Most Senior Design students are making a project selection of this magnitude for the first time and, therefore, do not have the experience to know which project they would like to pursue. This section is intended to provide you insight into what you are about to experience. This insight will aid you in making your first project selection. Here are some guidelines and suggestions to follow:

- You will be expected to make a significant individual contribution to the technical development of the project. It is helpful for you to have some familiarity with the components of the selected project. Choose a project that is related to courses or work experiences which you have successfully completed.

- Think carefully about what a project really involves and try to understand where the design focus of the project lies. In other words, a project may be described as addressing one topic, but the domain where the design team will spend most of their time and effort might be somewhere else; the latter is what you should use to gauge your interest in the project, so try to figure this out as best you can. If time allows, it may be helpful to ask questions of your instructors or others who might be in a position to assist you.
- Senior Design typically requires more effort per credit hour than other courses you have taken; you will need to dedicate many hours per week to the successful completion of your project. Choose a project that is interesting to you and/or which provides learning opportunities that might support your future career! Said differently, you will spend a lot of time learning about topics related to your project; try to direct this effort toward a subject that excites you and that will pay off for a lifetime.
- Interviewers like hearing about a candidate's project experience, in particular, Senior Design projects that have been successfully completed. Many students have landed a desired engineering position by impressing their interviewer with their achievements in Senior Design. Select a project which has enough challenge to be impressive or interesting to established engineers.
- Consider the fact that your project may put you in contact with people who may be mentors, supporters, or future employers! For example, if you do a good job and impress the people you are working with, your faculty advisor might help you figure out what to do after graduation and write you an outstanding letter of recommendation, and your project sponsor might connect you with a job opportunity or even become your future boss.
- The first point is so important; we will repeat it again here: Every individual student will need to make a significant technical contribution to the project in order to be an effective team member. This does not mean that you need to be an expert on every aspect of the project, as you will work with a team and learn as the project progresses. However, you should choose a project where you start out confident that you can make important technical contributions to at least one or two of the core aspects of the project and then learn and improve from there.

Avoiding Common Pit Falls 3.1 Selecting Your Senior Project

Students may use a strategy when selecting their Senior Design project that results in placement on a project they do not actually want. Here are some common mistakes.

- *Choosing your team rather than your project.*
  It is tempting to select a project in order to work with a friend or study partner. Doing so is not "wrong" but must be done so with care. Consider the following:
  - Do you both truly have the same interests? You should be independently interested in the project.
  - Do you have complementary or overlapping skills? If you essentially have the same skill set as your friend, you might want to consider working with some who can cover your weaknesses. Two strong programmers might not make the best team if the project requires for a balance between hardware and software.

Avoiding Common Pit Falls 3.1, continued

- *Choosing a project with a high coolness factor.*
  Again, this is not "wrong"; in fact selecting a project that is particularly interesting to you is one of the positive suggestions presented above. However, also consider the following:
  - What technical contributions can I make? If a particularly interesting project is related to power system design and you have not completed a power course, this might not be the best project for you.
- *Disregarding a technically challenging project.*
  Often some of the best projects are ones that are more challenging and will require additional effort. You may still be well suited for this project; consider the following:
  - Are you willing to put some additional time into the project in order to learn or improve necessary skills?
  - What does your advisor or instructor think about your ability to contribute to the project? These people can often give you a pretty good idea of what will be required.

3.3 Chapter Summary

In this chapter you learned about the Project Selection process. Here are the most important takeaways from the chapter:

- Successful companies spend a lot of time and consideration selecting the most beneficial project to pursue.
- You should choose a project which you will enjoy making a significant technical contribution toward.

3.4 Case Studies

Each of our case study teams were comprised of strong students who at the beginning of Senior Design had feelings of uncertainty and nervousness – perhaps just like you. A few students knew exactly which project they wanted to work on, but many questioned which of the projects offered that year would be the best experience for them personally. Let us look at how some of the students chose their projects and how their SWOT analysis supported their team development.

Augmented Reality Sandbox

Moe was adamant that he wanted to work on the AR sandbox project from the beginning. He strongly desired to work with an industry sponsor on a project which would provide the team with a lot of public visibility. Moe was an average engineering student but had extremely well-developed soft skills. He was excited to learn that not all aspects of engineering are technical and that he could use his natural skills in an engineering career.

- What went well: Moe was extremely motivated by this project and collaborated well with the Sponsor.
- What could have been improved: Moe occasionally neglected his technical responsibilities in favor of the soft aspects to the project. Remember, each student must contribute an approximately equal share of the technical workload as well as to the Project Management requirements.

James considered a number of software-based projects before committing to the AR sandbox. He desired a project that would challenge his already advanced coding skills. Ultimately, he chose this project due to the advanced image processing software that was going to be required.

- What went well: James was definitely challenged by the code required for this project. He learned a lot and developed advanced coding skills that he later applied to research position in an image-processing lab.
- What could have been improved: James did not delegate his responsibilities well. His "I can do it all" mentality worked for a while, but eventually, when he needed help on certain tasks, nobody was up-to-speed with the code. A better division of labor on this project may have reduced some of the bottlenecks James created.

Table 3.1 presents an excerpt from the AR sandbox team's SWOT analysis.

- What went well: Here we see that the personal and team strengths directly address a few of the identified weaknesses. For example, in the personal strength category, we see both Moe's (#1) and James's (#2) strengths as mentioned above. The team quickly realized that Moe's strength would help them address the concern about time management (#3) and James's strength would cover a few team member's coding weakness (#4).
- What could have been improved: Although item number 8 (lack of agricultural, et al., knowledge) was initially identified during the SWOT analysis, the team did not take steps to address this weakness until late into the PLANNING phase. They eventually

collaborated with an agricultural engineering research group to cover this weakness; however, doing so drastically impacted their initial design work. The team was forced to revisit some of the early PLANNING phase processes after gaining this additional information. Make a plan early in the project to address any weakness that the team cannot cover internally.

Table 3.1 Excerpt from AR sandbox SWOT

	Personal	Team
Strength	1. Work and collaborate effectively with team members with different skill. 2. Solid programing experience.	5. Joint EE/ME partnership allows variation of ideas coming from different perspective. 6. Ability to model prototype.
Weakness	3. Struggle with time management. 4. Not the best at coding.	7. Unfamiliar with the other engineering side. 8. Lack of sufficient knowledge on agriculture, geography, and hydrological concepts.

Smart Flow rate Valve

Tyler was a strong student, typically at the top of his class in most courses. He was motivated to work on the Smart Flow Rate Valve project partly because he had previous work experience with the Sponsor and was familiar with some of the technologies that would eventually be implemented. As such, he was assigned as the team leader at the bringing of the project.

- What went well: Tyler applied his technical skills effectively while designing this product. He also coordinated well with the Sponsor and led his team effectively; in fact, he was eventually offered a career with the Sponsor.
- What could have been improved: Tyler had little prior Project Management experience but was assigned to the team leader regardless. Since he was primarily motivated by the technical side of the project, he tended to focus on those aspects more than his leadership responsibilities. It took a little while for Tyler to develop into the strong leader he eventually became. This should have been recognized in the SWOT analysis process and addressed. Tyler could have used some additional support from his instructor early in the project development.

Gannon was a better than average student but not as strong as his two mechanical partners. However, he did bring significant modeling experience to the team – a skill that would be heavily required. He also was a "hands-on guy." As such, he was excited about the prospect of physically constructing the test apparatus required for the final deliverable.

- What went well: Gannon produced a number of excellent three-dimensional models that supported the team in a variety of ways.
- What could have been improved: Gannon eagerly applied his three-dimensional modeling skills to the project but initially did not integrate his modeling work with the rest of the team. This required the team to revisit Gannon's early work a number of times. Eventually, Gannon and his teammates worked out a process to ensure the necessary integrations occurred.

Table 3.2 presents an excerpt from the Smart Valve team's SWOT analysis.

Table 3.2 Smart Valve SWOT

SWOT Analysis – Team
Strengths: We are determined hardworking students. Half of us have taken a class on microcontrollers. Some of us have taken a class on thermofluids.
Weaknesses: Some of us have trouble staying organized. We do not have experience working on agricultural equipment. No one has experience in project management.
Personal SWOT Analysis – Tyler
Strengths: I have experience working with Raven. I have done some programming on the STM32F427 chip that we are constrained to use.
Weaknesses: Sometimes, I get too eager to dive in and start projects without planning enough.
Personal SWOT Analysis – Gannon
Strengths: I have extensive three-dimensional modeling experience.
Weaknesses: Time management can be a challenge for me.

- What went well: The team recognized the individual technical strengths brought to the team and effectively assigned responsibilities based on those strengths. The team also recognized that they, as a group, had a weakness understanding much of the agricultural equipment, terminology, and operations related to the project. The team mitigated this weakness by leveraging the sponsor's expertise during the INITIATING phase of the project.
- What could have been improved: We can observe that the team documented a weakness in Project Management. Gannon admitted to time management issues, Tyler recognized his tendency to skip necessary planning steps, and the team acknowledged those issues by stating "No one has experience in Project Management." However, the team did not mitigate this acknowledged issue until well into the PLANNING phase. It is not enough to document a weakness in the SWOT; it is important to take action so that the weakness does not negatively affect the project.

Robotic ESD Testing Apparatus

The members of the RESDTA team all selected this project as their top preference for similar reasons. Each student was personally motivated to contribute toward a technically challenging project. Each student recognized the multidisciplinary nature of the project and was excited about the novelty of that experience. And, each student desired to work on a project sponsored by a real customer who planned to integrate the students' project into a real-world system. The team members were already friends and had worked on previous projects together.

- **What went well:** The 2016/2017 RESDTA held the promise of developing into a performance team from the beginning (Sect. 4.1). The team was entirely comprised of capable and motivated students who were looking for a challenge and who had a deep commitment to success.
- **What could have been improved:** Working with close friends can be challenging. The RESDTA team experienced a few personality clashes. Each clash started with one team member censoring an opinion about a particular design topic in order to avoid a confrontation with their friends. In these cases, they prioritized their friend's feelings over their obligation to the project. Doing so led to less-than-optimal decision-making, and eventually resentment started to grow. Challenging a friend's opinions in a healthy and professional manner is a difficult skill to cultivate. Moreover, not allowing yourself to feel affronted when you are the one challenged by a friend or teammate is also difficult. However, when everyone on the team compassionately discusses technical issues and, at the same time, encourages constructive criticism, the team can potentially develop into a performance team that produces extraordinary results. Fortunately, the RESDTA team prevented their issues from derailing project success by openly talking with each other as the issues arose.

Table 3.3 presents an excerpt from the RESDTA team's SWOT analysis.

- **What went well #1:** Kaitlin recognized her leadership and communication strengths and helped Leah stay focused when discussing technical topics. Kaitlin would structure the team's documentation and assign small writing assignments to each individual. This gave Leah a narrowly focused topic to write about and helped her develop concise documentation. The team also recognized the complexity of the project stakeholders and the need to communicate well and constantly among everyone involved. Kaitlin was assigned to be the spokeswoman for the team to ensure that the project's communication management was accounted for at all times.
- **What went well #2:** While Kaitlin's natural communication abilities supported Leah, Leah's technical skill supported Kaitlin. Kaitlin wanted to contribute to the control subsystem of this project but was not particularly strong in that area. Leah excelled in Controls but probably could not have accomplished the design of such a large control system on her own. The team assigned portions of the Controls Design to Kaitlin, but Leah acted as a mentor and supported Kaitlin's efforts. Over the year, Kaitlin learned a substantial amount about Controls and developed a new technical strength, which is always a rewarding experience. At the same time, Leah was able to reduce her technical workload to a manageable level. She was able to contribute to the project without feeling constantly overwhelmed. When individual team members find creative ways of using their strengths to directly counteract another team member's weakness, the entire team benefits greatly.
- **What could have been improved:** The stakeholders for this team included the five students with three different majors, a mechanical advisor, and electrical advisor, a technical point-of-contact for the sponsor, the sponsor himself, and the course instructor. The team recognized the risks imposed by communicating with such a complex group of people and assigned Kaitlin as the team leader and spokeswoman. Managing communication, this complex would be challenging for anyone, particularly during the early stages of the project when the team is just learning to work together. Kaitlin ultimately performed this task as well as any developing Project Manager could, but she also could have used some help early on. Assigning sole responsibility of a high-risk issue to one resource may not always be the best idea. During their SWOT

analysis, Patrick claimed to have experience in both mechanical and electrical engineering, as well as his software background; however, this skill set was rarely leveraged throughout the project. Perhaps having Kaitlin take the commutation lead, but accept support from Patrick, much like the relationship Leah and Kaitlin developed on the Controls aspect, may have helped Kaitlin through the INITIALIZING phase.

Table 3.3 RESDTA SWOT

Team strengths
We have a diverse team with mechanical, electrical, and software skills. The team advisers have diverse backgrounds relevant to our project. We have a strong team leader.
Team weaknesses
The majority of students working on the project have part-time job commitments that may limit the amount of time that can be spent on project work. Technical communications within the unit will be challenging between the three disciplines. The unit has two advisors as well as two sponsor contacts which could lead to contending opinions about project direction.
Personal strengths
(Kaitlin) I have good communication skills. I am hardworking, and I (am interested willing) to act as the team leader. **(Leah)** I learn new programming languages quickly which will be helpful when trying to use G-code. I was good at linear controls and electronics.
Personal weaknesses
(Kaitlin) I enjoy controls but struggle with some of the concepts. I am not experienced with all of the documentations that will need to be done with this project. **(Leah)** my ability to condense information and present technical information clearly and concisely needs improvement; I often write more than needed for technical reports.

Chapter 4: **The Engineering Design Team**

The second process in the STARTING phase of the Design Life-Cycle is to form design teams. Once a particular project has been agreed upon, a team must be formed to work on that project. This is a critical step in the process. It is not enough to group individuals together and call them a "team." In order for the group to be considered a *team*, it will be important to select the *right* individuals. Then, in order to be successful, this group will need to be managed effectively.

An important part of the Senior Design experience is learning how to operate effectively within a team. This might be the single most important factor related to the enjoyment of your experience. An ineffective team will cause each individual to produce significantly more work than what is required of individual on an effective team. An ineffective team will produce lower-quality work even with the increased workload and that will result in more negative feedback from supervisors. This combination is sure to kill the morale of the team. Conversely, an effective team will support each other, increase productivity, and improve the final project results. These characteristics cannot help but increase the enjoyment of your experience.

A single individual can definitely cause issues within a team. However, the failure of a team to operate effectively is *never* the fault of one member. There are always things *you* can do personally to counteract the negative impact of an individual. When the majority of the team focuses on working well together, the team will overcome those negative impacts.

In this chapter you will learn the following:

- How engineering teams are defined.
- What is expected of you as a member of an effective team.
- How to complete a SWOT analysis to strengthen your team.
- How much time to allot toward your senior project to the project.
- How to run an effective team meeting.

C.J. Mettler, *Engineering Design*, https://doi.org/10.1007/978-3-031-23309-8_4

4.1 Definition of an Engineering Team

There are actually many ways of defining a "team," but one useful definition is as follows:

> A team is a small group of people with complementary skills who are committed to a common purpose for which they hold themselves mutually accountable [14].

Another excellent definition is as follows:

> Teams are made up of individuals whose collective competence and experience is greater than [what] any individual offers [15].

There are two important characteristics these definitions are attempting to succinctly attribute to teams; let us consider these characteristics in some detail.

- *Teams are comprised of effective individuals.*
  Teams are not a collection of random people. The members of a team must be committed to the outcome of the team. Each individual must offer a competency that contributes to the team's success. By this definition, the person who misses meetings, does not contribute individual thought, or is clearly not working toward the success of the team is not, in fact, a team member.
- *Teams are greater than the sum of the parts.*
  Teams are comprised of individuals who use their skill sets to complement the rest of the team's skills for the greater good. No single individual will have the skill set to "do it all." The team is most effective when those complementary skills are coordinated for the benefit of the group. An individual who is not using their skills to make a noticeable contribution is not, by definition, a member of the team.

A study performed in 1993 defined a categorization method of teams based on their effectiveness [14]. Using this method, teams may be separated into four groups as follows:

- Pseudo-teams
- Potential teams
- Real teams
- Performance teams

Figure 4.1 describes these four types of teams. Inevitably teams will be comprised of individuals of varying abilities ranging from weak to exceptional. The purpose of the team is to contribute more than the combined efforts of the individuals. This is shown in the first column of the figure. Here we see the individual contributions from an "exceptional" team member, two "normal" members, and one "weak" member. Teams who function well are called either real -teams or in special cases performance teams. These are teams whose contributions exceed the minimum benchmark set by the sum of individual contributions.

In this section, you will learn the common characteristics of the four types of teams and techniques to develop your team into a real team. By understanding what typical team members experience on each team type and how those experiences feel, you will be more capable of identifying the type of team you are working with. Once you identify the team type, you are better able to evaluate the team, and if necessary, take appropriate steps to improve your team into a higher-performing team.

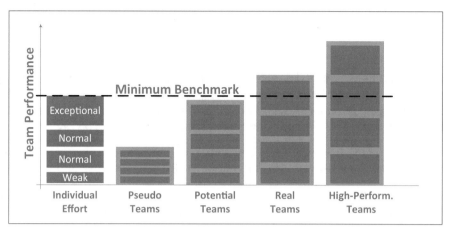

Figure 4.1 Types of teams and performance as related to individual performance

---

Pseudo Teams

Pseudo-teams are not in fact "teams" as defined above. This group's performance will be poor and well below expectations. Nobody benefits from being on a team like this and each individual would have likely produced more had they worked alone. Typically, each individual experiences frustration and will get little joy from the work. The project will almost certainly fail. Clearly, this type of experience is one to avoid at all costs.

Pseudo-teams are caused by apathetic individuals who wait, usually until it is too late, for someone else to take charge. The individual performance of even the "exceptional" and "normal" performers is lower than had those individuals worked alone. Typically, the lack of effort is not necessarily malicious (nobody wants to fail). Instead, the lack of effort can be the result of lack of leadership, procrastination, or a sense of being overwhelmed and not knowing where to start.

There are a number of things you can personally do to avoid working on this type of team:

- Take personal responsibility for the team's success. You are not responsible for doing all the work yourself, but you are responsible for pushing the team in the right direction if and when it stalls.
- Commit to making progress *every* week; in particular, you must make progress in the first week of the first semester to set a precedent for the entire project.
- If you are unsure what to do or how to make progress:
  - Voice those concerns with your teammates. Take responsibility to start the discussion.
  - Personally address the concern with someone in "management" (this might be a faculty advisor or course instructor), and make sure you follow up with the rest of your teammates. Better yet, organize a team meeting so that each team member becomes a stakeholder in problem-solving.

Potential Teams

Potential teams perform close to the same level as the sum of individual efforts. This is often the result of one exceptional team member carrying a large portion of the load. Project results will be underwhelming even if the overall project is a success. The weakest members of the team will underperform. The strongest team members will feel frustrated and used.

When recognized early, it is relatively easy to upgrade a potential team into a real team. However, the longer the issue persists, the harder it will be to make this improvement.

Here are some indications that your group is a potential team and what you might do about it. These solutions are meant to help team members and therefore the entire team improve.

- *Lack of progress:*
  There should be clear and documented progress from each team member every week. The lack of production from one team member in a single week can put a project in jeopardy. In upcoming sections of this textbook, you will learn multiple methods of tracking and documenting your production. What is important to realize now is that:

  1. Your team must carefully track and report everyone's progress.
  2. Your team must hold everyone accountable each week.

  If a team member did not make progress over the previous week, there is a concern that *must* be addressed. This might feel a little uncomfortable at first, but there is a very simple, nonconfrontational way of handling it:

    - Be understanding that team members all have busy lives, and everyone will at some point fall behind.
    - Realize that a "concern" is not a negative thing if it is corrected.
    - First occurrence: simply ask the teammate to explain how they will get caught up. Keep your tone of voice friendly and conversational, but make sure this is documented.
    - Second occurrence (within a short period of time): without accusing the person and still keeping a conversational tone, point out that this is "the second week within a [time period] that WE have not made progress in that area." Then, express that YOU personally are concerned about the situation. Ask what the TEAM can do to keep progress moving forward. Focus on the team, not the individual. Make sure you document this and consider voicing a concern to your instructor and/or advisor.
    - Third occurrence: Although "three times" might not seem like a major issue, three weeks is about 20% of your semester. It is not your responsibility to solve this problem, but you must react to it to save your project. Take this concern directly to your course instructor.

- *Unbalanced workloads.*
  Regularly assess the distribution of work; it should be evenly (approximately) distributed among team members. The workload is almost certainly off balance on a potential team. If you notice one team member with significantly more or less workload, a clear problem exists, affecting the team's ability to operate productively. Uneven workloads will often lead to resentment and team conflict. However, an appropriate reaction to this situation can often strengthen a team.

The best method of handling this situation somewhat depends on who is addressing the issues. Here are some possible scenarios to think about:

1. You notice that a high-achieving teammate is regularly carrying the bulk of the workload.

   Problem: When a teammate consistently works harder than the rest of the team, it is likely they will become resentful and possibly lose motivation to support the team.

   Impact: When a team's high-achieving members stop supporting the team, the team quickly collapses, team morale is reduced, and project success is jeopardized.

   Mitigation: Verbally acknowledge what you have noticed. Ask the high-performing team member how the team can help.

   Results: A slight redistribution of workload typically means other members will become more productive. The redistribution does not reduce the output of the high performer but rather results in the team's output increasing. Thus, productivity improves as well as working relationships.

2. You notice that a teammate is regularly not contributing at the same level as the rest of the team.

   Problem: Clearly a team member not producing their share of the workload means the rest of the team must work harder to compensate.

   Impact: Covering for an ineffective teammate requires your time and energy. Additionally, this person will often be unable, or unprepared, to effectively contribute to team project reports. This can reflect poorly on the entire team.

   Mitigation: Remember the learning objective for this section of the textbook is to improve your team's effectiveness from a potential team to a real team; this implies that everyone in the group is competent and committed to the mission. Mitigation strategies for a completely noncommitted or incompetent individual will be covered in Sect. 4.3; in this section, we are focusing on improving a relationship with a struggling *team member.*

   Remember, if you can help a weaker team member improve, even if it is "not your job," you are really helping yourself by improving your team. The key to solving this problem is first to determine why the person is not producing well, then specifically addressing that issue.

   One common reason a team member withdraws is that they feel that they "do not belong" in a group and that feel that they are inadequate or unwelcome; this is called the imposter syndrome. It is important to intentionally include this person into the group. This person will rarely volunteer an opinion, so regularly ask them what they think about team decisions (pro tip: do not accept one-word answers; ask follow-up questions such as "why do you think that" or "can you give us some details.") This person is not going to jump on opportunities to take the lead on difficult tasks but would appreciate being asked if they are comfortable taking that lead. It is relatively easy to help this person become an effective member of your team.

   Another possible reason a team member is underperforming is that they are struggling with their technical assignment and may simply be feeling overwhelmed. Offer this person a help or rearrange responsibilities. Sometimes the simple offer to help will give the person motivation to push on. Other times a small amount of effort from a supportive team member will get the struggling member back on track. Sometimes a more significant change is necessary and, when made, will allow everyone on the team to work up to their potential. Remember, the objective here is to develop the most effective team.

Results: Supporting and encouraging the weakest member of a team will assist that person in their professional development. Strengthening a weaker member and finding the optimal assignment of responsibilities will free up the stronger members to better apply their strengths. The entire team improves.

3. You notice that *you* have a disproportionate share of the team's responsibilities.

Problem: You are likely a high achiever who has always been able to "just handle it" in the past. But, in reality, you will not be able to do so long term on a complicated project like your senior project.

Impact: You will burn out and potentially become resentful. The team will not perform as well as it could.

Mitigation: Ask for help, even if you do not *really* need it! Be specific about what needs to be accomplished and related deadlines. Rather than correct someone's work yourself, point out issues that you notice, and ask the persons responsible what *they* think about your concerns; suggest they make any necessary changes. Make sure you document this exchange.

Results: A focus on team development, rather than simply completing the work, will improve your project results in the long run. Eventually, your personal stress will be reduced. Hopefully, you can evenly balance the workload, and the entire team becomes high achieving. In reality, you may end up doing slightly more than your "fair share," but this will be noticed by your superiors and you will stand out as the team leader.

If the attempt to evenly distribute the workload is ultimately unsuccessful, you will have sheltered yourself from poor reviews. You are not expected to produce more than your share. You *are* expected to be an effective teammate and actively work at strengthening your team. If you have done these things and the project still fails, you will not carry that responsibility. However, the key here is *regularly* documenting your efforts; it will be difficult for your superiors to *prove* you have attempted all that you could without your documents.

- *Repeated feedback:*
  It is normal and expected that you regularly receive criticism from experienced engineers involved with your project. Certainly, this will come from course instructors and possibly from faculty advisors, sponsors, or Sponsors. It is extremely important to respond positively to this feedback in a timely manner. This is part of the design process and is in no way a negative thing. However, potential teams often require repeated critiquing whereas real teams correct the issue, learn from it, and improve. Here are some characteristics of potential teams that cause issues:

  1. Members of potential teams often assume "someone" is addressing the feedback. When everyone assumes someone else is responsible, the feedback remains unaddressed.
  2. Responsibilities on potential teams are often not well defined and therefore are in a state of flux. A person responsible for a task one week may receive feedback but then pass that responsibility to a teammate the following week who is unaware of that feedback.

3. Potential teams typically do not make effective use of documentation. This can cause problems in at least two different ways. First, feedback may come to you in a variety of sources. For example, if you submit a report, the report might be edited, and a rubric might be filled out; both will have valuable feedback. Potential teams are apt to miss some of the feedbacks by only reviewing a single source. Second, the team itself typically does a poor job recording the feedback. For example, there will certainly be a lot of discussions after a review session. It is rare for potential teams to be seen taking notes during this discussion, maybe assuming they will remember it all.

Avoid repeated feedback by being diligent about responding to feedback. Here are a few techniques that potential teams could use to develop into real teams:

1. Assign a secretary for each review to record all verbal feedback.
2. Be sure to review *all* feedback from *all* sources provided to you.
3. Openly discuss all feedback with the team. This includes feedback provided to you as an individual.
4. Compile all feedback into a public location. The Engineering Notebook (0) would be a good option for this.
5. Assign follow-up action items to individuals so that the team is clear who is responsible for addressing the feedback.
6. If any feedback is not completely understood, ask the reviewer for more information.

Real Teams

Your goal for Senior Design should be for your team to operate as a real team. The risk of project failure on a pseudo- and potential teams is high, so it is important for your team to reach the status of a real team. On the other hand, forming a performance team during the Senior Design sequence is difficult and not necessary; this requires skills which can only be developed from experience and it takes time. If your Senior Design group can operate as a real team, the probability of project success is extremely high. Fortunately, achieving this level is relatively easy so long as you are intentional about being a good team member.

Real teams exhibit the following characteristics:

- The team focuses on a clear and singular goal.
- The team effectively makes group decisions.
- Responsibilities are effectively communicated and distributed.
- Each person takes responsibility for his/her individual work.
- Each person is committed to the success of the team.
- Each team member uses effective communication.

Often, real teams assign clear roles to individual members. This is something to consider for *your* Senior Design team; however, it is not necessarily the best method for *every* team. Some teams perform well when all team members equally share all responsibilities. Other teams perform better when at least some of those responsibilities are clearly assigned to individuals. Here are some roles you may wish to consider on your team:

- *Team leader – responsible for team scheduling:*
  There are many assignments and due dates to track throughout the Senior Design sequence. Sometimes it is beneficial to assign this duty to one person. The team leader's responsibility is to verify work is completed on time, that all assignments are formatted correctly, and to formally submit those assignments. This role should *not* require an individual to actually *complete* the assignments on his/her own; everyone must contribute to the technical work evenly. Rather, the team leader is overseeing the team's formatting and submission of that work.
- *Team secretary – responsible for notes taking during all meetings and reviews:*
  In (0), you will learn about the Engineering Notebook. One aspect of this document is to record notes for all meetings and project reviews. Some teams choose to rotate this responsibility among members while other teams prefer to permanently assign a secretary role to one individual. Teams rarely have much of an issue with recording their regularly scheduled team meetings, but sometimes recording help sessions, impromptu meetings, and project reviews are forgotten.
- *Team treasurer – responsible for maintaining the project budget:*
  Most Senior Design projects are allotted a limited budget. Carefully tracking the money being spent is extremely important. Often there are no additional funds available if you accidently overspend. This can result in the students being responsible for any unapproved overages. A treasurer is often helpful to ensure all expenditures are tracked and accounted for.

Your team has the option of assigning these roles or others you may wish to implement. However, if you do so, remember to be flexible once the roles are assigned. If an assignment is not working, change it sooner than later. For example, if your team assigned the role of team leader to an individual, but that person proves to be disorganized, it might be better to reassign or eliminate that role. Also, remember that accepting one of these roles does not necessarily mean that your share of the technical work will be reduced. Typically, on most Senior Design projects, each individual must share an equal portion of the technical work.

Another option is for your team to rotate these roles among the team members (e.g., weekly or monthly). This is only an effective strategy if it is absolutely clear who has each role at any given point in time, and everyone fully understands the responsibilities and expectations of each role. As with any of the role assignments discussed here, be prepared to adjust your strategy if it is not working well and remember that taking on one of these roles does not reduce a team member's responsibility to contribute technical work.

Performance Teams

Performance teams have all the characteristics of real teams, but simply perform at a higher level. The dedication and passion of each individual on the team must be extraordinary. All team members are exceptional in their area of expertise. Each member has highly developed communication and teamwork skills. The team is led by a competent and motivational leader.

However, a group of exceptional individuals does not necessarily produce a performance team. The distinction of performance teams is that each team member feels a deep commitment to the other members and to the success of the team's mission. Certainly, a group of exceptional individuals may be grouped onto a Senior Design team. This group will effortlessly perform as a real team. However, the jump to becoming a performance team takes time, training, and experience that most individuals simply have not developed by their senior year. It has happened, and some Senior Design teams have performed at unbelievably high levels. But those teams are rare.

Examples of performance teams include Gene Kranz's mission control team (discussed in the Part II intro), the 1980 US Olympic Hockey Champions, and the Boeing Engineers who designed the 747 Airplane. Kranz essentially employed young men fresh out of college in 1970, the Hockey Champions were college kids playing against professionals, and the Boeing Engineers changed the way the world thought about air travel. These teams were comprised of highly capable individuals, but what set them apart and allowed them to accomplish amazing feats was the dedication to the team and the unified goal.

Each individual of mission control was committed to being "Tough and Competent." They trained hard to know every aspect of their responsibilities. But they did not stop there. They also knew every detail of their immediate team member's responsibilities too. This allowed them to effectively support and cover for each other when the crisis occurred in 1970.

The 1980 US Olympic Hockey team exemplified what it means to work as a team. No single individual on that team was better than their competition's counterpart. Their team practices were legendary. They simply outworked their competition. Of course, they honed their individual skills. But to make up for their lack of talent and experience, they spent extra practice time learning to play as a team.

During the late 1960s, "average" people started to demand an affordable means of air travel. Boeing's answer was the 747. Designed from scratch in roughly 16 months, the 747 dwarfed all other commercial planes; the plane is so large that the air required to pressurize the cabin during flight weighs nearly a full ton! Newly developed engine technology significantly improved fuel efficiency. The result was a drastic reduction of the cost-per-passenger to travel which completely changed the travel industry. The engineering feats accomplished during this short period of time were so impressive that Boeing nicknamed the design team "The Incredibles." A real team could have produced similar results but not within the 16 months allotted.

You are not expected to develop a performance team for a Senior Design project. However, it is useful to understand what the characteristics are for such a team. Perhaps, with some additional effort, your team could achieve these high standards at least in some areas of your design project.

Industry Point of View 4.1: Teamwork Makes the Dream Work

I work for an international computer company. At the time, I was a Project Manager for a new product developed in the UK. My responsibility was to coordinate with a counterpart in the UK to ensure that a quality product was released to our Sponsors.

A few months into this new product release, it became evident that there was a major issue with the internal power supply of our product. Working with the team in the UK, we determined that all of the products coming into the USA had to flow through my site where each power supply would be rebuilt. The problem was that we did not have the equipment, resources, or staff to complete the task. I was starting from scratch – no staff, no lab space, no equipment, no rework process, etc. Worst of all, the ability to rebuild these supplies had to be ready within two weeks to reduce the risk of not meeting our financial goals for the fiscal year.

I knew a performance team would be necessary if we had any chance of accomplishing the task.

The first step to building my team was to determine which set of skills would be needed for the task. I compiled a list of skillsets I believed would be necessary and approached many of the managers in my division who supervised individuals with at least one of these skills. Then, we held a brainstorming session with these managers. The managers agreed to share resources. A dedicated engineer was identified for each skill, and I soon had a list of people that I wanted to be part of my team.

On Day 2, I met with this group of engineers and presented the problem, the financial importance to the company, and the impossible schedule. Then, I asked for their help. I did not ask anyone directly to apply the skill which landed them on their managers list. But, in turn, each individual recognized a project need for which they could personally take responsibility.

I think that was important. I did not attempt to dictate roles or force responsibilities on anyone. Instead, I asked a group of dedicated people to be part of this team and allowed them to personally buy into the challenge. I needed them to take charge of their area of expertise; this was not something I could micromanage.

By the end of Day 5, extraordinary results began to occur. Lab space and equipment were identified and prepared. We coordinated with shipping managers to ensure all packages were rerouted to our lab as they entered the USA. A procedure was developed to quickly, but carefully, rebuild these power supplies. A quality control process was established. Technicians were reassigned roles to make sure we had the manpower to complete the volume of work in a timely manner.

We held very short daily meetings to stay organized and make team decisions, but mostly I met one-on-one with anyone who needed support so that the rest of the team remained focused. Everyone had their responsibilities. As the Project Manager, my responsibility was to organize a group of capable people and to provide a clear vision of the problem. Then, empower those people to do their job. After that, my role was to support their roles and then get out of their way.

Normally, in a large company, getting so many people to work together to design a complete process like this would require weeks; my team did it in five days. We were ready to begin rebuilding power supplies when the first one arrived on Day 7, a full week ahead of schedule. This team continued to support each other for another six months until the power supply vendor could permanently solve the issue. During that time, not a single supply failed on a Sponsor.

I think the real key to our success was a team, not necessarily made up of exceptionally talented people but of exceptionally dedicated people who knew their jobs and were willing to make sacrifices for the team goal. I will never forget working with that group of people.

4.2 SWOT Analysis

As defined in Sect. 4.1, a team is comprised of dedicated individuals with complementary skill sets. In order to best use these skills, the team must understand the strengths of each individual which the team can leverage to its advantage. Understanding an individual's weakness can also prevent some of the shortcomings which may arise. One common technique for doing this is called a SWOT analysis. SWOT stands for strengths, weaknesses, opportunities, and threats:

> By analyzing yourself as an individual and then developing a team SWOT, you will better understand how to divide your resources as the project develops.

The first step is to consider **strengths** that you can apply to the benefit of the team. What technical areas (classes) have you enjoyed or excelled at? Maybe you enjoyed a thermodynamics course but struggled with structures. Maybe you have completed an internship experience analyzing the power grid but did not love your programming course. What interpersonal skills do you have? Are you an exceptionally organized person? Do you have experience in a leadership role? Do you enjoy public speaking or writing?

Also consider what personal **weaknesses** you might have. Maybe there were specific courses you struggled with more than others. Maybe you have preferred to work as an individual during previous courses, and teamwork has been less effective for you. By articulating each individual's weakness, the team can better support that individual. Sometimes individual weaknesses can be completely avoided; if you are not an effective programmer, maybe you can assign the programming aspect of your project to a teammate that enjoys it. Other times weaknesses cannot be avoided, but they can be mitigated. You are required to work effectively as part of team regardless of your previous preferences; however, by talking about what has not worked for you in the past, you might find an acceptable accommodation with this team.

Next, it is important to discuss **opportunities**. There are two aspects to consider here: personal opportunities and technical ones. Personal opportunities might include a chance to develop a new skill set or gain experience that you can leverage when seeking permanent employment after graduation. Technical opportunities focus more on what impact the project's solution will have if successfully designed and implemented. Maybe your sponsor is looking at improving a testing process or hoping to develop a profitable product. It is not important to specifically identify whether the opportunities you identify are personal or technical, and they might be both. For example, a renewable energy project may produce a technical solution to a power generation concern while also providing a personal sense of achievement.

Finally, you should consider any **threats** to the project. These also can be personal or technical. Do you have a particularly busy schedule and are concerned about finding common meeting times? Does the technology exist that is required to solve your problem? Does the technology cost fit into your allotted budget? Is your team going to be dependent upon an outside source for support? A business might also consider the likelihood of a competitor solving the same problem faster and capturing the market before you are able to release your product.

The SWOT analysis is an excellent icebreaker at your first team meeting. Privately consider how you would answer each category of the analysis prior to meeting with your group. Then, as a team icebreaker, share your thoughts in order. By discussing strengths and opportunities, you are sharing a piece of yourself for the benefit of the team. When discussing weaknesses and threats, the discussion should evolve into potential solutions, which allows the team to show its support for you. Start by discussing each member's strengths. Then, discuss weaknesses, but focus on strategies to mitigate some or all of those weaknesses. Continue by sharing what you personally see as opportunities; this will provide your team with an insight to your motivation and personality and will get to know you better. Finally, communicate what you perceive as threats. Again, focus on solutions in the hope of preventing the threat from occurring.

Example 4.1 presents a personal SWOT analysis for one individual student. Example 4.2 presents how that student, along with her team, incorporated her personal SWOT analysis into their team's SWOT analysis.

EXAMPLE 4.1: Personal SWOT Analysis

The following SWOT analysis was developed by a student designing an automated switching control system to optimize the amount of capacitance on a distribution powerline for a newly installed hydroelectric power plant on the Red Rock Reservoir Dam near Pella, IA. The project was sponsored by the Missouri River Energy Services (MRES).

Strengths:
I have work experience in the power industry and am familiar with communication devices such as Real-Time Automation Control (RTAC) devices. I personally know the engineers responsible for installing the capacitor bank.

Weaknesses:
I have never witnessed this extensive of a project, an end-to-end design. I struggle with some of the power system control schemes that I have been exposed to.

Opportunities:
I am looking forward to cultivating a professional relationship with the MRES field engineers and hope to gain employment with them upon graduation.

Threats:
I have a heavy course load this semester due to my second minor. I am also an officer in the Eta Kappa Nu (HKN) and Institute of Electronic and Electrical Engineers (IEEE). I am concerned with my ability to dedicate adequate time to this project.

EXAMPLE 4.2: Team SWOT Analysis

The previous example presented one individual's personal SWOT analysis. After each team member analyzed his/her own situations, the team combined input from each individual SWOT analysis to produce the following team SWOT analysis:

Strengths: Together we have the following:

- Familiar with Real-Time Automation Control (RTAC) devices.
- Personal relationships with the field engineers involved with the project.
- Familiarity with the university's power lab equipment.
- Experience delivering technical content in professional presentations.

Weaknesses:

- Neither of us have ever witnessed this extensive end-to-end design. We will need to work closely with our advisor and Sponsor on this aspect.
- We will attempt to avoid certain power system control schemes during our design efforts. If it appears that one of these schemes is the only scheme that will work, we will need technical assistance from our advisor.

Opportunities:

- I am looking forward to cultivating a professional relationship with the MRES field engineers and hope to gain employment with them upon graduations.
- The experience with the OPAL-RT real-time simulator will be a unique experience to put on resumes.

Threats:

- We both have a variety of outside commitments and scheduling with each other will be challenging. We will hold meetings on Sunday evenings to avoid other conflicts.
- We will be unable to implement our design on the regional power grid. This will leave some uncertainty in our final design and possibly make it difficult to fulfill all the course requirements when it comes to system validation. We will work with the course instructor to avoid as many concerns as possible.

Analysis:

This two-student design team produced a thoughtful SWOT which guided many decisions throughout the design process. Each student contributed to strengths, opportunities, and threats. One student did not contribute to weaknesses in a meaningful way. The team might have benefited had this student been more thoughtful regarding this category.

Notice how the strengths and opportunities are simply a summary of the individual's thoughts. Conversely, the weakness and threats do not focus on the individual's concerns but rather present a solution to overcome each issue. This is the proper focus of a team SWOT analysis.

4.3 Expectations of Effective Teammates

It is quite possible to develop a real-team without excessive training or skill. A group of people cognizant of the importance of developing a cohesive team and who are willing to put a minimal amount of effort into being effective team members will likely form a real team naturally. This section provides a set of minimum expectations that when followed by each individual will dramatically improve your team's ability to function together as a real team and successfully complete your Senior Design project.

There are five areas addressed here. They include personal conduct, communication, conflict resolution, team building, and addressing an ineffective team member.

Personal Conduct

Each individual should commit to the following codes of conduct:

- Attend all meetings:
  - Be on time.
  - Be prepared with something to contribute.
  - Be engaged by providing opinions, voicing concerns, and asking questions.

  What if you cannot: In reality, it is likely that daily life will require that you miss a meeting occasionally. If this happens, you have a few responsibilities:

  1. Promptly inform the rest of the team of your situation.
  2. Attempt to reschedule the meeting if possible.
  3. If it is not possible to reschedule, read the meeting minutes (Sect. 5.3) and then *follow up with a team member.*

- Adhere to the schedule:
  - Carefully monitor the schedule using a Gantt Chart (Sect. 8.2).
  - Know what your action items are and when they are due.
  - Complete your tasks *on time.*

  What if you cannot: Project schedules will inevitably slip because you do not have the experience to know exactly how much a time is required for a particular task. This is tolerable as long as you respond to the slippage appropriately:

  1. Inform your team about the slippage as early as possible (this means you must start your task long before it is due).
  2. Update the schedule to reflect the delay.
  3. Create a plan on how to make up the lost time.

- Support your teammates:
  - Speak positively to each other about their ideas, the project, and the team.
  - Focus on solutions not failures; if something is not working, help solve the problem rather than simply point out the failure.
  - Engage quiet teammates by asking respectful questions.
  - Be enthusiastic about participating; take responsibility as if the entire success of the project depends on your contribution.

What if you cannot: Sometimes it is challenging to support everyone on your team, but it is critically important that you *do* support all real-team members. If you find that you really cannot accomplish this on your own, seek help from an instructor.

Effective Communication

Effective communication is the number one factor in project success at all levels. This section is focused on intra-team communications, specifically communication skills that facilitate productive team meetings and work sessions.

- Be an active listener:
  - *Do not interrupt* a teammate unless it is to ask a clarification question.
    By interrupting a teammate, you are inadvertently communicating to that person that their thoughts do not matter to you. They will be less likely to be involved, and over time you may completely lose their support.
  - Make eye contact with the speaker. This demonstrates that you are listening and care what they have to say.
  - After the speaker has expressed their thoughts, respond directly to their comments before providing your own. Specifically, you might do the following:

    1. Thank the speaker for their contribution.
    2. Praise the idea (or at least a portion that has merit) even if you do not fully agree.
    3. Ask a clarifying question in order to procure more information.
    4. "Parrot" their comments. This is a technique that ensures proper understanding. Start by stating "What I understand you said was. . . ." or "I think I understand, you mean. . . ."

What if you cannot: Challenge yourself to regularly address what someone says before contributing your own ideas. This will force you to listen to the speaker more closely and then show the speaker that you paid attention.
- Challenge yourself to use proper technical language:
  - Use proper terminology. Current "travels through an element"; it is not "in an element." Voltage is "divided across elements"; it does not "go" anywhere. Often seemingly insignificant misuse of terminology leads to significant misunderstandings.
  - Use proper units. Probably every student has complained at least once about losing points on something as insignificant as forgetting to include units in an answer. However, the misunderstanding between inches and centimeters caused the first satellite sent to Mars be destroyed when it skipped off the Mars atmosphere.
  - Reduce or eliminate the use of pronouns; be specific. Rather than state "We have a problem with communications," state "we are unable to transmit the data." If a teammate uses pronouns, ask what they mean by "it", "she" or "an."
    What if you cannot: Practice makes perfect. This is a hard skill to master; keep working on it.
- Structure your discussion effectively:
  - Remember that no one will be as familiar with your part of the project as you are. You need to put your discussion into context before getting to your main point.
  - Start with the big picture. Make sure everyone knows what part of the project you are about to refer to.

- Explain what you were trying to accomplish before explaining the problem.
- Provide what you know to be correct or completed before discussing the new concern.

What if you cannot: Every discussion will start with an assumption of what the group already knows. It is not possible to perfectly accomplish this category. However, simply putting yourself in your listener's point of view will make your discussions more efficient and save you time and energy.

Conflict Resolution

Not all conflict is bad. Different opinions and viewpoints will cause disagreements. Constructively working out those disagreements brings a team closer together, but improper management of them can derail a team.

- Challenge ideas; do not criticize people.
- Use "I" statements, rather than "you" statements. "I think there might be a better way" is much more productive than "Your idea is not going to work."
- Brainstorm solutions; do not focus on problems.
- Take time to clarify the situation and/or concern so that everyone understands the conflict before attempting to reach a resolution.
- Be willing to compromise.
- Privately document the conflict, discussion, and resolution in case the issue escalates.

Team Building

It is extremely likely that your project will be unsuccessful if your team does not perform well together. Therefore, it is your responsibility to strengthen the team as much as it is to perform the technical work you are assigned:

- Actively encourage participation from everyone on the team:
  - Intentionally include everyone in all team discussions.
  - Ask for input from someone you notice is not fully engaged.
  - Spend time together; when possible, meet in person.
- Hold each other accountable:
  (After following the tips above to hold yourself accountable)
  - Understand everyone's assignments.
  - Expect teammates to produce, document, and report their own work.
- Celebrate each other's successes; downplay failures.
- Have fun!

Addressing an ineffective, noncommitted, or incompetent "teammate"

Unfortunately, it *is* your responsibility to properly address an ineffective person on your team. First of all, remember that an ineffective person assigned to work with you is not automatically a "teammate." As defined above, "teammates" are effective people committed to the team and the project goal. This section will provide you strategies to deal with someone who has repeatedly failed to commit to the project or participate effectively; let us refer to this type of person as a superfluous member.

> Your responsibilities are to your team, the project goal, and the sponsor; you do not have a responsibility to cover for a superfluous member.

- Attempt to make everyone an effective teammate:
  - Follow the advice provided in the subsections above. Doing so will avoid the chance that a person assigned to your team will become a superfluous member.
  - All action on your part should be constructive, not punitive. If you can transition a struggling person into a teammate, rather than a superfluous member, your team wins.
- Double check to see if you have *really* followed all the advice provided above:
  - This is so important; it is worth mentioning twice.
- Document your concerns:
  - Accurate records will help you explain your concerns if the need arises.
  - Records will also make it clear that *you* are not the problem.
  - These records will ensure that *you* are assessed based on your work.
- Do not cover for this person:
  - It is tempting to pick up this person's slack, but do not do this.
  - If work is completed even though one person rarely contributed, it is difficult to demonstrate that there was a problem.
  - Remember, the documentation you keep shows you completed your work effectively and attempted to foster a real team.
  - *Be careful*: there is a big difference between helping a teammate and covering for a superfluous member.
- If you feel comfortable, discuss your concerns with this person directly:
  - Be open, straightforward, and honest.
  - Focus the discussion on how the team might help this person be more effective.
  - Be supportive.
- If all of your efforts fail, report your concerns to your instructor:
  - Do not wait too long. The sooner the issue is addressed, the more likely a positive solution can be reached, which in turn will positively impact your workload.
  - Provide specific detail using your documentation.
  - Remember that your instructor's first response will also be to correct the situation in hopes of helping this individual, you, and your team. The first steps will be corrective, not punitive.
  - Remember that you have an obligation to foster a *real team* atmosphere. A superfluous member fails to produce value to the team, reduces team morale, wastes team resources, and endangers project success. Although impossible to prosecute, this is in all practical purposes a form of *embezzlement*. Yes, that is a bold statement, but consider for a moment the negative affects this person can have. This person is directly reducing the team's resources.
  - At this point, it is not your responsibility to correct the problem. Allow your instructor to do so. Regardless of the outcome, it will benefit your team and you will actually be helping this person in the long run.

4.4 Project Time Commitments

The project you are about to undertake is going to require a significant amount of time from each individual on the team.

> It is beneficial to understand how much time is expected before getting started so that you can plan accordingly.

A typical Senior Design course consists of a two semester sequences of either two or three credit courses. Most universities expect between two and three hours of work per credit hour outside of lecture to pass the course; additional effort is expected to earn an A. This means that you should schedule between 4 and 9 hours of work each week depending on your university's guidelines *to pass the course* and possibly slightly more to achieve a desirable grade.

For the sake of discussion here, let us assume your course is a two-credit course and your university expectations are three hours per credit. Your project should require about 6 *effective* hours per week.

There is a difference between "time spent on the project" and "effective hours." Effective hours are time spent either mastering the course objectives or efficiently advancing your project goal. Effective hours might include the following:

- Reading about real team characteristics
- Investigating a new technology
- Designing a new circuit

Additional time that might necessarily be spent on the project but would not count as effective hours might include the following:

- Redeveloping skills that were supposed to be mastered in other courses
- Correcting mistakes

Effort has been made to scope your project so that it requires the appropriate number of hours per week from each student. It is important to understand that it is an *individual* effort. This means that if three of you attend a team meeting for an hour, you did not contribute an hour of effort each. Rather, you contributed 20 minutes of effort (one-third of the total effort). For this reason, you will want to efficiently balance the amount of time spent working individually versus collaboratively. Sometimes working collaboratively is much more effective than working alone. However, if you are able to complete the work as an individual, it is more efficient for the team that you do so.

EXAMPLE 4.3: *Allotting Time to the Project*

This example provides guidelines of how to allot work time to certain project activities.

Writing and proper documenting: full credit hours:
This counts directly toward your overall workload. During weeks which you are writing reports, this effort should be accounted for by a slight reduction in your technical output. Assuming you followed best practices and documented while you were working, this should be a trivial amount of work but should be counted.

EXAMPLE 4.3, continued

Individual technical work: full credit hours for correct work:

Obviously, any time you spend making technical progress on the project directly counts. The design process is often iterative. If you have to make changes, you should ethically consider whether to count that extra time or not. If the iteration is due to your mistake, then you should not count it. If the iteration was unavoidable or due to someone else's mistake, then that is a legitimate contribution to your team.

Team meetings and group work sessions: Hours/attendees:

While it is important that you attend meetings, you will not be contributing during the entire meeting. If you are working a technical problem with a partner, you will only be contributing a fraction of the overall effort. On a *real* team when everyone is contributing equally, you should only count a fraction of this time to the project.

Meetings with advisors and/or instructors: No credit hours:

This is analogous to going to an instructor's office hours; they are help sessions for you. These people are available to provide guidance, but this time does not directly add value to the project.

Lectures: No additional credit hours:

Remember, the guidelines provided above are after you have already attended lecture.

Retroactive documentations, revisions, and/or editing: No credit hours:

If you forget to document something and have to retroactively record it, you were not as efficient as you should have been; this does not count toward your project time. Follow best practices and avoid this. Similarly, when you submit a document, it should be polished and well written. If you need to make corrections, you should complete that on your own time. Avoid this by proofreading your work and by double-checking each other's material.

4.5 Running an Efficient Team Meeting

Meetings are often called a "necessary waste of time." Time spent talking about the project does not directly produce results, yet without regular meetings, a project team would quickly disintegrate into a group of individuals. Effective teams hold short and efficient meetings only as often as necessary. This section will provide guidelines to make your meetings more efficient and effective.

You will conduct a few types of meetings throughout your design process, including the following:

- Regular team meetings
- Weekly advisor meetings
- Technical help sessions
- Sponsor/Sponsor meetings

This section includes descriptions of each of these meetings and guidelines for getting the most out of those meetings. Before discussing the specifics of these meetings, we will look at some universal tips for all meetings.

Universal Tips for Conducting Effective Meetings

Regardless of the type or purpose of the meeting, there are a few practices which should regularly be employed to help your team conduct meetings effectively. These include the following:

Know what the purpose of the meeting is:
Are you meeting to provide status updates, get technical help on a particular problem, or accomplish some specific tasks? Come to the meeting focused on the issue at hand.

Be prepared to deliver on your responsibilities:
One of the most frustratingly wasteful types of meetings is when nothing gets accomplished because one team member was unable to complete a task necessary for the group to make progress. This might include reporting on an item you were assigned to research or delivering a piece of technical progress. It would be better to postpone the meeting if you are unable to deliver your piece rather than assemble unproductively.

Assign a "chair" or "host" for the meeting:
Another common waste of time is when a group of people sit around chatting waiting for the meeting to start. By assigning a person to be "in charge" for the meeting, it will be clear whose responsibility it is to get started. This person is also responsible for making sure the meeting stays on schedule, that the discussion does not ramble, and that all topics are addressed. If the team has previously assigned a team leader, it is often this person who chairs meetings. However, the assignment can rotate among members so long as it is clear who has the responsivity for each meeting. It is rarely the project advisor's responsibility; do not expect your advisor to run your meetings for you. Not only is this a learning experience for you, but you will also be better able to handle this responsibility than your advisor because you will be more familiar with the project.

Ensure everyone contributes:
When appropriate, everyone should contribute something to the meeting. If a particular member is not required to contribute, you should question whether their time might be better spent doing something else. It is relatively easy for everyone to contribute status reports on their responsibilities, but if the meeting is a technical help session, the content might only apply to a subset of the team. Include only those who are needed, and allow the others to make progress on another aspect of the project.

Create and distribute an agenda for every meeting:
Ensure that all members are aware of the meeting's purpose and what they are expected to contribute by providing an agenda. Typically, this should be done 24 hours in advance by the meeting chair. This is also a chance to ask if anyone else needs to put something on the agenda in case the chair missed something important. Your team should create an agenda template to save time in future weeks. Example 4.4 provides a suggested template for a general meeting agenda.

Assign action items:

Ensure that everyone knows what is expected of them before adjourning the meeting. Tasks assigned to an individual are called "Action Items." As a general rule, each individual should be assigned at least one action item every week. Action Items may include both technical and nontechnical items. If an Action Item needs to be assigned to someone, but it is unclear who the task should be assigned to, ask for volunteers.

Start and end on time:

To ensure that you are not late, show up 5 minutes early. Give yourself time to unpack your things, open any necessary software applications, and review the agenda. A few minutes preparing yourself will make the meeting run more efficiently. If someone does show up late, do not review what has already been covered; to do so is disrespectful of everyone else's time; catch this person up after the meeting.

EXAMPLE 4.4: *Recommended Agenda (Template)*

Each team should consider how to run their own meetings. This template is a generic example that will work for most teams, but you might want to modify it slightly so that you are more effective. Remove the curly brackets and enter your information.

Team Name: *{your team's name}*

Agenda for: *{type of meeting, or name of meeting, and date}*

Focus of the meeting: *{provide 1 or 2 sentences to inform everyone of the meeting's purpose}*

Required material: *{provide a bulleted list of who should be prepared with what material}*

Call to Order by *{meeting chair's name}* at *{time and date}*

Review of Gantt Chart:

Item one (*{name}*):

Item two (*{name}*):

Item three (*{name}*):

Review of action items:

Meeting Summary:

Notice: In Sect. 5.3, you will learn that recording Meeting Minutes is an important step in the project documentation process. You *could* use this template to format that recording by simply adding a few bullet points below each topic.

Regular Team meetings

Recall that real teams spend time together and meet in-person when possible. Meeting with your team regularly will foster a healthy working relationship with the other members. However, if you are not careful, this can be an inefficient time sink. Remember, you should only account for a fraction of the time spent in a meeting relative to how many people contributed.

Your weekly team meetings are typically more informal than the other meetings discussed in this section. You may want to consider whether it is necessary to create formal agendas or assign meeting chairs for regularly scheduled team meetings. However, the principles previously discussed still apply: know why you are meeting, be prepared, get started on time and stay focused, and end with action items.

There are many reasons to conduct team meetings that will reduce the team's effort in the long run. These reasons include the following:

- Get to know each other.
- Develop team understanding of goals and objectives.
- Brainstorm project ideas.
- Develop project schedules.
- Get technical help for a teammate.
- Practice presentations.
- System-level testing.

Teams often get together for other reasons as well:

- Completing technical work that could be accomplished individually
- Writing reports
- Status reports (this will be redundant since you will do this during advisor meetings)
- Individual subsystem testing

Ideally, all of your project activity will be as efficient as possible. However, in reality, students often feel the need for support from teammates. Even though some time spent collaborating might be less than efficient, it still might be preferred at a personal level. In fact, some teams intentionally choose to conduct meetings many more meetings than other teams. This is your choice; it is not *wrong* to meet often! Maybe your team is willing to be less efficient during team meetings because it is more enjoyable to work together than to work in isolation. Perhaps reviewing status reports prior to an advisor meeting makes the advisor meeting more efficient. Often meeting to debug a problem to implement the "two heads are better than one" concept may actually prove more efficient in the long run. How you allot your time is your decision; just be sure that you are making an informed one. Relying on teammates to assist in completing individual responsibilities may produce a better *result* but will increase the *time* dedicated to the project.

> In general, if a task can be completed by one individual, involving the entire team
> is an inefficient way of completing the task.

You should conduct your first team meeting as soon as possible to get acquainted with your teammates and to set up the structure of the team. Example 4.5 provides a suggested agenda for your first meeting.

Before concluding the meeting, be sure to internally address and document any concerns related to a member not completing an Action Item or producing a quality result. The primary purpose of doing this is to be intentional about expectations. The secondary reason for doing this is to create a record if a problem persists.

After the first meeting, the format, material, and focus of team meetings will vary. No specific example agenda is provided here due to this wide variety. If you wish to create agendas for these meetings, use the basic structure provided in Example 4.4, but adapt it to the current focus of the team. There is no *correct* format for conducting your team meetings. If you are attempting to stay focused and efficient, and addressing the following items each week, then you are "doing it right:"

- Review the project schedule
- Assignment of Action Items
- Meeting summary

EXAMPLE 4.5: Suggested Agenda for First Team Meeting

Let us get your project team off to a good start. Plan on meeting your team as soon as possible, ideally immediately after team assignments are made. The goal of this meeting is to get acquainted with your team and your project.

Team Name: *TBD*

Agenda for: *{Date} Team Meeting*

Focus of the meeting:

    1.   *Get to know each other*
    2.   *Determine team structure*
    3.   *Define the Problem Statement (1ˢᵗ draft)*

Required Material:

    1.   Individual SWOT analysis
    2.   Individual Problem Statements
    3.   Work schedules

Call to Order at *{time and date}*

Get to know each other:

    1.   Name
         a.   Background/Degree
         b.   Interest in project

> Get everyone talking. This is a chance to begin the team building process. Have FUN!

    2.   Complete SWOT analysis

Determine team structure (this can change at any time)

    1.   Who is the team advisor and sponsor?
    2.   When/Where will the team regularly meet?
    3.   Does anyone want the role of team leader?
    4.   Does anyone want the role of team Secretary?
    5.   Does anyone want the role of team Treasurer?

> These are personal choices and do not have to be made right now.

Develop draft of our problem statement:

    1.   Review individual problem statements
    2.   Complete a team problem statement

> Everyone should contribute. Practice asking for contributions from each person in an encouraging way.

Review of action items:

> Make sure to list specifically who was assigned to and when it must be completed.

Meeting Summary:

> As a team, openly discuss whether everyone contributed equally. The point of doing this is not to point fingers, but to help the team bond.

Weekly Advisor Meeting

Meeting with your advisor regularly is important. Weekly meetings are typically necessary for most teams early in the design sequence. As a team transitions from the PLANNING phase to the EXECUTING phase and has built a strong foundation for the project, it may be better to reduce time spent meeting with your advisor and increase time spent executing the project. However, even then, meeting every 2–3 weeks will be necessary.

During your first advisor meeting, you should plan on the following:

- Introducing your team and getting to know your advisor. You might even take a few minutes to share some personal backgrounds or ask your advisor about their primary research interest.
- Settling on a mutually agreed upon, reoccurring meeting time. Set a schedule for your regularly occurring meetings, at least for the first semester. It is okay to change this when schedules change at the start of second semester.
- Discussing the project's Problem Statement (Chap. 6). It will be important that everyone has a mutual understanding of the Problem Statement so that the entire team pulls in the same direction throughout the duration of the project. Often teams, accustomed to "normal" college coursework, expect the advisor to define the Problem Statement. But this expectation is not valid in Senior Design. **You** have an equally valid opinion when defining the Problem Statement as any other team member, including the advisor. [An exception to this rule exists if the advisor is also the project Sponsor or Sponsor.] The most effective way to approach this discussion is to have an open discussion where everyone on the team (including the advisor) are equals. Do not be afraid to share your thoughts and opinions!

Example 4.6 provides a suggested agenda for your first advisor meeting.

After the first advisor meeting, the purpose of subsequent advisor meetings is to synchronize work among the teams, first by determining whether you are on schedule and then by determining a plan of action if your schedule is slipping. By meeting regularly, schedule slippage can be recognized before negatively impacting the project. At a minimum, the team should review the project Gantt Chart (Sect. 8.2), report on the progress made in the past week, and assign action items for the coming week. It is probable that the Gantt Chart will not be created until after the first couple of meetings. If that is the case, then during those first few meetings, replace this discussion with a review from the previous action items using the recorded meeting minutes.

It is important to be as efficient as possible during these meetings so that all necessary topics are covered each week. A typical meeting should include the following:

- A quick introduction. The meeting chair should call the meeting to order and quickly review what is on the agenda. If there are any significant issues to be addressed during the meeting, it is a good practice to mention those at this time; this will help the chair keep the meeting on track and on time. Notice: This is not a "personal" introduction but rather an introduction to the meeting.
- An overview of the project status. This includes each member discussing their assigned items, whether those assignments were accomplished and whether they have any

pressing concerns to discuss during the meeting. This discussion should directly refer to the Gantt Chart (assuming it has been developed). Allot approximately 2–3 minutes per student for this.

EXAMPLE 4.6: Suggested Agenda for First Advisor Meeting

Shortly after your first team meeting, you should meet with your advisor. This first advisor meeting is primarily an introduction meeting. Here is a suggested agenda for that meeting.

Team Name: *{your team's name if you have one yet}*

Agenda for: *{date}* Advisor Meeting

Focus of the meeting: *First introduction and review of Problem Statement*

Required Material: Draft of Problem Statement, work/class schedules

Call to Order by *{meeting chair's name}* at *{time and date}*

    Team introduction:

    Discussion of a mutually convenient
      weekly meeting schedule:

    Discussion of project solicitation and
      problem statement:

> Take time to get input from everyone. Consider all aspects of the problem. Take notes.

    Improvement of Problem Statement Draft:

> Compile all thoughts into a cohesive statement.

Review of action items:

- Technical discussion. The exact content during the technical discussion will change from week to week. For each new topic, make sure to provide a "top-down" discussion, including the following:
  - How the current topic fits into the big picture. Pro-tip: use the Block Diagram (0) or Requirements Document to do this (0).
  - Any related Action Items and/or due dates.
  - Technical *details*. For example, do not simply tell your advisor "We completed the design of the power circuit" but present the power circuit on a slide (or similar media) and discuss how it works in *detail*. The more detail you can cover during these advisor meetings, the better your results will be when you submit graded assignments.

- Review of action items. End every meeting with a review of action items and verify that each action item is portrayed in the Gantt Chart. Ensure that any course assignments are included in the discussion; remember your advisor does not always know the details of what is expected in the course, it is your responsibility to inform your advisor of these details.

Example 4.7 provides a suggested agenda that can be modified for use during weekly advisor meetings. Consult with your advisor to determine their personal preferences as well; there might be items your advisor wishes to regularly address beyond what is provided here.

EXAMPLE 4.7: *Suggested Agenda for Weekly Advisor Meetings*

After the first advisor meeting, you should focus your meetings on the project schedule and action items. If you need additional time for technical help, schedule follow-up meetings with only the individuals who are working on those technical issues.

Team Name: *{your team's name}*

Agenda for: *{date}* Advisor Meeting

Focus of the meeting: *Status report, help on topic X*

Required Material:

- Each team member should update their portion of the Gantt Chart.
- *{Name}* will provide a few slides on topic X

Call to Order by *{meeting chair's name}* at *{time and date}*

Review of Gantt Chart (or action items if the meeting is held before the team creates the Gantt Chart)

1. *{Name}*
   - Recently tasked to:
   - Recently completed:
   - Still working on:

   > It is not quite time to discuss details of your progress. The team is just providing an overview of the schedule.

2. Repeat for each student

Summary of Technical Progress

1. *{Name}*
   - Progress on *{item}*
   - Need help on *{item}*
   - Plan to *{item}* next

   > Now each person should share details of their technical work. This should eventually include flow charts, circuits, data, etc.

2. Repeat for each student

Review of action items:

Technical Help Sessions

It is probable that you will run into technical issues multiple times throughout the project. This is normal and expected. You may rely on any source, within reason, for help, so long as you document and/or cite that source. Your advisor will be the most common source that you turn to for help, but there are a number of other sources to keep in mind, including the following:

- Instructors of related courses, including your Senior Design course.
- Other students in your Senior Design class.
- Graduate students working in a shared lab space.
- Sponsors: however, you should procure permission prior to using this source.

You may use weekly advisor meetings to get technical help when possible; this likely will be the most time-effect method of solving technical issues. However, remember that the *primary* purpose of the advisor meeting is to monitor the project schedule. A difficult technical concern should not overshadow this purpose. In the case where you run out of time to discuss both the project schedule and the technical issue, a follow-up meeting should be scheduled to resolve the technical issue.

When you are asking for help on a technical issue, make sure to thoroughly explain the problem before expecting any valuable advice. This will likely require the use of figures; bring your schematics, flowcharts, or diagrams to this meeting. Students often forget that the advisor is not nearly as familiar with your project as you are. Rather than jumping directly into the problem, it is important to explain the entire situation. Start with the big picture; maybe this includes a discussion of the project Block Diagram (0) or the Requirements Document (0). Discuss any related components that are working. Then, explain what is not working. Be specific. Rather than say "the microcontroller does not work," explain that "when code is downloaded to the microcontroller it does not seem to run" or "when the code is running, it seems to constantly reset itself." These are two different ways the microcontroller does not work, but the specificity will focus the discussion so that the problem can be effectively identified.

Technical help sessions vary widely in content and format; therefore, it is difficult to provide a useful example agenda. In fact, most technical help sessions should be specifically focused on a single topic and so do not require an agenda at all. Use your judgment to determine whether an agenda is necessary. If so, modify Example 4.4.

Avoiding Common Pit Falls 4.1: Discussing Technical Issues

Occasionally, students will feel that an advisor is "running them in circles" by providing different answers to similar questions. While that feeling is understandable, certainly the advisor is not intentionally doing this.

What is most likely happening is that the student presents a limited or incomplete picture of the issue. The advisor probably does not recognize that there is a broader context and attempts to support the student with an answer.

When the issue arises a second time, the advisor may understand the context differently, or more fully, and provide a different answer. This situation is compounded if the student is asking multiple people the same question in isolation, rather than collaborating with everyone.

To avoid "the runaround," attempt to do the following:

- Provide a full accurate description of the problem.
- As the project advances, ensure you are considering the broader context.
- Use figures to aid your discussion.
- Update everyone involved as the situation evolves.

If your advisor is also your Sponsor, your weekly advisor meetings may replace dedicated sponsor meetings. Assuming your Sponsor is someone else, you will need to regularly conduct meetings with this person as well. These meetings are more formal compared to the other types of meetings presented in this section.

Your team should definitely assign someone to chair this type of meeting; however, the student chairperson should be ready to give up control of the meeting to the Sponsor if that person "takes charge" of the meeting. This is perfectly acceptable, and somewhat normal, but you should be prepared to conduct the meeting yourself rather than wait for the Sponsor to do so.

Double-check your agenda and send out a polished version at least 24 hours prior to the meeting. Make sure to incorporate the sponsor's concerns in the agenda, and ask if they have any additional topics to add. It is also polite to refer to your agenda as a proposal or draft and verify it is acceptable to the sponsor, rather than dictate that you have finalized the agenda without the Sponsor's input.

One of the most important meetings of the entire project will be the first meeting with the sponsor. The primary purpose of this meeting is to get a better understanding of the project's Problem Statement (Chap. 6) and Project Requirements; it is part of a technique called "Requirements Gathering" (Sect. 7.5). The more you understand about the problem and the Sponsor's motivation during this first meeting, the more efficient your team will be during the INITIATING phase.

During this meeting, you will want to address the following issues:

- Personal introductions: **Take a moment to introduce more than your name. Maybe include your degree, background, why you chose this project, or hometown. The specifics of what you share do not matter much, but an intentional effort to be personal will set the team on a good foundation with the sponsor. Record your names and the piece of information that you plan to share directly on the agenda. This will provide the sponsor with a reference for future meetings. Be sure to express an interest and excitement about the project.**
- General project description of the project: **The solicitation which you used to select this project will have provided you with important information about the project. However, it is highly probable that this information was incomplete, unintentionally misleading, or out-of-date. The primary focus of the first sponsor meeting is to clarify exactly what the sponsor desires from the project.**
  It is important that you come to the meeting prepared. To do this, prepare a *draft* of what you *think* the Problem Statement is. Then, be willing to modify, correct, or even eliminate your draft based on your discussions with the sponsor.

Do not initially share your Problem Statement Draft.

Instead, after your personal introductions, transition the conversation to the project by briefly mentioning the solicitation to show that you have come prepared, requesting that the Sponsor describe their goals in their own words. Then, request that the Sponsor describe the project in their own words. Listen to this description, take notes, and consider which portions of your Problem Statement draft need to be modified.

- Problem Motivation: This is a part of an official Problem Statement. The project solicitation and verbal discussion with the sponsor should enlighten you with reasons *why* the sponsor wants the project completed. You will have previously considered this with your team and advisor when completing the Problem Statement draft. Were you correct? Did you really capture the *primary* motivation? Understanding the *why*, at this point, is more important than the *what*.

  After the sponsor has described the project, you should verify that you have understood everything correctly. This step will require skilled communication by one of your teammates. Consider everything you thought you understood about the project prior to the meeting, incorporate what you have just learned from speaking with the sponsor, and summarize what you now understand in your own words. You might be correct, but more likely, the sponsor will modify or correct at least part of your statement. Continue this discussion until you are sure you have correctly understood *why* the project is desired.

- Project goal: This is the second part to an official Problem Statement. After you are sure you understand why the project is desired, focus on what the sponsor desires as the outcome of the project. Use the same technique described for the Problem Motivation. Integrate everything you know so far and describe in your own words what you under-stand to be the project goal. Continue the discussion until you are sure the sponsor and the team are in agreement.

- Project requirements and constraints. It is not quite time to formalize a Requirements Document, but you will need some information during this meeting to get started on the next step. In 0, you will learn that a Requirements Document is organized by Goal → Objectives → Requirements and Constraints → Specifications. Now is not the time to try to categorize all the information correctly, but rather you should simply gather as much information as possible during this meeting and organize it later.
  During the sponsor's discussion of the project description, they likely provided many technical details. Make sure you record these, but to not lose focus on the motivation and goal until those are fully understood. Then, review the technical details you have recorded, and ask whether there are any other concerns to be aware of.

- Plan for team interaction: This is a chance to verify with the sponsor how best to connect with them. Would they prefer to communicate by email, group chats, or a message board? Would they like you to connect with them regularly or just if you have questions or concerns? Sometimes teams prefer to funnel all correspondence through one individual on the team; other times a completely open style of communication is more fitting; if you choose to have one individual handle all correspondence, make sure to copy *everyone* on all communications. Similarly, if you have more than one sponsor, know who to whom you should direct questions. There may be other logistics to cover at this time as well.

Example 4.8 provides a template for the first sponsor meeting. Feel free to modify this as much as necessary so that it is appropriate to you, your project, and your Sponsor. Ensure that you conclude the meeting on time and have a plan for the next meeting.

After the meeting, send a well-written summary of the meeting to the Sponsor and make sure you thank them for their time.

Industry Point of View 4.2: Running Efficient Meetings

I work in the Triangle Research Park in Raleigh, NC. My company recently recognized how inefficient our meetings were and fortunately has attempted to do something about it. Managers rarely had a formal agenda nor not we have and official structure for the meetings. In fact, managers would often arrive late to their own meetings. When the meeting did start, the discussion was unfocused, often the person with important information came unprepared, and we rarely accomplished anything actionable.

I would spend many hours each week sitting in pointless meetings which meant my real work was not being accomplished, and I would have to work evenings and weekends to make up for that time and I can think of few things more demoralizing to a workforce.

Fortunately, our company leaders recognized the impact this was having on productivity. To resolve this issue, a number of new polices were enacted. These included RSVPing to all meeting invites, attending the meeting prepared and showing up on time. Additionally, in order to call a meeting, managers were required to publish an agenda with the meeting invite.

The company put such a high priority on this that they actually included these issues on each employee's (including the manager's) personal evaluation. Evaluations affect promotions and pay raises, so we all started paying more attention to what was coined "meeting etiquette."

Now, our meetings are shorter and more productive. Also, I am not required to attend meetings that do not directly involve my responsibilities. I really value a manager who respects my time!

EXAMPLE 4.8: *Suggested Agenda for First Sponsor Meeting*

It is important to foster a solid working relationship with the sponsor during the first meeting and to generate a firm understanding of the project's Problem Statement.

Team Name: *{your team's name}*

Agenda for: *{date}* Sponsor Meeting

Focus of the meeting: To improve the understanding of the Problem Statement

Required Material:

- Review of the project solicitation
- Problem statement draft

Call to Order by *{meeting chair's name}* at *{time and date}*

Personal Introduction:

  Student team –

    1. *{Name, background}*
    2. *{Name, background}*

  Sponsor –

> Documenting this will provide the Sponsor a reminder for future reference.

Discussion of Problem Statement

  General description *{Sponsor}*

  Review of Project Motivation *{Name}*

  Review of Project Goal *{Name}*

> To transition from the introductions to the Problem Statement, you might say:
>
> "We have read the project solicitation and have developed a draft of the problem statement, but would you mind describing the project in your own words?"

Summary of Project Requirements and Constraints:

Plan for team interaction:

Review of action items:

> Create a list of all technical details to be organized at a later date/time.

4.6 Chapter Summary

In this chapter, you learned about the engineering design team. In particular, you learned what is expected of you in order to contribute to an effective design team. You also learned how to conduct many of the important meetings that will be necessary during Senior Design. Here are the most important takeaways from the chapter:

- A real team is comprised of highly motivated team members who are each willing to contribute to the success of a project.
- Your responsibilities include both a technical contribution to the project development and a personal contribution to the team development.
- Before concluding a team meeting, the team should have an open and honest discussion about each individual's contribution during the meeting summary. In particular, any extraordinary effort or lack of effort by an individual should be recorded. If a pattern develops, steps should immediately be taken to rebalance the team's workload.
- Meetings are necessary to coordinate project efforts. The most important technique you can contribute to making meetings effective and efficient is to come prepared to each specific meeting – this means understanding the purpose of each meeting and adapting accordingly.

Chapter 5: **Project Documentation**

Often students feel that documentation requirements in Senior Design are academic exercises that are time-consuming and pointless "course assignments." This could *not* be further from the truth – a quick reference search on "importance of documentation" will produce literally millions of links to professional sites attesting to the necessity of proper documentation.

You will soon experience one particular example of this. In a few short months, you will have nearly completed your senior project, all that stands between you and graduation will be writing a Final Project Report (FPR). The *last* thing you will be interested in doing during the final weeks of your collegiate career will be digging up scattered information or worse recreating results for this report. Imagine how devastating it would be, if during the last days of the Design Life-Cycle, you are informed that "these design decisions have not been validated well enough" or "that conclusion is not supported by sufficient data!" You may be required to address those shortcomings in the very short time you have left. One of the advantages to consistent and proper documentation is to provide a record of everything you have done so that when you write the report, you will have all the information you need to successfully complete the project.

The Engineering Notebook is used to organize and maintain your design decisions, data, and analysis throughout the project and will be of great use when writing the final report. The notebook will provide a record of good practice, research integrity, and product development that can be used in legal proceedings, patent claims, and knowledge transfer between engineering teams [16].

In this chapter you will learn the following:

- How to maintain a proper Engineering Notebook.
- The types of entries should be included in an Engineering Notebook and how to format them.

C.J. Mettler, *Engineering Design*, https://doi.org/10.1007/978-3-031-23309-8_5

5.1 Project Notebook Background

Engineering Notebooks are *legal* documents similar to a driver's license or real estate contract – really, they are *that* important! While there is some variability between different issuers, they all have a few common characteristics that must be maintained. For example, Montana driver licenses do not look like South Dakota licenses, but they both include the owner's picture, date of birth, and height. Engineering Notebooks may be formatted differently, but they all essentially record the same content.

Common Characteristics of Engineering Notebooks

There is no one "correct" style of notebook (a common theme of the Design Life-Cycle we have already encountered), so it will be important for you to adapt to different standards based on the preferences of your employer or, if you are self-employed, generate your own standard and meticulously adhere to that.

There are, however, common characteristics of Engineering Notebooks that should be considered regardless of the specific standards:

- Engineering notebooks provide a *complete* history of project activities by including regularly spaced entries without any significant time gaps.
- Entries are numbered and/or dated sequentially.
- Entries must include a *topic* and *reference* to Project Objectives.
- Entries provide a clear and detailed description of project activities.

Rules for Maintaining Handwritten Journals [16]

Historically, before electronic records became standardized, there were very strict rules for maintaining handwritten journals. While these rules are becoming obsolete, they are mentioned here because many references still assume you will be using handwritten journals rather than electronics notebooks; it is useful to understand what those references are referring to. Also, the rules provide some insight into the formality, and hopefully the gravity, of engineering record keeping:

- Journals were required to be written in permanently bound notebooks with printed page numbers so that no new pages could be added.
- Entries were required to be written in ink so entries could not be modified after they were created.
- Mistakes could be crossed out but still needed to be readable.
- Any unused/blank spaces were required to be crossed out so nothing could be added at a later date.
- External resources, such as graphs, could be pasted onto the notebook's pages, but all four edges had to be initialed so that the letters appeared on both the original page and the pasted page so that pasted resources could not be exchanged.
- Each entry was signed by the engineer attesting to the validity of that entry.

Example 5.1 provides an example of a handwritten journal entry.

EXAMPLE 5.1: *Example of Handwritten Journal Page [16]*

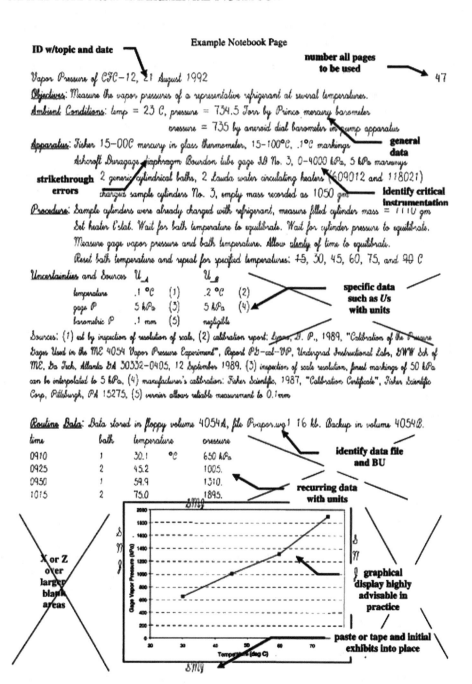

SAMPLE PAGE FROM EXPERIMENTAL NOTEBOOK

In this example page from a laboratory notebook, note that page number, general information, and representative data are entered.

5.2 Electronic Notebook Guidelines

With the standardization of electronic record keeping, many companies migrated from permanently bound handwritten journals to electronic versions. You will need to be able to adjust to whatever standard your company uses. Some companies have even gone so far as to eliminate a corporate policy on notebooks but will still expect that you have personally maintained a consistent record of your work – it is highly advisable to do so using techniques similar to those presented in this textbook.

There are many advantages to electronic record keeping platforms, including the following:

- Electronic copies are easier to share among team members.
- Electronic copies tend to be easier to read.
- Digital data can easily be inserted into the notebook.
- Information can easily be copied out of the notebook and into reports.

However, do not assume that the electronic version is less formal or that it reduces the need for consistent and meticulous record keeping. The migration to electronic versions was meant to increase the effectiveness of records and not eliminate the need for them.

Rules for Maintaining Electronic Notebooks

Here are some general guidelines for maintaining an electronic notebook:

- Create a new notebook entry *every* time you start working on a new project activity – even if this occurs during the same sitting. For example, if you run two tests during the same lab session, it is often useful to have two different, sequential entries.
- Although this is likely an automated process, verify that the entry is time-stamped and include both the date and time you started the activity.
- Record the project objective which the current activity relates to.
- Regularly record *something* about what you are doing. You may want to set a timer for 5 minutes, and when it goes off, enter one or two sentences about what you are currently doing.

5.3 Primary Types of Notebook Entries

Different project activities require slightly different notebook entries. For example, you will need to record design decisions during team meetings and engineering data during technical sessions. Also, some of that information will need to be summarized so that tracking the team's progress is more manageable. The three primary types of entries are the following:

- Meeting Minutes
- Technical Notes
- Action Item Reports

This section will discuss each of these entries and provide a number of examples.

Avoiding Common Pitfalls 5.1: Doubling Documentation Efforts

Many students will record data on paper with the intent of later copying the work into the electronic version. This is setting yourself up for failure in two different ways:

- You are forcing yourself to spend twice as much time record keeping, once for the original and once for the copy.
- It is extremely likely that you will forget to copy the work or lose the hardcopy before you have a chance to create the permanent record.

Follow these steps to reduce the time and effort required for appropriate recording keeping:

- Create the notebook entry as soon as you start work – this should literally be the very first thing you do *every* time so that it becomes a habit.
- Enter your work directly into the electronic version in real time – do not count on updating the document later.
- If you must work out parts of a problem by hand, include one line of text in the electronic record related to the work, take a picture of the hand calculations, and upload the image before the end of the lab session.

Meeting Minutes

Meeting Minutes are used to record design decisions. The minutes also provide a record of who presented specific information, of who had questions and the related answers, and of the team's progress. Most importantly, Meeting Minutes document team-level decisions and assignments (Action Items).

Meeting Minutes are detailed records of the conversation held during a meeting.

Entry Title Format
Each entry should be labeled with a formatted title that includes the date, type of meeting, and who recorded the Meeting Minutes.

EXAMPLE 5.2: Proper Titles for Meeting Minutes

Here are some examples of common types of meetings:

- 10.10.2015 – Team Meeting Notes (Larson)
- 04.22.2016 – Advisor Meeting Notes (Halverson)
- 01.16.2018 – Sponsor Meeting Notes (Machado)

Entry Content
Every entry should include the following contents:

- The purpose of the meeting
- Who attended the meeting
- Detailed notes on the discussion during the meeting (Sect. 4.5)
- Action items, including to whom they were assigned and when they are due

Documenting the exact purpose of the meeting at the top of the Minutes will reduce your time searching for information later. You can scan the meeting's purpose to quickly determine if what you are looking for might be found in that particular document.

EXAMPLE 5.3: *Possible Purposes of Meetings*

Here are some examples of common purposes of meetings you might experience:

- Weekly team meeting
- Regularly scheduled advisor meeting
- Requirement gathering
- Procuring feedback on block diagram
- Updating Sponsor on current progress

It is a common practice to document who attend the meeting; if, under normal circumstances, the entire team attends, you may simply state "Entire Team" rather than list each name. You may also consider only documenting who missed the meeting. Make sure to send an email to anyone who missed the meeting and verify that they plan to review the Minutes.

The bulk of a Meeting Minutes entry should include detailed records of what was discussed. In particular, any decision decisions made during the meeting should be included. Ensure that the record not only records the final decisions but also records the justification for why the decision was made. Often, the justification is more important to record then the decision itself.

Action items are an important way of making sure tasks get accomplished in a timely manner, that responsibilities are spread fairly among team members, and for monitoring project progress.

It is a good practice to immediately assign **Action Items** to an individual as soon as the need is recognized.

Remember, while the Senior Design course instructor will be providing you guidance on what needs to be accomplished for your course grade, you need to independently determine project-specific Action Items in order to complete your project. The assignments provided by the instructor are only the starting point of what needs to be accomplished. These assignments are designed to focus your efforts as appropriate during different stages of the Design Life-Cycle. However, it is not possible for the instructor to specify exactly what you must accomplish for your specific project; only your team, with the assistance of your advisor, can determine that.

EXAMPLE 5.4: *Proper Meeting Minutes #1*

9.29.2020 Weekly Advisor Meeting (Jesse)
Tuesday, September 29, 2020   4:34 PM

Participants:
Advisor, Jesse, Russell, Stephanie

Purpose: Weekly meeting
        Discuss access to technical loading data

Discussed:
- Updated Advisor on initial project research
- Prepared for meeting with Administrate support staff
  - Need access to campus's building load data
  - Need information on rooftop areas
- Next step is to select loads to design for
  - Each person picks top choices,
  - Be able to defend it, pros and cons
  - As a group pick three
- Must figure out how best to administer agendas for future meetings
- First time looking at Gantt Chart will be on October 6th

Action Items:
- Finish filling out Smartsheets - Jesse, Russell, Stephanie
- Schedule a meeting with Admin staff - Stephanie
- Decide on *simulation* program - Jesse, Russell
- Make initial design selections - Stephanie, Jesse, Russell

Summary:
- Everyone was prepared for this meeting.
- Slight concern regarding how responsive the Admin rep might be.
  We will need to evaluate this next week.

Analysis

This is an example of Meeting Minutes taken early in a project. Normally, there should be more technical information in an entry. In this case, the lack of technical information is understandable due to the infancy of the project.

*The **Purpose** could have been left as "Weekly Meeting," but including the primary concern or discussion point improves the quality of this document. This is something the team did well throughout the project.*

*Few of the items **Discussed** included sufficient detail. For example, the first entry is very generic. Recording what topics were researched would have improved this document.*

***Action Items** should be assigned to an individual; however, during this meeting, most of the action items were assigned to a group. This is an example of inefficient use of resources. The team improved this aspect of their team management over the course of the project.*

EXAMPLE 5.4, continued

*The **Summary** provided a record of a concern related to an important stakeholder. Communication with this person continued to be a challenge. But the team's documentation showed a track record of their attempts to improve the situation and were not penalized for their inability to procure some important information.*

EXAMPLE 5.5: Proper Meeting Minutes #2

03.05.2021 Advisor Meeting (Shawn)

Friday, March 5, 2021   10:30 AM

Attendance: Brooke, Dylan, Shawn, Zeke, Advisor

Meeting Time: 10:30am - 11am

Purpose: Current status and questions for prototyping and testing

Meeting:
- **Housing:** Dylan will be able to start printing the housing this weekend and the design is finished/finalized.
- **Transceiver:** Had some issues with the transceiver but those were fixed, had to do with soldering issues and not ideal breakout board layout.
- **PCB:** Putting last parts of the layout together and Brooke will finalize with Shawn and Jon for electrical layout and Zeke and Dylan for checking dimensions. Possible bottleneck with Mettler, either order Monday morning or Wednesday morning.
- **Meeting with Sponsors:** Want to wait until we have something to show but soon enough to allow for any changes they may have. This could fall into the week of March 22nd -26th earlier in the week to allow time for the PCB to arrive. Plan to invite the sponsors to our weekly meeting on the 26th. Start meeting at 10:00 a.m. and invite the sponsors to meet at 10:30 a.m.
- **In-Circuit Programming:** Asked about the different ways of programming the chip and what option would be best for the sponsors. Can program the chip using the cheaper option since this is a prototype.
- **Next Week Meeting:** Not planning to meet with Advisor next week unless something urgent comes up.

Action Items:
- ☐ Jon: Write in the report the options for the in-circuit programming.
- ☐ Brooke: Finalize the PCB design and order with Mettler next week.
- ☐ Shawn: Finish up the programming of the display
- ☐ Zeke and Dylan: Incorporate new circuitry topology into housing development

Summary:
- Jon was unable to attend the meeting, Brooke will follow up with him on his action item.
- Shawn was unable to complete last week's action on display programming. He made progress and the team in not concern about this effort.
- Zeke and Dylan continue to work together, but are attempting to divide future work for efficiency

---

Analysis

This example comes from a meeting which occurred well into the technical development of the project. Sufficient details were recorded in an easy-to-follow format. Action Items were appropriately assigned and a Meeting Summary was provided. It might be a little concerning that two team members have continued to share Action Items this far into the project. But the team discussed that situation and did not seem to be overly concerned.

Technical Notes

The majority of the entries in an Engineering Notebook are technical notes. This is where all technical decisions are recorded, designs are documented, data is collected, and conclusions are drawn.

Technical Notes are records of individual contributions toward project development.

Entry Title Format

The title of a Technical Note is formatted similar to the Meeting Minutes' titles. Each title should include the date, a short phrase to describe the work, and who was primarily responsible for the entry. Each title should have **one** name even if a group of people worked on that task, credit for the work can be given in the entry itself – the title records who was primarily responsible for the entry.

EXAMPLE 5.6: *Proper Technical Entry Titles*

Here are some examples of Technical Note titles:

- 11.07.2015 – Power conversion design (Larson)
- 03.21.2016 – Forward movement code (Halverson)
- 02.10.2018 – PID control of system rotation (Machado)

Relationship to Project Activities

Since the primary audience of these entries is not you, it is important to inform the reader of how the activities recorded on a particular page fit into the broad scope of the project. This should primarily be a direct reference to the Requirements Document (0). If this is difficult for you to articulate, then you should take a moment and question why effort is being spent on that particular task. There are two common reasons for this struggle as follows:

- You forgot to consider how the work you are currently focused on relates to the project requirements.
- You are actually not focused on a project priority and so are likely not being efficient with your time.

Obviously, the first reasons can easily be addressed. The second reason, however, presents a greater concern. You should stop and ask yourself, *why* are you conducting time on an effort that is not a priority? Certainly, you are not doing so intentionally – so, what caused the situation? A lack of clear priorities, failure to monitor the schedule, weak or missing Action Items, or some other form of miscommunication? Once the source of the problem has been identified, refocusing the effort should not be all that difficult.

EXAMPLE 5.7: *Project Relationship References*

Here are some examples of possible project relationships you might consider:

* Requirement 3.1: System power must be supplied from wall voltage.
* Specification 2.1.3: System must move forward at least 2 m/minute.
* Objective 1: System must move hard drives into position to be kitted.

In some cases, entries may not relate *directly* to the Requirements Document. It is acceptable to create additional categories of work. Here are some examples:

* Market research
* Cost analysis

These may not *seem* to exactly fit in the "Technical" Entry category. However, they *are* documenting engineering efforts and so they do, in fact, belong here.

Technical Progress Documentation

The content of each entry should be put into context with some simple phrases or headings and then the details should be recorded. Do not just simply record a bunch of technical details. Deciphering what you were doing months after the entry was created will be cumbersome if you do not provide the context. The details should be clear and easily understood. It is not necessary to write a full report in each entry, but you should provide enough structure so that a reviewer can understand the documentation.

EXAMPLE 5.8: *Potential Technical Content*

Here are some examples of Technical Note titles:

* Pros/cons list of possible components in an alternative selection (which include links to data sheets)
* Flowcharts for sections of code that was developed
* Circuit schematics or three-dimensional graphics of structures
* Calculations
* Tables of test data

Conclusions

Briefly summarize your session in one or two sentences. Include a list of Action Items for yourself so it is clear what you need to do next. Remember, sometimes you might put a lot of work into the project in a short amount of time and then focus on other course work for many days before coming back to the project. It is convenient to quickly look back at the previous entry's summary to remind yourself where you were and what needs to be done next.

Example 5.9 displays a technical entry which could have been improved and Example 5.10 presents a proper entry.

EXAMPLE 5.9: Poor Technical Entry

04.02.2019 Final Testing Program for the MSP430 (redacted)

Relationship to overall Project:
This is the final program needed to operate the mMSP430 prototype vehicle and show it off at the design fair. This is a required component of the course.

Requirement:
　　Requirement 4.1

Technical Progress:

```
#include <Servo.h>
#define LED RED_LED
const int buzzerPin = 8; //identify which pin is connected to the buzzer
int adPin = A2;
Servo servoleft;
Servo servoright;
int LedPin = 11;
void setup() {
pinMode(adPin,INPUT); //IR Receiver

  servoleft.attach(2); //attach the left servo to pin 2
  servoright.attach(3); //attach the right servo to pin 3
  pinMode(LED,OUTPUT);
  digitalWrite(LED,HIGH); //board power indicator
  halt();
```

Analysis

This is a poor example of a technical entry, so the student's name was redacted.

*First, notice that the **Relationship** does not match the title. The title suggests this entry would record an important test. The **Relationship** accurately states that this is just a record of the final code.*

It is unnecessary to copy/paste large sections of code into the Notebook. A link or reference to the latest version of code would suffice. What should be recorded is a high-level explanation of the code, perhaps in the form of a flowchart.

Furthermore, without further documentation, we cannot tell what Requirement 4.1 is. Simply listing the requirement number is not sufficient when we look back months or years after the document was created. Also, if this is the final code, it must surely relate to more than a single Requirement.

Unfortunately, there is a larger problem with this entry. The title correctly conveys the intent of this entry. The team did in fact perform their final test on this day. What was the test set up? What data was collected? What were the results?

These issues strongly suggest that the team was simply trying to create a Notebook entry but was not serious about documentation. This type of effort is sometimes called "bean-counting," meaning they were more concerned about the appearance of the Notebook than they were about the documentation of the project. They were more focused on their grade than their results.

EXAMPLE 5.9, continued

At the end of this project, there were some questions about the team's final results, and they were asked to produce the data they used to derive their conclusions. Since they had not recorded the data, they were unable to support their conclusions. This directly impacted their project evaluation and final grade.

EXAMPLE 5.10: *Proper Technical Entry*

3.4.2021 Wiring Corrections - Elinore

Thursday, March 4, 2021   10:39 AM

Project Relationship: Recalculate current specifications to ensure a functioning circuit

Requirement 3.1.2: Storage system allows for utilization of 16 hours per week (9.5%)
Requirement 3.2.1: Remain at 80 % efficiency of storage for at least 7 years
Requirement 3.2.2: Mechanical and electrical integrity to last for at least 7 years

Action Items:
- Wires were getting hot during the circuit tests – verify we have the correct gauge wire, determine the correct gauge if we not
- Verify Calculations with our Advisor
- Verify calculations for final nanobubble cord choice with Maryam and Barry
- Email Mettler about purchasing additional length of wire

Below is a document with my calculations for the battery module.

Batter
Current Fe...

> Note: the original entry is continued on the next page.

Actions Taken:
- I have redone the current calculations for the battery modules based on our run time and have submitted them to our Advisor for verification. I have run into one snag which is that we are not sure how long our batteries will take to charge since the power from the panels will vary with sun exposure. Since our current calculations will vary with the length of time that they are run, I cannot pin down a certain current value. I am currently talking to our Advisor about this and we have a meeting with a PV-expert tomorrow morning and I plan to ask him as well.
- I have discussed with Sponsor our plan for purchasing – hopefully I can sort out my problem with the current today or tomorrow so that we can make a trip to Lowe's either Saturday or Sunday.
- I am waiting for our Advisor's approval before emailing Barry.

EXAMPLE 5.10, continued

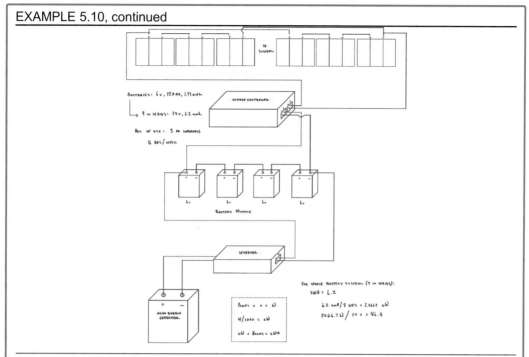

Analysis

A quick glance at this technical entry tells us the primary focus of the work (wiring corrections), who was responsible (Elinore), and how this effort is related to the project (the list of requirements).

This student went further than the basic requirements and included which Action Items she was addressing by this work and provided a summary of her accomplishments. This is an interesting and useful technique to provide context to the work.

The point of this entry was to determine the wire gauge required on Elinore's project. She included a PDF link to her calculations. This entry could have been improved had she actually displayed the calculations here. Or, at the very least, documenting the results here would have been helpful. A reviewer is not going to enjoy tracking additional documentation (such as the PDF) to find basic information.

Although the full set of calculations would have been useful, Elinore did provide a very clear wiring diagram and summary information. This diagram is neatly labeled and color coded. Elinore's effort in documenting this work became very useful as a reference when the system was installed at the Sponsor's test site months after the entry was created.

Action Item Reports

At the end of each week, it is useful to summarize the weekly progress with an Action Item Report. This will provide a quick reference for the following:

- What was supposed to be done during the previous week.
- What was actually completed during the previous week.
- What was assigned to be completed in the upcoming weeks.

Entry Title
Provide each report with a date, the phrase "Action Report," and your name.

> ## EXAMPLE 5.11: Action Item Report Titles
>
> Here are some examples of Technical Note titles:
>
> - 11.09.2015 – Action Report (Larson)
> - 03.27.2016 – Action Report (Halverson)
> - 02.14.2018 – Action Report (Machado)

Entry Content
There are three simple categories to include the following:

- Last week I was assigned to do the following:
  - Copy and paste the Action Items at the end of last week's Action Item report.
  - Use a numbered list to keep organized.
- My progress this week includes the following:
  - List the same numbered items from the previous category.
  - For each item, record either the following:
    Completed, along with a link to the technical entry page in which that material was recorded.
    Not completed, along with a brief explanation of why
  - If any tasks were addressed which "Last week I was assigned" category, include them at the bottom of this list.
- My action items for the coming week are the following:
  - Create a new numbered list.
  - The first items on this list should be any actions from the previous category for which you recorded *not completed*.
  - Copy all your action items from any of the Technical Notes and/or Meeting Minutes; make sure to include the due dates.
  - This is the category which you will directly copy and paste into next week's first category.

Action Items reports, while important and useful, are one of those "small items" which are easy to forget. Many students find it useful to create a calendar reminder to complete the report at a convenient time toward the end of each week.

During the first few weeks of a project, there is not be much technical progress made, so the initial Action Items will primarily address "project-based" concerns. However, as the project develops, the primary focus of the Action Item Report should be on the *technical* progress.

Example 5.12 displays a set of properly formatted Action Item Reports. Notice how little effort an Action Item Report should require. The first category (Last week I was assigned) is a direct copy/paste from the previous week's Action Items. The second category (My progress this week includes) starts with a copy/paste from the first category and then primarily includes links to preexisting documentation. The final category is a copy/paste of Action Items already documented in other sources. The only real effort required is a quick justification of why any Action Item was not completed and how that issue will be handled.

5.4 Chapter Summary

In this chapter, you learned about project documentation. Proper and thorough documentation is a critical component of a large effort like your Senior Design project. The documentation substantiates the effort you put into the project and should be used as a reference during presentations and within reports.

> In fact, making claims about project progress that are not substantiated by project documentation is an ethical violation equivalent to plagiarism!

Proper documentation provides a record of events which will make writing the final report months after design decisions were made significantly easier. It also provides an undisputable record of the effort you put into the project and evidence that you used best practices and made reasonable choices while developing the product; this can be absolutely critical if the project is not 100% successful; proper documentation can be used as evidence that you did all you could do given the time and resources available. Conversely, poor documentation substantiates the perception that you personally were at fault for project shortcomings either due to making invalid design decisions or through a lack of effort.

Here are the most important takeaways from the chapter:

- Electronic documentation reduces the effort required to document your project.
- To further reduce your documentation efforts, document little bits of work often and consistently.
- Start every project session by opening your Engineering Notebook and creating the appropriate entry.
- There are three types of entries you should regularly contribute:
  - Meeting Minutes
  - Technical Notes
  - Action Item Reports
- Each entry should be properly titled and formatted.

EXAMPLE 5.12: *Proper Action Item Reports*

<u>01.26.2020: Action Report (Joe)</u>

Last week I was assigned:
1. Begin setup of single laser system
2. Write verification plan
3. Make sure parts are being ordered

My progress this week includes:
4. Basic setup of single laser system
5. Wrote verification plan
6. Ordered more parts

My action items for this week are:
7. Execute polarization
8. Incorporate polarizers
9. Test polarization of light channels

<u>02.02.2020: Action Report (Joe)</u>

Last week I was assigned:
A. Execute polarization
B. Incorporate polarizers
C. Test polarization of light channels

My progress this week includes:
D. CDR presentation Thursday 01.28
E. Reorganize Gantt Chart
F. Order polarizers

My action items for this week are:
G. Execute polarization
H. Test polarization of light channels
I. Begin image processing coding

Analysis

Here we see the Action Item Reports from a student in two consecutive weeks.

Consider the report on the left first. Joe was productive this week.

- *During the previous week, he was assigned three action items to complete during the week of 01.26.2020 (items 1–3).*
- *We can see that he did exactly that from his "My progress" list (items 4–6). However, the numbering in the second category should have matched the numbering in the first category. That way we can easily see that Action Item #1 was assigned one week and completed in the next.*

Now consider the report on the right. Here there is an issue that must be addressed.

- *Joe appropriately copied and pasted items 7–9 into items A–C.*
- *Ideally, under the "My progress" category, we should see items A–C repeated with a* **complete/not complete** *comment. Instead, we see new tasks.*
- *Joe did not make any technical progress this week. He worked on the Critical Design Review (CDR) and Gantt Chart and purchased some equipment.*
- *What Joe recorded as tasks D–F are perfectly fine to have included, but they should have been cascaded below the report on items A–C.*

Not completing a previous week's Action Items is not a negative concern if:

- *It happens occasionally.*
- *The team agrees Joe's priorities should be focus elsewhere.*
- *Joe has a documented plan to make up the lost time.*

5.5 Case Studies

Let us look at a few notebook entries developed by our case study teams.

Augmented Reality Sandbox

Figure 5.1 presents a technical entry from the EXECUTING phase of the AR Sandbox project. Students occasionally struggle to determine what to include in a technical entry when the work was primarily software-based. This is a good example to follow:

- What went well: James provided a succinct summary of what was accomplished during this work session. The "primary issue" is well stated and provides the reader context to understand the remaining points. James did not provide the details of the

---

Integrate 3D Corn Model (James)

Thursday, January 19, 2017   11:36 AM

Primary Issue: Corn models have been rendered in VRUI, now that code must be merged with the source code.

Solution:
- Created a new class called CornRenderer
- Copied most of the code from the VruiCorn application into the source file for the new class.
- Worked out compatibility issues and investigating positioning of the newly rendered corn.
- Finally got the corn model to show up in the sandbox, fig. 1 (black dots in center).

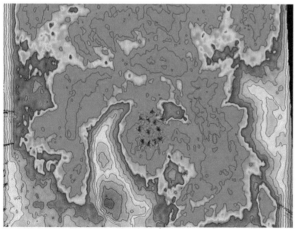

Figure 1: Corn Rendered in the ARsandbox

Action Items:

- Investigate –fpv switch and positioning problems.

Figure 5.1 Technical Notebook Entry from AR Sandbox

code itself, rather he explained what the code accomplished. He also provided results, in this case the image.

- What could have been improved: There are two issues that James could have improved. First, the "project relationship" is not addressed here. James should have included a Requirements Document item (Spec or Req number) at the top of this entry. Second, maintaining version control on a software project is important. James should have included filenames of the newly developed (or improved) code with version-control numbers in this entry.

Smart Flow rate Valve

Figure 5.2 presents a software-related technical entry from the Smart Valve team:

Compensator Code (Keith)

Saturday, February 24, 2018   3:55 PM

Relationship to project:
   Objective 2 - System must accurately dispense liquid
      Req2.1 - Control system must be responsive

      (note: the related specs define the parameters of the control system)

Project Context:
   The theoretical controller design must be implemented in code
   We plan to use MATLAB/Simulink

Technical Entry:

Started writing code for PI compensator

   Defined gain parameters w/ #define for ease of change for development or sponsor use

   Moved motor control to be ran each time through main code, will eventually be checking position feedback from encode at the start of motor control section

   Plan to figure out how to make timers work for necessary timing delays on motor driver

      Should be able to start a timer that will allow coding without wasteful delays

Added slight modifications to Simulink diagram for better accuracy

   Changed feedback path to be actual flow output from valve rather than linearized flow output

   Move gain scalar from feedback path so that input to system was actually based on flow rather than input being position scaled by linear relationship to flow

   Discovered important evaluation factor for the compensator while expanding test vector to test nonlinear region response

      Changing the gains so that the system response to a positive increase in desired flow causes the valve position to overshoot on a decrease of desired flow

      The gains can be changed so that valve position does not overshoot on decreasing flow but the system takes longer to settle on an increase in desired flow

   Tested effects of adding a nonlinear disturbance (sin(t))

Figure 5.2 Technical Notebook Entry from Smart Valve

- **What went well:** The heading of Keith's entry puts the material into context. The Requirements Document items related to this entry are included. Keith also included a "context" segment in most of his entries to further clarify what he was doing. His explanation of the work was very detailed.
- **What could have been improved:** Keith's entry does not provide any results nor action items. We cannot know whether Keith's work was successful or not. In this case, typical control-system graphs (which relate desired inputs to actual outputs) for a variety of commands were produced by the software. These would have been useful results to include here.

Robotic ESD Testing Apparatus

Figure 5.3 presents a debugging-related technical entry from the RESDTA team:

Motor Performance (Leah)

Wednesday, March 22, 2017   1:33 PM

Relationship to project: Requirement 2.2 - RESDTA must move in and out of test envelope quickly

Motors were meeting specifications, but when system was relocated motors performance became erratic.

- Measured both signal to driver and one pole of X motor (blue = control, Red = motor pole):
  - Motor signal was erratic and noisy coming from driver (left picture below)
    - Signal was measured from driver output directly (should not be noise)
  - Some noise on control signal- potential issue?

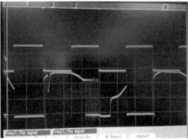

- Connected different driver (ST-4045-A1) to one of the Nema 23 motors
  - Requires 5V signals to run (would need level-shifters if we changed drivers)
  - Has optically isolated inputs (would be nice to reduce potential noise)
  - Pins wired up (SW6 ON, all others off to set current to 0.6A and single step mode):

  - Coils connected for Nema23 and power supply connected to motor supply

- Results: Motor ran smoothly in both directions at both slow and fast speeds

- Conclusion:
  - Drivers damaged due to movement of motors when unpowered or powered but not enabled
  - If we continue to use these drivers, protection must be included or we have to state that the machine can't be moved without power unless of emergency at which point drivers may need to be replaced

Figure 5.3 Technical Notebook Entry from RESDTA

- What went well: Leah provides us with a clear understanding of her work. She starts this entry by explaining the problem she is attempting to address (the motors quit working). She explains what she did to investigate the issue (measured signals) and included a screenshot of data. She drew conclusions from the data (noise was present) and explains her next step (connected to a different driver). She again documents results (motors ran smoothly) and – most notably – draws conclusions from the results (manually moving the motors must have caused damage). The entry is concise but informative. Leah had an excellent habit of explaining what she was doing in a bulletized list but also providing context or her personal thoughts in parenthesis after each bullet.
- What could have been improved: A schematic or picture of the test step-up would have been useful. Also, Leah presented a screenshot of data from the first test but failed to do so for the improved test. Documenting the comparison would have been useful.

PART III: THE INITIATING PHASE

"Would you tell me which way to go from here?" asked Alice.
"That depends a good deal on where you want go," responded the Cat.
"I don't care where, I just do not want to be here," said Alice.
"Then, it doesn't matter which way you go," retorted the Cat.
– Alice's Adventures in Wonderland, by Lewis Carroll

"I had ambition not only to go further than any man had ever been before; but as far as it was possible for any man to go."
– Captain James Cook

Captain James Cook, a Commander in the British navy, traveled further during his three exploration voyages between 1769 and 1779 than any human being had ever traveled, or would ever travel again, until the age of space exploration [17]. He spent years crisscrossing the world's oceans on circuitous routes that led to the discovery[1] of more land than any other human in history. He is credited with discovering Australia – but he was in search of a theoretical place, referred to at that time as Terra Australis Incognita (unknown land of the South), what we now know as Antarctica [18]. He made a second, much longer and more expensive voyage to again attempt to find Terra Australis Incognita. On that voyage, he discovered New Zealand, South Georgia Island, and a number of smaller islands in both the Pacific and Antarctic Oceans but failed to achieve his primary objective. In fact, he circled Antarctica three times on that voyage and came within a few miles of the land mass, but each time turned back at the last moment. He is credited with discovering the Hawaiian Archipelago on his third voyage – but, he was in search of the fabled Northwest Passage [19]. Maps of his voyages portray that he fulfilled his true ambition, *traveling as far as it was possible*, and made significant contributions to navigation and cartography, but they also show that he never accomplished any of his primary objectives.

Compare this to another of history's great adventure stories – that of Ernest Shackleton [20]. After his ship, the *Endurance*, was trapped and crushed by ice floes off the Antarctic coast in January 1915, he and his crew spent over a year trying to survive on the ice. In April of 1916, the ice floe upon which they were stranded started to break up as it moved north. With few options left, Shackleton ordered his men to sail three small lifeboats a short distance to Elephant Island. From there he and the best of his crew navigated one of the lifeboats, the *James Caird*, 1300 km through the world's most violent waters using precise mathematics and navigation techniques to affect rescue at a whaling station on South Georgia Island. South Georgia Island is 165 km long, meaning that a miscalculation of $\pm 3.6°$ over the course of the month-long voyage would send the small boat past the island and into the South Atlantic Ocean. The next opportunity to find land would have been South Africa, nearly 5400 miles further. With no hope of turning around in those unforgiving seas their only chance of survival was to land on that island on their first try. Their course was straight and true and is still considered one of the greatest boat journeys of all time [21]. They landed on South Georgia Island on May 10th, 1916, returned to Elephant Island to rescue the remainder of the crew, and returned to England without losing a single life in their 3-year journey.

---

[1] It should be noted that the peoples indigenous to the land Cpt. Cook "discovered" were well aware of the existence of that land long before his arrival.

Compare the stories of these two master-navigators. Both accomplished great things and earned their place in history. Cook's charter from the British Admiralty did not provide him with a specific scope for his voyage, nor did he have a limitation on time. His resources were essentially unlimited as he scavenged what he needed from the bountiful lands he explored. He was free to adventure and explore as far as his ambition took him. Shackleton's mission was quite different. His scope was unmistakable: South Georgia. Margin of error: minuscule. Consequences of failure: certain death. His time and resources were tightly bound by what the *James Caird* could carry: stores and supplies for 1 month's voyage.

Cook was a master scientist – discovering the unknown, adding a wealth of information to the accepted body of knowledge. Shackleton was the consummate engineer – efficiently achieving great things with limited resources. With all due respect to scientists, the purpose of *this* textbook is to teach *engineering design*.

When the results matter, when the resources are limited, or when the timeline is constrained, there is no room for guess work. Engineers cannot wonder. The destination must be clearly understood, the route carefully charted, and results meticulously monitored for any unexpected deviation to the plan. During the INITIATING phase of a project, the Project Motivation and Goal will be identified. These high-level concepts will be supported with Objective Statements and further defined with detailed Requirements and Specifications. The final solution will be constrained with any necessary limitations and a project schedule will be outlined. During the PLANNING phase, the focus of the next part of this textbook, the design plans will be carefully charted. During the EXECUTING phase, covered in Part V of the textbook, the plan will be enacted, and the results carefully monitored.

There are four primary processes that must be addressed during the INITIATING phase, each of which has a formal structure used to present the outcome. The following table lists these processes and structures. The chapters in this section will aid you in completing each process of the INITIATING phase.

Processes of the INITIATING phase	Presentation Structure
The problem must be carefully investigated and full understood.	Problem Statement
Acceptable solutions to the problem must be carefully defined.	Requirements Document
All Stakeholders must be convinced that the engineering team has the knowledge and ability to solve the problem	Project Kickoff Meeting
A legal agreement or contract must be produced To document the project definition and protect The interests of all stakeholders.	Project Charter

Chapter 6: **The Problem Statement**

The first real step of any project is to carefully articulate the **Problem Statement**, which is comprised of two parts, the **Project Motivation** and the **Project Goal** – in layman's terms "what is the issue and what needs to be done about it."

If pressed to define the *Endurance* crew's "Problem Statement," Ernest Shackleton might have stated, "There is no chance of rescue on Elephant Island. Therefore, we must make a voyage to the inhabited South Georgia Island." Motivation: no help is coming; Goal – navigate to South Georgia Island. Obvious, right? But, consider for a moment the tremendous struggle which Shackleton must have endured before steadfastly arriving at this Problem Statement.

The party's motivation could not have been an easy one to choose. They had just successfully spent a year on the ice pack – and they actually thrived, food was plentiful, there was no shortage of fuel, fresh water was readily available, and now Elephant Island provided an abundance of resources compared to the ice pack. Compare that to the uncertainty of putting back to sea! There must have been at least some votes for a motivation statement "Let's survive here first."

Now consider the Goal Statement. If the vote for the "Let's survive" motivation statement had carried the day, the goal of navigating to South Georgia Island would have made no sense. Even after the motivation was "self-rescue," the goal still was not necessarily clear – South Georgia was not the closest inhabited land; it was just the easiest to navigate to.

The point of this analogy is this: decisions flow in a top-down structure. Without carefully considering *which* challenge to tackle, and specifically articulating your final decision, you will not know where to start your design process nor be able to determine whether you successfully finished the design. Once the motivation is clearly defined, we must specify what it is we are going to do about it. A clear goal is necessary to weigh all future design decisions against.

© The Author(s), under exclusive license to Springer Nature Switzerland AG 2023
C.J. Mettler, *Engineering Design*, https://doi.org/10.1007/978-3-031-23309-8_6

Avoiding Common Pitfalls 6.1 Focus on the Problem

Students often struggle to write their Problem Statement because they are too focused on the *solution*, rather than the *problem*.

> It is vital that all efforts during this initial Design Life-Cycle processes be focused on defining the *problem* as opposed to the *solution*!

We must understand the problem before the appropriate solution can be designed. This is so important that it will be covered again at the end of the chapter in Avoiding Common Pitfalls 6.4.

In this chapter you will learn the following:

- How to define a proper Motivation Statement.
- How to define a proper Goal Statement.
- How to define proper Objective Statements.

6.1 Project Motivation Statements

A clear understanding of the Sponsor's *real* concern is critical if you are to select the optimal design choices. As students, it is easy, and reasonable, to assume that what you are told is exactly what you are meant to do – literally every academic problem you have ever encountered has been provided to you with the intent to describe exactly what you were supposed to accomplish. But those problems are not real-world problems; instructors have spent time carefully creating a specific problem in order to specifically evaluate your answer. Real Sponsors will not have been so careful in soliciting a project even if they could. Certainly, they do not mean to deceive you, but the limitations of our language skills lead to imprecise communication, and limited time or resources limit the amount of thought dedicated to developing the problem. It is the engineer's job to investigate the problem thoroughly so that the *real* problem is understood.

EXAMPLE 6.1 Clear Motivations

Shane, an engineering manager at Raven, has a long history of supporting Senior Design projects of the highest caliber. He is an experienced engineer and wholeheartedly supports student education – he is an *ideal* Sponsor for a student team to work for.

In 2016, Shane pitched the following project description (paraphrased for brevity):

We need a robotic arm developed that will carry a set of sensors over the top of a crop canopy and measure the height of the crop and, simultaneously, measure the distance between the arm and the top of the canopy. This will then be implemented in sprayer-implements to automatically adjust the height of a sprayer while traveling across a crop-field. It is necessary to select a set of possible sensors and then test those sensors to determine which is the optimal type to use in the final product.

Notice, there is no Motivation Statement in this description (most initial project solicitations do not include one). The Motivation might reasonably be assumed to be:

It is beneficial to maintain a consistent distance between a sprayer's arm and the top of crop canopies to provide even coverage when dispensing pesticides.

While this was, in fact, the company's overarching Motivation, it was not the Motivation for this particular student project. A more accurate representation of Shane's Motivation was as follows:

Simultaneously detecting the crop canopy and the ground directly underneath is a difficult task for most sensors, the optimal method of doing this is unclear.

Both of these Motivation Statements can reasonably be derived from the same project description but will lead to very different projects. We will continue to look at this situation and how the two Motivations might drive different Project Objectives in the next section.

Developing Accurate Motivation Statements

Your initial responsibility as a design engineer is to develop an accurate and meaningful understanding of the Sponsor's problem. But how do you do this this if you cannot completely rely on what the Sponsor tells you? Here are some tips:

- Carefully read the project solicitation but keep an open mind – there is a lot of work to do before any of it is official.
- Generate a mental picture of the *problem*, as opposed to the *solution*, and be ready to modify it as you develop further understanding.
- Meet with the Sponsor and focus the meeting on the problem. Start the meeting off with "would you start by explaining the challenge you would like us to address." Notice, this is very different from starting with "what we can do for you" or "what would you like us to design." These questions skip over the problem and jump to solutions.
- Ask a lot of questions and, in particular, ask "why?"
- Recognize that the information you initially collect will be partly Motivation, Background, Description, and Solution (upcoming sections will define each of these categories). You will need time to digest and categorize the disjointed pieces of information.
- Summarize and verify your understanding. Your Motivation Statement should be summarized in one or two sentences. You may need to provide more information to put the Motivation into context eventually, but the Motivation itself should be very concise.

Avoiding Common Pitfalls 6.2 Accurately Determine Motives

Here are the most common mistakes when writing Motivation Statements:

- The statement is more related to a solution than a problem.
  Prevent this mistake by avoiding action verbs such as build or design.
- The statement does not completely represent the Sponsor's concerns.
  Prevent this by verifying the statement with the Sponsor. Try asking "So, your primary Motivation for needing this product is. . . ."
- The statements are background information rather than Motivation.
  Prevent this by forcing yourself to phrase the Motivation Statement in one or two at most sentences. If you need more text than that, you are likely providing background information.
- The statement describes what needs to be accomplished, as opposed to why the work needs to be accomplished. Prevent this mistake by focusing on the environment where the solution will eventually be implemented, rather than the solution itself.

Industry Point of View 6.1 Project Motivations Dictate Solutions

As a Product Manager for a large computer company, there often were times where two divisions of the company had conflicting requirements for the next generation of disk drives. The disk drive industry was in the process of converting from 5 1/4″ drives to 3 1/2″ drives. One of our divisions wanted the 3 1/2″ drives as quickly as possible. Their priority was to get the fastest and most cost-effective drive possible (the 3 1/2″ drive). The other division was more worried about their existing customers who continued to want the older drive to fill up empty slots in their storage racks. Our disk drive manufacturing division shut down the 5 1/4″ line and it was not possible to restart producing the 5 1/4″ disk.

I was tasked with "finding a way to produce more 5 1/4 drives," but it quickly became evident that the cost of doing so would never be recouped by sales to the limited 5 1/4″ drive customers remaining in business. On the other hand, losing those clients completely would have equated to a huge profit loss for the company.

When we understood that producing more 5 1/4″ drives was impossible, we were forced to reevaluate the situation. I started visiting directly with clients and realized that the customer did not specifically care about the size of the drive. These high-end clients often buy large storage racks but do not necessarily fill each slot with a hard drive at first; this allows for future expansion. Most clients had hundreds, if not thousands, of unused 5 1/4″ slots. They wanted to fill those slots as their companies grew with 5 1/4″ drives – they did not want to purchase new racks.

Once I understood the real Project Motivation, we were able to rephrase the project definition and develop a creative solution. The width of a 3 1/2″ drive's case is less than half the length of a 5 1/4″ drive's case. A mechanical packaging solution was easily developed to hold two small drives in the slot originally designed for one large drive.

The customers who originally thought they wanted 5 1/4″ drives were ecstatic because this solution allowed them to fully utilize their existing racks while at the same time allowing for a much denser storage system (meaning more storage in less physical space). By understanding the real motivation, we were able to look past the preconceived solution and develop a solution that met the needs of *both* sets of customers.

6.2 Project Goal Statements

Now that you have a clear understanding of the challenge which you are trying to address and have condensed that understanding to a concise Motivation Statement, you must decide what you are going to do about it. This is necessary to ensure all project activities are focused on solving this challenge.

EXAMPLE 6.2 Motivations Drive the Goal Statement

Consider how the two interpretations of Raven's Motivation Statement in Example 6.1 would result in different Project Goals.

Scenario 1: The overarching company's Motivation:

It is beneficial to maintain a consistent distance between a sprayer's arm and the top of crop canopies to provide even coverage when dispensing pesticides.

The following Goal Statement might reasonably develop from this Motivation:

A system is required that will monitor its own distance above the crop canopy and automatically adjust itself to maintain a preset distance.

Scenario 2: The specific student project's Motivation:

Simultaneously detecting the crop canopy and the ground directly underneath is a difficult task for most sensors, the optimal method of doing this is unclear.

The following Goal Statement was the intent of the solicitation:

A test system will be designed to analyze the performance of different sensors.

Notice, neither scenario is "right" nor "wrong," but only one correctly represented immediate concerns. It is true that Shane's ultimate Goal was stated correctly by scenario 1 but had the students attempted *that* project with reasonable Triple Point Constraint limitations placed upon them; they would not have had the resources, time, or budget to successfully produce a quality project. Instead, the team worked together to properly define a Project Statement which correctly represented what Raven desired of the student project. Developing this proper statement took a couple of iterations before the final draft was worded in such a way that all Stakeholders understood and agreed upon the intent. The effort applied during the INITIATION phase resulted in a successful project. In the end, the student team delivered a high-quality and useful test apparatus to their Sponsor.

The Goal Statement clearly describes the singular focus of all future project activities. Each project has exactly one Goal. If you think you have more than one Goal Statement, you likely have not fully developed your understanding of the project. If you *actually* have more than one Goal Statement, you have more than one project and will need to manage multiple project teams – this case is definitely possible in industry but is not appropriate for most Senior Design projects.

The Goal Statement should be a single sentence which clearly articulates what is required to address the Motivation Statement.

Here are the most common mistakes when writing Goal Statements:

- The statement does not directly address the Motivation.
  Prevent this mistake by checking for a direct causality between the two statements. If the two statements do not naturally flow together, one or both probably needs to be reworded.
- The statement focuses on solutions rather than the Project Goal.
  Prevent this avoiding any mention of technology. The use of technology is *rarely* the Project Goal, rather it is the potential solution for the problem.
  (There are some rare exceptions to the previous statement, particularly in the academic world. These exceptions typically relate to the desire for some sort of demonstration apparatus. However, even then, a better Project Goal can usually be identified by asking "why is the demonstration desired.)

6.3 Project Objectives

The final part of an effective Problem Statement is Project Objectives. Objectives further define the Project Goal by providing high-level details. They also create categories of work. A Project Goal stated in a single sentence is often too ambiguous to develop project plans around. The Objectives add context and depth to the understanding.

A Project Goal stated in a single sentence is often too ambiguous to develop project plans around. The Objectives add context and depth to the understanding.

EXAMPLE 6.3 Objectives Further Define the Goal Statement

Let us continue investigating Raven sensor project. Recall, we have identified the Motivation and Goal as such:

> *Simultaneously detecting the crop canopy and the ground directly underneath is a difficult task for most sensors; the optimal method of doing this is unclear. Therefore, a test system will be designed to analyze the performance of different sensors.*

However, at this point, it is unclear exactly what is required of this test system. A few Objective Statements will clarify what needs to be developed.

Here are the Objectives defined for this project:

- This system must interface with a variety of sensor types.
- This system must contain plant life with a clearly distinguishable canopy.
- This system must easily adjust the distance between the canopy and sensors.
- This system must clearly display the sensors' output.

We still do not have a clear picture of the solution – that comes much later, but we are starting to get a clear picture of the design challenge.

Each of these Objective Statements still requires more detail before we could be sure the engineer and the Sponsor are in complete agreement, but we are getting closer.

6.4 Summarizing the Problem Statement

The project Problem Statement is an excellent tool to use when presenting your project. In fact, nearly every verbal presentation you deliver throughout the Design Life-Cycle will start with this statement. The Problem Statement simply consolidates the Motivation, Goal, and Objective Statements into one cohesive paragraph.

This section will provide a generalized framework to help you get started. However, it would be a mistake to force all Problem Statements into this exact framework. In the end, formulating a perfect Problem Statement requires a little ingenuity and some effective technical writing skills.

A general framework for effective Problem Statements:

- Start by stating the Motivation.
- Tie in the Goal using the word *therefore*.
- Support the Goal with Objectives following the phrase *in order to do this*.
- Organize the Objectives in a logical order, probably from input to output.
- Consider whether the Problem Statement is understandable by itself or an additional sentence of background information is warranted.

EXAMPLE 6.4 Finalizing the Problem Statement

Summarizing Example 6.3 into a cohesive Problem Statement might look like:

*Simultaneously detecting the crop canopy and the ground directly underneath is a difficult task for most sensors; the optimal method of doing this is unclear. **Therefore**, a test system will be designed to analyze the performance of different sensors. **In order to** be effective, this system must deploy a variety of sensors over a crop canopy. The distance between the sensors and the canopy and the actual height of the canopy must be independently adjustable. Finally, the output of the sensors must be clearly displayed and recorded for analysis.*

Notice the exact phrasing is not identical to the sentences produced during the development of the Problem Statement, but all the pieces are there.

The last step in the list above is to question whether this statement is sufficient for the average person to understand the project. In this case, a sentence of context would greatly improve the message.

*Raven plans to introduce an auto-leveling feature to a chemical-spraying implement but must develop a method to detect the height of a crop's canopy and the distance between the implement and that canopy. Simultaneously detecting the crop canopy and the ground directly underneath is a difficult task for most sensors; the optimal method of doing this is unclear. **Therefore**, a test system will be designed to analyze the performance of different sensors. **In order to** be effective, this system must deploy a variety of sensors over a crop canopy. The distance between the sensors*

EXAMPLE 6.4, continued

and the canopy and the actual height of the canopy must be independently adjustable. Finally, the output of the sensors must be clearly displayed and recorded for analysis.

Industry Point of View 6.2 Problem Statements

Project Management (PM) is not necessarily an engineering discipline. As a certified PM (Project Manager), I have managed a wide range of projects that include consultations in the medical and construction fields, as well as in engineering.

Regardless of who is conducting the project, my first order of business as a project consultant is to determine the organization's Problem Statement. A clear understanding of the problem will dictate which solutions are acceptable and which must be avoided.

My most recent consulting contract was for a therapy clinic for children with special needs located in Bozeman, MT. The clinic outgrew their current location and had leased a new space. The new building was wonderfully located but required significant interior remodeling. The clinic had a six-month window to design the new space, procure the necessary permits, complete construction, and relocate. Realistically, there was no chance the work could be completed before they had to vacate their old building.

Obviously, the Project Motivation for the clinic was "our lease is up and we have to relocate." The Project Goal required more thought.

One responsible Goal may have been to "reduce the financial impact of the move." The remodeling process could then be completed as efficiently as possible but not before the clinic had to vacate the original building. They considered temporarily moving their services into the patents' homes – typical operating model for many similar clinics. There would be the cheapest solution to the Motivation; the clinic would be able to limit the cost of the remodeling while continuing to bill the patents at regular rates. However, this solution carried a risk. A change to some patents' treatments could case setbacks in their therapy.

A number of alternative goals were considered, and ultimately the clinic determined that their Project Goal was to "expedite the move without reducing the quality of care provided to patents." This changed the solution space. Initially, a portion of the new building's remodeled was completed and then was sectioned off so that services could commence immediately in that area while construction was completed throughout the remainder of the building. Patents were transported to the new building during the construction to gradually introduce them to the new space. Once the majority of the remodel was completed, construction workers were paid overtime to remove the partition over a weekend. This Project Goal required a significantly larger financial investment than the original Goal but allowed the clinic to find creative solutions which allowed them to maintain their high quality of care throughout the transition.

The Project Management "rule" to avoid considering solutions during the INITIATING phase must be appropriately understood. Considering the solution space a particular

---

Industry Point of View 6.2, continued

Goal Statement will force you into is not the same thing as predefining solutions. While developing this clinic's Goal, we had to consider the impact each Goal Statement would have on the solution. We considered the impact of in-home services and temporarily subdividing the new building. These are the solution spaces which would be defined by particular Goal Statements. We did not consider which patents would be seen in-home or which service providers would perform particular tasks. We did not consider where the subdivision would be placed or how the details of the transition would occur. Those would be the "solutions" you are instructed to avoid.

Today, the clinic is successfully up and running in the new building. The transition was affordable and did not negatively impact their patents.

6.5 Concept Check

The following two examples provide a variety of valid statements that might have been discussed during the first sponsor meeting for two different senior projects. It is important to categorize each statement correctly and then develop a cohesive Problem Statement.

EXAMPLE 6.5 Problem Statement for a Nameplate Reading System

Identify each statement below as either a Motivation, Goal, Objective, or Other:

1. A method to remotely read nameplate information of industrial exhaust fans will be developed.
2. Industrial exhaust fans generate a lot of excess heat and therefore are installed on top of buildings to promote natural cooling.
3. A quadcopter will be used to fly a GoPro camera to the top of the structure and image nameplates.
4. Reading industrial exhaust fan nameplates is necessary for correct maintenance but is often difficult when fans are installed in hard-to-access locations.
5. This solution will need to remotely locate the fan, locate the nameplate, and record a high-resolution image of the that nameplate.

Analysis:

1. *This statement is the primary "thing" the following statements are trying to explain. Therefore, it is a Goal Statement.*
2. *This statement explains an important aspect of the environment which your product will need to operate within. It is a Background Statement, so the correct choice would be Other.*
3. *This statement explains the solution. So, it is not part of the Problem Statement either; the correct choice would be Other.*
4. *This statement explains why the project is necessary. Therefore, it is a Motivation Statement.*
5. *This statement lists a number of high-level tasks which the solution must accomplish. Therefore, it is an Objective Statement, or rather three Objective Statements merged into one sentence.*

EXAMPLE 6.5, continued

Summary:

A proper Problem Statement for this project might read as follows:

(Supporting background) Industrial exhaust fans generate a lot of excess heat and therefore are installed on top of buildings to promote natural cooling. *(Motivation)* Reading industrial exhaust fan nameplates is necessary for correct maintenance but is often difficult when fans are installed in hard-to-access locations. *(Goal)* Therefore, a method to remotely read nameplate information of industrial exhaust fans will be developed. *(Objectives)* This solution will need to remotely locate the fan, locate the nameplate, and record a high-resolution image of the nameplate.

EXAMPLE 6.6 Problem Statement for a Mars Rover Exhibit

Identify each statement below as either a Motivation, Goal, Objective, or Other:

1. The Kirby Science and Discovery Center has a Mars Rover exhibit which they would like to modernize.
2. The new Mars Rover exhibit must allow children to interact with a Mars Rover replica.
3. The Mars Rover replica must communicate with a "ground station" in a similar manner to how scientists actually communicate with the real Mars Rover.
4. The existing exhibit is out of date and a number of features are no longer operational.
5. Nearly 50,000 children visit the exhibit each year.

Analysis:

This example is less straightforward than Example 6.5. This is because this Problem Statement does not follow the exact pattern presented above. Rather, this is an example of how a performance team was able to modify the standard format of a Problem Statement to effectively deliver their message:

1. *This is the overall Goal of the project. Although until the other statements are assessed, this could be confused with the Motivation.*
2. *This is an objective that is further defining what the term "exhibit" means. This is clarifying that the exhibit is not simply pictures and wall hangings, but rather some sort of interactive experience.*
3. *This is further defining the exhibit, so it is also an objective statement. Not all Objective Statements are required to be merged into a single sentence like the previous example.*
4. *This is the Motivation Statement; it directly describes what is wrong with the existing situation.*
5. *This is background information. Unlike the background information in the previous example, this statement does not directly support the Problem Statement and should not be included here.*

EXAMPLE 6.6, continued

Summary:

Although this Problem Statement does not exactly follow the standard flow presented above, when statements 1–4 are put together, the problem is very clear.

(Goal) The Kirby Science and Discovery Center has a Mars Rover exhibit which they would like to modernize. *(Motivation)* The existing exhibit is out of date and a number of features are no longer operational. *(Objectives)* The new Mars Rover exhibit must allow children to interact with a Mars Rover replica. The Mars Rover replica must communicate with a "ground station" in a similar manner to how scientists actually communicate with the real Mars Rover.

Avoiding Common Pitfalls 6.4 Focus on the Problem!

Avoiding Common Pitfalls 6.1 stated:

> It is vital that all efforts during this initial Design Life-Cycle processes be focused on defining the *problem* as opposed to the *solution*!

But what does that mean and how do you avoid it?

Consider this project scenario:

Bruce and Anne, co-owners of the Joy of Hunting, desired a dog collar that operated like most invisible fence containment solutions. However, they wanted the "fence" to move with a hunter rather than be stationary around a yard. So the Goal of their project was to "develop a containment solution to control a hunting dog while in the field."

One of the original Project Objectives was stated as follows:

> *The system will use ultrasound technology to determine the distance between a dog collar and a human operator.*

Analysis:

Consider why this Objective Statement will eventually cause problems.

The technical Project Management answer:

This is not an Objective statement! It does not define the problem; it attempts to define how the problem will be solved. But, so what? Why does that matter?

A more effective answer:

Ultrasound technology is an excellent method to accurately determine the distance between two objects and can fulfill the Objective as it is written. However, ultrasound technology, when operated at battery-supplied power levels, is limited to a range of a few hundred inches and is utterly unable to accomplish the Goal of the project (limiting the range of a dog to reasonable hunting distances).

Avoiding Common Pitfalls 6.4, continued

Summary:

To avoid writing a Problem Statement that will cause you issues down the road, keep the following concepts in mind throughout the INITIATING phase.

* The use of technology is **rarely** the true Goal/Objective of the project. The use of technology is the solution to the Goal/Objective.
* The focus of the Problem Statement process is to define the **problem**, not the solution.
* Defining the solution at this stage (before the problem is fully defined) limits the solution to preconceived ideas rather than allowing best engineering practices to develop the optimal solution.
* If the client starts discussing the solution or technology as part of the Problem Statement:
  – *Do not* correct the client.
  – Take notes.
  – Recategorize the concerns later.
  (You will learn in Sect. 7.3 that this statement might be categorized as a "constraint.")
* If a team member (Student, Instructor, or Advisor) starts discussing the solution or technology as part of the problem,
  – *Do* correct the team member – respectfully.
  – Simply state:
  "I believe that topic is more related to the solution. Let's take note of that idea, but for now, can we get back to defining the problem?"
  or
  "I am not sure that is really one of the real Objectives of this project, can we reword that as an Objective Statement?"

The Joy of Hunting Resolution:

Bruce and Anne knew that the application of GPS technology for this application was already protected by a US patent. They could not legally implement a GPS-based solution without infringing on someone's intellectual property. Their concern was not necessarily that ultrasound technology be implemented but rather that GPS technology was *not* used. Therefore, the Problem Statement was modified to include the following Objective:

> *The system will determine the distance between a dog collar and a human operator.*

And a constraint was included in the project documentation that stated:

> *The solution cannot include GPS technology.*

Now, the design team was able to investigate other wireless technologies that held more promise than ultrasonic signals, while the Sponsor's concern of patent infringement was all addressed.

6.6 Chapter Summary

In this chapter, you learned about Problem Statements. Having a clear understanding of the project's Motivation, Goal, and Objectives is necessary for engineers to produce the optimal solution. We do not want to produce a solution to the wrong problem. All Stakeholders must agree on the precise Motivation so that the Project Goal can be properly formatted. A succinct Project Goal helps us narrow our focus and ensure that all project efforts are directed appropriately. The Project Objectives are used to further define the Project Goal so that there is no ambiguity in the Problem Statement.

Here are the most important takeaways from the chapter:

- The Problem Statement consists of the following:
  - Project Motivation (one sentence)
  - Project Goal (1–2 sentences)
  - Project Objectives (one phrase per Objective, typically 3–4 on a senior project)
  - If necessary, one or two critically important background statements
- The Problem Statement should be developed through a top-down hierarchy, allowing the Motivation to drive the Goal and the goal to be further defined by Objectives.
- The Problem Statement should be developed with a focus on the problem, not the solution. Discussing potential solutions is exciting and tempting but should be avoided to prevent preconceived ideas limiting effective engineering later in the process.

6.7 Case Studies

Let us look at each of our case study teams' Problem Statements as they were initially developed early in the project and then observe how the statements were improved in the final drafts.

Augmented Reality Sandbox

The first draft of the AR sandbox team's Problem Statement was as follows:

> *The goal of this project is to successfully project a clear, real-time image over the sand. An initial prototype will be completed consisting of a temporary structure holding the sand, the required Kinect sensor and BenQ projector, and a custom-built PC this fall semester with initial development on the additional features. The final aluminum structure will be built and the three unique features will be fully implemented during the spring semester.*

- What needs to be improved:
    - This team was focused on the solution throughout the INITIATING phase rather than the problem. As such, a number of solution-based statements were incorporated into the Problem Statement.
    - The first sentence of their initial Problem Statement starts with "the goal of this project..." rather than discussing the motivation.
    - They never specify what the "real-time" image is nor what the "unique features" will be.

By the time this team presented their Project Kickoff Meeting, the Problem Statement read like this:

> *The Kirby Science and Discovery Center (KSDC) desires to install an Augmented-Reality (AR) sandbox to teach children about the agriculture of Eastern South Dakota. There are a number of AR sandboxes commercially available but none that are designed to teach children about agriculture. Therefore, an AR sandbox will be designed that includes agricultural-related features. The sandbox structure will be designed to safely withstand child's play in a public area. An image of a topographical map will be projected upon the sand. Crop growth models will be generated and predictive results will be depicted over the topographical map.*

- What went well:
    - This updated version of the Problem Statement is properly formatted with Motivation, Goal, and Objectives (in order).
    - The Problem Statement clearly specifies the problem and what needs to be done about it.
    - The statement provides a background sentence explaining why the KSDC should not simply purchase a commercially available product.
- What could to be improved:
    - The term "AR sandbox" is not likely to be understood by all audiences. A background statement might have been included to define this term. However, the statement already includes a more important background statement

as mentioned above, so the team chose to avoid increasing the length of the Problem Statement any further. Instead, they carefully defined the term "AR sandbox" immediately after presenting the Problem Statement.

Smart Flow Rate Valve

The first draft of the Smart Valve team's Problem Statement was as follows:

> *Agronomists often develop a "prescription map" that details an exact amount of each chemical needed on specific parts of a field. These maps, along with precise control of the application's flow rate, allows for greater efficiency in applying chemicals. Raven currently uses a flow meter and a pressure sensor to monitor an application's flow rate, and an adjustable value is used to modify the rate if necessary. These are three distinct components and are connected to each other with cables that are expensive to manufacture and install and have a risk of becoming damaged while the sprayer is in operation. A product is needed that combines these separate components into a single, water-, and dust-resistant housing. Combining these components will decrease the number of potential failure points by reducing the amount of cabling in the system. The product will implement a closed-loop control system to monitor and control the liquid product flow. The control system will use the valve position and the current flow rate from the flow meter to adjust the position of the valve, which regulates the flow rate. The product will output the measured flow rate, valve position, current pressure, and GPS speed onto a field computer or universal terminal to allow a user to monitor the system or manually input a desired flow rate if a prescription map is not available.*

- What needs to be improved: Overall, this was a good first draft. However, the Problem Statement should be concise. This version is longer than necessary.

By the time this team presented their Project Kickoff Meeting, the Problem Statement read like this:

> *Agronomists often develop a "prescription map" that details an exact amount of each chemical needed on specific parts of a field. These maps, along with precise control of the application's flow rate, allows for greater efficiency in applying chemicals. Raven's existing solution to controlling an application's flow rate implements three distinct products which are expensive to manufacture and install. A product is needed that combines these components into a single, easy-to-install package. The product will implement a control system to monitor and control the flow rate. The product will output the measured specific parameters to a terminal to allow a user to monitor the system. The user must be able to select the type of input control between a prescription map and a manual override option.*

- What went well:
  - The final draft of the Problem Statement is much more concise than the original draft while still clearly defining the problem.
  - This statement starts with two necessary background statements and then follows the standard structure of one Motivation Statement, one Goal Statement, and a few Objective Statements.
  - All of the information eliminated from the original statement was included in the Background section of any documentation produced.

Robotic ESD Testing Apparatus

The RESDTA team's first (and only) draft of the Problem Statement was as follows:

> *Our Sponsor advertises their hard drives as the world's most reliable memory storage devices. To make that claim, this company tests every possible aspect of every drive before selling it to a customer. Unfortunately, the resources required to rigorously perform their electrostatic discharge (ESD) testing have become unacceptable. A robotic ESD testing apparatus was desired to automate this specific test. The apparatus must accommodate a variety of devices for testing, each with a different physical size and ESD test locations. The solution must perform an entire test matrix without interaction from the operator. Test data must be recorded and displayed to the operator throughout the duration of the test.*

- **What went well:** This was an excellent Problem Statement. It is concise but informative. The RESDTA team summarized a complicated problem with the first three sentences, two background and one motivation sentence. The Project Goal is clear (*ESD testing must be automated to reduce the resources necessary to complete the test*). The Objective Statements, contained in the last three sentences, further clarify the Goal. The Objective Statements provided in this Problem Statement are only a summary of the *nine* Objectives defined for this project. The RESDTA team wisely realized that they could not precisely follow the rules for constructing a Problem Statement (which state that each Objective should be mentioned) and at the same time adhere to the purpose of the Problem Statement (which is to provide a short introduction to a larger discussion). The team summarized the list of Objectives with a few important statements here and then explained the remainder of the Objectives when they presented their Requirements Document.
- **What could be improved:** One of the most challenging aspects to this project was the fact that the solution was constrained to use the existing ESD delivery device. This is what is called a Project Constraint (Sect. 1.2). While a Problem Statement is not required to present any or all of the Project Constraints, this particular constraint was important enough that it probably should have been mentioned here.

Chapter 7: **Requirement Documents**

The Requirements Document is the most important output of the INITIATING phase. It will definitively define exactly what the project should accomplish, and as such, it will be used to assess the project results during the EXECUTING and CLOSING phases. As with all project communication, it is a top-down thought process, Fig. 7.1, where the Motivation, Goal, and Objectives from the Problem Statement generate Requirements (Req) and Specifications (S).

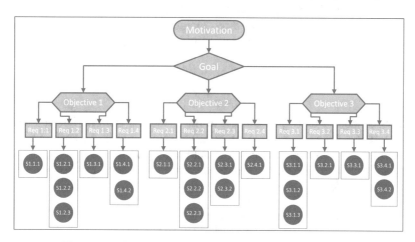

Figure 7.1   Structural flow of the Requirements Document

Students usually are able to define the Project Motivation and Goal without too much trouble but are often challenged to organize the rest of the Requirements Document correctly. This is partly due to being lax in our common vernacular and carelessly interchanging similar words. In fact, we are *taught* to use creative synonyms early in our education. This makes creative writing more interesting to read but causes a problem in technical writing. Precise language must be used in technical writing to avoid confusion (much more on this in Sect. 10.1). For now, focus on terms related to the Requirements Document's categories, Goal, Objective, Requirement, and Specification – can you distinctly define each of these? Probably not – in our day-to-day lives – these terms mean almost exactly the same thing and therefore are interchangeable. In Project Management, each has a precise definition.

In this chapter, you will learn:

- How to define proper Requirements
- How to define proper Specifications
- How to define proper Constraints
- How to format a Requirements Document
- How to gather and organize the information required for the Requirements Document

C.J. Mettler, *Engineering Design*, https://doi.org/10.1007/978-3-031-23309-8_7

7.1 Project Requirements

In this section, you will learn about project "**Requirements**." Unfortunately, the terms, "Goal," "Objective," and "Requirement," can often lead to confusion because they are often interchanged in common vernacular language. Even though we have already been using the terms "Goal" and "Objective," this section will formally define these terms so that the definition of "Requirement" will be fully understood; we will then define "Specification" in the next section.

Unfortunately, the Oxford Dictionary uses a circular definition to define the first two terms that we are trying to uniquely identify ("Goal" and "Objective"), so it is understandable that these two terms are easily confused. Oxford defines these terms as follows:

"**Goal** – the object of a person's ambition or effort."

"**Objective** – a thing aimed or sought; a goal."

Admittedly, it is difficult to distinguish these terms using the common vernacular definitions alone. To properly use these terms, it is important to remember that a project has exactly one Goal and that there are multiple Objectives. Although PMI (Project Management Institute) does not formally define "Goal" other than to say there is only one per project, its definition of "Objective" is helpful [22]:

"**Objective** – a result toward which effort is directed; a result to be obtained, a service to be produced."

So what is the difference between these terms in a practical sense? To answer that, it is useful to replace the word "result" in PMI's definition of Objectives with the word "feature." Now, if we assume there is one ultimate "Goal" by which a Motivation will be satisfied, the "Objectives" can be interpreted as the features or services which that Goal will exhibit. It is in this way that the "Objectives further define the Goal" as explained in Sect. 6.3.

Requirements are used to further define the Objectives in a similar fashion. PMI goes on to define "Requirement" as [22] follows:

"**Requirement** – a documented condition or capability."

When put into context, this means "Requirements" are used to document exactly what is meant by the related Objective Statement. Example 7.1 provides a list of common features which will require Requirement Statements.

EXAMPLE 7.1 Common Requirements

This is a short list of product characteristics, or features, which you should consider when trying to identify your project's Requirements.

- Physical limitations
- Power limitations
- Speed limitations
- Specific functionality
- Communication procedures
- Software features

Example 7.2 demonstrates that the terminology used in Objectives Statements is not sufficient to precisely define the project and that Requirements are necessary.

EXAMPLE 7.2 Requirements Define Objectives

Many engineering projects contain the Objective "must be mobile." However, taken out of context, that Objective carries very different connotations. Here are some examples:

- Goal – build a "tiny" home
  - *Objective – must be mobile:*
    Requirement – must be towable by light weight pickup truck
    Requirement – all interior features must be stowable
- Goal – design a mobile phone for children
  - *Objective – must be mobile*
    Requirement – must be carriable by small children
    Requirement – must survive daily abuse

Both projects contain the same Objective, but the context is very different. The Requirements derived from the respective Goal/Objective flow defines what exactly is meant by the term "mobile."

On the surface, maybe this seems ridiculous – how would anyone attempt to build a tiny home using the mobile phone's definition of "mobile"?

Well, first off, the Problem Statement will become part of a legal document (Chap. 6), and nothing can be "left to interpretation." But there is a more important answer here. Would it have been inherently obvious that the sponsor of the tiny home wanted "all interior features to be stowable" by simply stating they wanted the system mobile? Not necessarily, that feature could easily have been overlooked. Similarly, would the Requirement of a "mobile phone" meant the same thing if the phone was not intended for small children but rather for teenagers?

The Requirements that support a particular Objective are necessary to properly define the Objective in the appropriate context.

7.2 Project Specifications

The sequence of Goal/Objective/Requirement is used repeatedly to accurately describe the necessary features of a design. However, while a full set of those sequences will *qualitatively* describe the necessary attributes of a solution, the information these sequences provide will not be sufficient to develop *and validate* an optimal solution. For that, **Specifications** are required – these are the values by which success is measured during the development of a project (*before* final testing occurs).

PMI defines "Specifications" as [22] follows:

"**Specification** – a precise statement of the needs to be satisfied."

The Oxford Dictionary provides a complementary definition:

"**Specification** – a detailed description of the design."

Notice that these definitions use the phrases "precise statement" and "detailed description." Putting that into an engineering context, it means that the statement or description can be *measured*. A Requirement or Objective may often be validated qualitatively; a Specification will *always* be validated quantitatively (i.e., with measured data).

All Requirements must be defined by at least one Specification. However, most often a Requirement should have more than one Specification. Example 7.3 provides a few common Specifications to consider for each of the Requirements provided in Example 7.1.

EXAMPLE 7.3 Common Specifications

Example 7.1 provided a list of common Requirements. Let us now observe how each of those Requirements might be further defined with quantifiable values:

- Physical limitations:
  - Weight, in pounds
  - Height, in meters
  - Volume, in liters
- Power limitations:
  - Voltage, in volts
  - Current, in milliamps
  - Power, in horsepower
- Speed limitations:
  - Time to collect data, in microseconds
  - Time required to report data, in milliseconds
  - How fast/slow something must move, in minutes
- Specific functionality:
  - Distance, in inches
  - Tolerance, in percentage
  - Repeatability, according to a ratio
- Communication procedures:
  - Accuracy, in bytes dropped per transmission
  - Data packet size, in gigabytes

Each Specification should be measurable; they are *not yes/no* characteristics. Ask yourself when developing Specifications: "What tool (or instrument) will I use to verify this Specification?" If you cannot answer this, you should reconsider whether the statement is a Specification or not. Example 7.4 provides an example of why a detailed Specification is necessary to define a Requirement during the engineering design process.

EXAMPLE 7.4 Specifications Quantitatively Define Requirements

Consider the Tiny Home project presented in Example 7.2. The Goal/Objective/Requirement sequences have begun to describe a number of design challenges that must be met. However, exact values are required to produce an optimal design.

Dissecting the first sequence in Example 7.2 will illustrate this. Recall that sequence was as follows:

- **Goal – Build a tiny home**
  - **Objective – must be mobile**
    Requirement – must be towable by a light pickup truck

This particular Requirement produces a relatively clear mental picture of one aspect of the design challenge. We have all seen pickup trucks towing loads on the interstate. But *exactly* how heavy can you make your tiny home?

You could build a potential solution, hook it up to a truck, and take it for a test drive. But consider the potential outcomes of that test:

1. The truck cannot pull the tiny house so the entire design must be scrapped.
2. The truck can just barely pull the tiny house, but the load has exceeded towing capacity and it eventually damages the truck.
3. The truck can pull the tiny house, but the load is drastically under towing capacity, and although the design meets Requirements, unnecessary compromises may have been made in other areas (e.g., the house might be much smaller than what it might have been).

Based on the results of that test, the design could be improved iteratively until the results show a truck can pull the tiny home, the load is slightly under towing capacity, and the design meets all other Requirements.

But then you take the design for one last test drive and a cop pulls you over because the load surpasses the maximum width of a legal load (because we have yet to define that aspect as well), and the process starts all over. This is the "explorer's" circuitous route which engineers try to avoid – it is called *development* rather than *design*.

A proper Requirements document must include detailed Specifications:

- Goal – build a tiny home
  - Objective – must be mobile
    Requirement – must be towable by light weight pickup truck
    - Spec – weight must not exceed 7500 lb.
    - Spec – width must not exceed 8.5 ft.
    - Spec – height must not exceed 8.5 ft.

EXAMPLE 7.4, continued

With this information, the design can be verified using engineering data before the first test drive. The weight, width, and height can be estimated during the PLANNING (design) phase and verified during the EXECUTING (build) phase. A test drive is still necessary to verify the design was fully specified and that something did not get missed in the early stages, but this list of Specifications drastically improves the likelihood of achieving the Requirements on the first test.

Avoiding Common Pitfalls 7.1 Not Enough Specifications

In Sect. 7.4, you will learn how to compile all your Objectives, Requirements, and Specifications into a formal Requirements Document. Usually this is accomplished in outline format, but sometimes the Requirements Document is produced graphically like Fig. 7.1.

The graphical representation forms an inverted tree structure where there are multiple branches. In the outline representation, the old adage if there is an A, there must be a B holds true.

The point is this: in either case, a single phrase does not typically fully define a category. The Goal is supported by multiple Objectives, each Objectives should be defined by multiple Requirements, and each Requirement should be clarified with multiple Specifications.

Although this guideline will not be perfectly true in all cases, if you notice that you have *many* "straight branches" where each Requirement only has one Specification, you probably have one of the following issues to address:

- You have not thoroughly defined your Requirements. Put some more effort into developing Specifications to more completely define the Requirements.
- Your Requirements Document should be organized more effectively. Consider how you might regroup some of concepts into categories which allow for the proper structure.
- You have incorrectly labeled some of the concepts. For example, a concept you have listed as a Requirement might actually be a Specification. Review the definitions of each term, and verify you have properly placed each concept into the correct category.

Exception to the Rule for Software Requirements

The Requirements Document should be presented as a standard "tree" format, meaning that you start with a singular Goal which produces a couple of Objectives. Each Objective is defined by multiple Requirements. Each Requirement is further defined by one (rare) or more Specifications.

There is an exception to this standard. It is sometimes not possible to define Specifications for software-related Requirements. You should take care when implementing this exception. First, are you *sure* the statement is not a Constraint (Sect. 7.3)? "Requirement" statements which are difficult to Specify are often mischaracterized Constraints. Second, have you thoroughly considered any numerical metrics that should be applied to the Requirement. Example 7.5 provides examples of common software Requirements that, at times, can be difficult to define with a Specification.

EXAMPLE 7.5 Software Exceptions to the Requirements Rule

Some software Requirements do not require further definition or explanation and therefore do not require Specifications. Some examples include the following:

- Req: Software must display the processing time.

  This is a YES/NO statement – either the software displays the processing time, or it does not – which is a good indicator that the statement is a Constraint. Yet, this particular statement is not a Constraint; the processing time is a feature the product must have, not a limitation on the design.

  Now, consider whether this Requirement needs Specifications:

  - Does it matter how quickly the processing time is displayed after the time has been calculated?
  - Has the required accuracy of the time been defined elsewhere?
  - Does the font size of the display matter?

  This is an example where *some* project teams may decide not to include a Specification. Whereas other teams, when faced with the same Requirement, may have Specifications which must be incorporated.

- Req: The application must include a splash page with Sponsor's Logo.

  Displaying the logo is something the application must do; therefore, it is not a Constraint. But what Specifications could be used to further define this? Consider the following:

  - How long should the splash page be visible?
  - How much area of the total screen should the logo encompass?

EXAMPLE 7.5, continued

This Requirement is an example of a software Requirement that may initially appear to stand alone without the need of Specifications. However, upon further consideration, there are probably a few Specifications that should be included. Make sure you dig deep if you think you do not need Specifications.

- Req: The Graphical User Interface must be programmed in MATLAB.

This is a *yes/no* statement. It is a Constraint, not a Requirement. This statement is a limitation on how the problem can be solved. The statement is not a feature which the product must exhibit – the software does not program itself. If you believe you have a software Requirement, that does not need a Specification; double-check that the statement is not actually a Constraint.

7.3 Project Constraints

The last term that needs to be defined before building a Requirements Document is "Constraints." This term is the concept which students tend to use incorrectly most often in Project Management. This section will define the term and provide examples of Constraints you should consider when defining your project. It will also explain why Constraints are often mischaracterized and provide guidelines how to avoid this mistake.

Project Constraints

PMI defines "Constraints" as [22] follows:

"**Constraints** – A limiting factor that affects the <u>execution of the project</u>."

In other words, a Constraint is a limitation related to the *project*, whereas Requirements are related to *product*. A Constraint puts bounds on what the *engineer* can do to solve the problem, whereas Specifications put bounds on the *deliverable*. Example 7.6 provides a list of common Constraints.

EXAMPLE 7.6 Common Constraints

Here are three examples of common Constraints Senior Design students often must consider for their projects:

- Constraint: Solution must be coded in MATLAB:
  This constraint limits the engineer's ability to choose the coding language. Maybe the engineer is more familiar coding in C++, but the Sponsor has employees familiar with MATLAB.

EXAMPLE 7.6, continued

- Constraint: Components must communicate using CAN protocol:
  This constraint limits the component selection. Maybe SPI-protocol components are cheaper, but the solicited project is meant to be embedded into a larger system which already uses CAN.
- Constraint: System must multiplex four sensors:
  This is a tough one to identify because it contains a number. It may seem like this is a Specification. Additionally, it might be assumed this is a description of what the solution must do and, therefore, mistaken as a Requirement.

  However, this is a Constraint. The designer may wish to save space using only two sensors, but that is not an option. The design must use exactly four sensors.

Avoid Mischaracterizing Constraints

Constraints often are mischaracterized as Specifications or Requirements. There are three reasons why this commonly happens. The good news is that understanding those reasons makes it easy to avoid mischaracterization.

Perhaps the primary reason why Constraints are mischaracterized is that the term "Constraint" seems negative – something that *cannot* be done, rather than something that *must* be done. This is another example of how common vernacular interferes with technical understanding. The truth is that a Constraint can be either positive or negative; something that must be done or something that cannot be done. Example 7.7 lists two Requirements and provides examples of positive and negative Constraints and Specifications for each.

The most common mischaracterization students make is to label a Constraint as a Specification. This is caused by the fact that identifying Specifications in order to define a particular Requirement can be difficult. At the same time, a particular Constraint might seem to directly relate to this challenging Requirement. Mistaking a Constraint for a Specification is a common occurrence since both Specifications and Constraints are used to define a Requirement; therefore they often seem interchangeable. The temptation is to slap the Constraint under the Requirement as a Specification. While this is the most common mischaracterization, it is also the easiest to avoid! Keep in mind the following:

1. Constraints can be verified with *pass/fail* criteria.
2. Specifications require equipment and data to verify.

Consider the statements provided in Example 7.7. Start with the Specifications. All the Specifications would require equipment to measure and verify. A stopwatch could be used to time how long is required to collapse a system, and a scale could be used to determine the weight. Voltage and current could be monitored with an oscilloscope to determine power consumption. Contrast this with the Constraint Statements. What piece of equipment would you use to verify the programming language used? That is a silly question, is it not? This is simply a *yes/no* test. How would you collect data to determine whether your package can be shipped by the USPS (US Postal Service) or not? You do not need data to verify that; you simply need to take your package to the local post office and ask if it can be shipped.

Another reason why Constraints are mischaracterized is that both Constraints and Requirements are used to define objectives and therefore are often identified simultaneously. This can result in a concept that is actually a Constraint to be characterized as a Requirement. However, remember that a Constraint limits the *designer*, whereas a Requirement defines the *product*. This is the fundamental difference between the two terms, but applying this technical definition can be tricky. Consider the examples provided in Example 7.6; two out of the three examples could easily be misinterpreted as statements about what the product must do. For example, the second Constraint provided, "components must communicate using CAN," certainly sounds like something the system is doing. However, the end-user will not care *how* the internal components communicate; the user will care about *what* is communicated. This statement is not defining *what* the system must do; it is limiting the design of how the system performs. The way to avoid this mistake is to remember that Requirements can be further defined with Specifications and Constraints cannot. Whether the system is using CAN or SPI can be verified with a simple *yes/no* test and no further information is required to define the statement. Therefore, it is a Constraint.

EXAMPLE 7.7 Constraints Versus Requirements/Specifications

Here are some common examples of Requirements, Specifications, and Constraints. While there may be both positive and negative versions of Specifications and Constraints, it is uncommon to have a negatively written Requirement:

Requirement: The test system must be portable.

- Positive Specification: The system must be collapsible in under 10 seconds.
- Negative Specification: The system cannot weigh more than 10 pounds.
- Positive Constraint: The system must be packaged in a US Postal Service approved container.
- Negative Constraint: The system cannot contain hazardous shipping material.

Requirement: The photovoltaic system must include a monitoring system.

- Positive Specification: The system must report generated power to within $\pm 1\%$ of actual generation.
- Negative Specification: The monitoring feature cannot draw more than 3% of total generated power.
- Positive Constraint: Monitoring system must be easy to interpret.
- Negative Constraint: System cannot be developed in Visual Basic.

Caution: the terms "positive" and "negative" were used here to illustrate that not all Constraints are negative nor all Specifications are positive. However, it is not necessary to actually identify any of these statements as positive or negative. Simply listing "Constraints" and "Specifications" correctly will suffice for any documentation purposes.

7.4 Building the Requirements Document

Now that you understand the general terminology of the Requirements Document, you can start to build one for your project. Keep in mind, this is the most important deliverable of the INITIATING phase. It will require time and effort to get it "right," so it is unlikely that the first draft of the document will be sufficient. In a way, the Requirements Document is like a big puzzle; you have all these disjointed pieces of knowledge about your project, and you need to organize them in a way that precisely communicates your project.

Formatting the Requirements Document

Remember that Specifications add precision to Requirements, Requirements put Objectives into context, and Objectives define the Goal of the Project. One descriptive representation of this thought process would be an organizational tree similar to Fig. 7.1. However, while this structure is sometimes used, it can be difficult to fit this image into a document with enough resolution for the audience. Another representation commonly used is a standard numbered outline. This provides a hierarchical numbering system for each important item in the Requirements Document, which will be useful later.

EXAMPLE 7.8 Format of a Requirements Document

A Requirements Document should look like the standard outline shown in this example:

Obj 1 State Objective 1 directly from Problem Statement. This is probably the first (or most important) input.

> Req 1.1 State the most important Requirement that defines this Objective
>> Spec 1.1.1 State the most important Specification for this Requirement
>> Spec 1.1.2 If the statement is particularly long and, therefore, requires an additional line of text, it should be properly indented.
> Req 1.2 State the second most important Requirement of this Objective
>> Spec 1.2.1 List the specifications
>> Spec 1.2.2

Obj 2 Repeat for the second Objective

> Req 2.1
> Spec 2.1.1

Organize the list of Objectives starting with the inputs, continuing through the processing, and ending with the outputs.

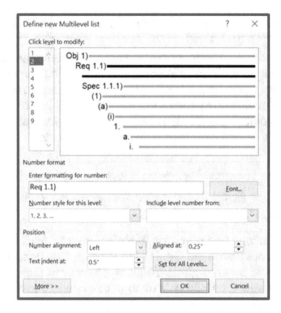

Figure 7.2 Pop-up window to format Req Doc

It is possible to automate this outline structure in MS Word and other editing softwares. Follow following steps to create the necessary outline structure:

1. Open a new MS Word document.
2. On the toolbar, navigate to Home/Paragraph Settings/Multilevel list, Fig. 7.2.
3. Select level "1" in the upper-left corner:
   (a) In the "Enter formatting for number" field, type *"Obj"* before the default "1)."
   (b) In the "Text indent at:" field, enter "0.5."
4. Select level "2" in the upper-left corner:
   (a) In the "Enter formatting for number" field, type *"Req"* before the default "1)."
   (b) After "Req," enter a *space,* and then use the "Include level number from:" dropdown menu to select *"Level 1"* and add a period (.) after the new number.
   (c) Use the "Number style for this level:" drop-down menu to select *"1,2,3. . ."*
   (d) Now the "Enter formatting for number:" field should read *"Req 1.1)."*
   (e) In the "Text indent at:" field, enter *"1.0."*
5. Select level "3" in the upper-left corner:
   (a) In the "Enter formatting for number" field, type *"Spec"* before the default "1)."
   (b) Repeat the formatting steps described for the Req level.
   (c) **You will need to include Level 1 first and then Level 2, in two separate steps.**
   (d) In the "Text indent at:" field, enter *"1.5."*

If you like to modify any of the formatting after creating the list do the following:

1. Click directly on the first item number (*"Obj 1"*) in the MS Word document.
2. Once the item number is highlighted, right-click on that same number.
3. Select *"Adjust list indents."*

Now you are ready to build your own Requirements Document. A typical Senior Design project should have between three and five Objectives. Each Objective should have more than one Requirement. Most of the Requirements should have at least two Specifications. If your Requirements Document does not follow those guidelines, you likely have not developed the document far enough.

Avoiding Common Pitfalls 7.2 Completing Your Requirements Document

After your Requirements Document is nearly complete, double-check the following:

- Does the outline follow the organizational tree structure? Each Objective should have more than one Requirement, and almost all Requirements should have more than one specification.
  - A few exceptions are acceptable, but the majority of the outline should follow the guideline.
  - Corrective Action: Consider reorganizing and/or redefining your Objectives or Requirements – your document might flow better.
- Do all your Requirements have valid specifications?
  - Corrective Action #1: Reconsider whether that statement is a constraint.
  - Corrective Action #2: If that statement is actually a Requirement, brainstorm how you will measure the results – this will likely generate ideas about the necessary specifications.
- Can all your specifications be measured with equipment or is the validation of that statement a *yes*/*no* question?
  - Corrective Action: Reconsider whether that statement is a constraint.

7.5 Requirements Gathering

The Problem Statement and Requirements Document are created through a process called "Requirements Gathering." This process starts during first sponsor meeting (Sect. 4.5) and continues through an iterative process as the team develops more knowledge about the project. A draft of the Requirements Document is presented during the Project Kickoff Meeting (Chap. 9) to the Sponsor. The term "draft" does not imply "incomplete." This "draft" should be as comprehensive and accurate as possible, but it will not be considered "final" until the Sponsor has reviewed the document. The Sponsor typically has a few corrections, modifications, or requests to add to the "draft" during the Project Kickoff Meeting. The project team will need to incorporate those requests into the final version of the Requirements Document which is presented in a written contract call the Project Charter (Chap. 10).

> It is imperative that the Requirements Document be developed accurately and thoroughly by the time the Charter is written.

This document will guide all project design choices. It will also be the litmus test used to evaluate the success of the team's work at the end of the project.

> The output of the Requirements Gathering process includes a valid Problem Statement and an organized Requirements Document.

This section will discuss techniques to improve your Requirements gathering process.

Initial Preparations

There are a few tasks you can complete before meeting with your Sponsor which will help your Requirements gathering efforts.

- Review the Project Solicitation as submitted by the Sponsor and develop a draft of the Problem Statement in your own words.

  You have probably not understood the exact desire of the Sponsor perfectly, but that is okay. The purpose of this draft is to formulate your own ideas about the project so that the first meeting with the Sponsor will be more efficient.

- Review the Project Solicitation again and identify all technical information.

  It is not important just yet to correctly identify whether a specific piece of information is an Objective, Requirement, Specification, or Constraint. For now, just compile a list of all information provided to you.

- Briefly research the topics you find in the solicitation.

  This effort is not to make you an expert on the topic nor is it to develop any solutions just yet. Rather, the purpose of this effort is to become familiar with the key concepts that will be discussed during the first meeting with the Sponsor.

- Meet with your team and discuss the project prior to the sponsor meeting.

  Make sure the team, including your Advisor, has a unified understanding of the project. If there are any differences of opinions, make sure to address these on the meeting agenda.

Once you have a clear picture of the information you were initially provided, you are ready to meet with the Sponsor. You will want to meet with the Sponsor as soon as possible, so do not take too long preparing. Depending on the project, most teams are able to accomplish this within a few days after the project team is formed. It is imperative that you meet with the Sponsor as soon as possible.

First Sponsor Meeting

The format of the first sponsor meeting was discussed in Sect. 4.5. This section will discuss methods which can be used during that meeting to effectively gather the Project Requirements.

Do not wait until you have completed your preparation work to start scheduling a meeting with your Sponsor. Get the meeting on the calendar as soon as possible. Use the time between scheduling the meeting and the actual event to complete the preparation work mentioned above. You should coordinate with your course instructor on how best to contact a nonuniversity Sponsor – there may be university regulations to be aware of.

Here are some ideas to keep in mind about the first meeting with your Sponsor:

- This meeting will often set the tone for the working relationship with your Sponsor.

  So, treat this as a professional event. That includes arriving on time, dressing appropriately, and conducting yourself well.

- Procure all of the information you will need during the next few weeks.

  As much as possible, "listen – don't talk." It is tempting to share your ideas in hopes of demonstrating that you are a competent engineer. Doing so often derails the conversation, probably shifting the focus from the *problem* to the *solution.* Instead, ask questions, listen, take notes, and get as much information as you can.

- The highest priority of this meeting should be to fully define the *problem.*

  It is tempting to discuss *how* you will solve the engineering problem. However, at this stage, you must focus on defining the *problem* to be solved. Due to time limitations, it is not always possible to fully flush out the entire Requirements Document and discuss potential solutions. Your priority should be developing a fully agreed upon Problem Statement. Then, collect as much information as you can about the Requirements Document.

- Do not attempt to categorize information into the Requirements Document.

  The Sponsor is not going to care whether a statement is a Constraint or Specification; they "just want the statement considered." This means that after you have gathered all the information, you have complete freedom to organize this information; however, it makes the most sense to you. It is not worth the time during the meeting to try to organize everything on the spot.

- End your meeting on time.

  As the scheduled time for the meeting is nearly up, make sure you save time at the end to review Action Items and thank the Sponsor for their time. You may also consider scheduling a follow-up meeting in the near future.

Techniques to Gather Information Efficiently

Requirements Gathering is a skill that takes practice to develop. It is hard to ensure that you have captured *all* the pertinent information and even more difficult to identify whether you have included superfluous information. Developing an effective set of Requirements Gathering skills will improve the efficiency of this process.

Here are some tips to remember during the Requirements Gathering process.

- Maintain focus on the problem.

  (Have we said this already? If so, that is okay – it is that important.) If you feel the conversation is starting to devolve into a discussion about solutions, try politely refocusing on the problem. Here are a few phrases that might help:

  - "I feel like that is one *possible* solution, are we allowed to consider others?"
  - "I understand that you have a *preference* for that component, but are we constrained to it?"
  - "Is there a reason that technique is necessary?"

Sponsors will often give you permission to explore other possibilities, and you should. However, occasionally they do have a justification for why you are constrained to their solution-based ideas; if so, include that information in the Constraint list. Your Instructor will probably ask you to defend this Constraint; make sure you understand *why* the Sponsor limited you.

- Ask as many follow-up questions as you can.

Conversations typically start at the surface level of an issue. We do this naturally because providing too much detail too fast will confuse our listeners. By asking follow-up questions, you encourage the speaker to give you the details you are going to need. Do not take anything at face value, dig deeper. The more you understand about the problem, the more effective your team will be come. In fact, challenge your team during the first sponsor meeting to ask *at least* one follow-up question for every new piece of information you are given. Examples 7.9 and 7.10 provide examples where student teams could have navigated the Requirements Gathering process more effectively by asking follow-up questions.

EXAMPLE 7.9 Requirements Gathering Process #1

A 2014 Senior Design team was provided with a Project Objective of "determining the distance between two objects (up to 100) feet apart using ultrasonic technology."

Eventually, it was determined that ultrasonic technology could not reliably be used at this distance. The team questioned "**why** was it necessary to use ultrasonic technology?" The reasoning was that the Sponsor had previously used ultrasound for an unrelated application but had no real reason that it had to be used again. The ultrasonic Constraint was removed from the Objective statement, and the team developed a very successful product using GPS technology instead.

Analysis:

This team ran into nearly the exact situation described in the Joy of Hunting example discussed in Avoiding Common Pitfalls 6.4. *Both teams should have identified the phrase "using ultrasonic technology" in an Objective Statement as superfluous during the Requirements Gathering process. The phrase is too focused on the technology that would be used to solve the problem, not the problem itself – this is the **how**, not the **what**. These types of phrases should be eliminated during the Requirements Gathering process. Without that preconceived idea tainting the Problem Statement, the team would have identified GPS technology as the optimal solution during the "Alternatives Selection" process (Chap. 12), the first process in the PLANNING phase, and saved the team a lot of time in the long run.*

EXAMPLE 7.10 Requirements Gathering Process #2

A 2013 Senior Design team was tasked to produce high-resolution underwater imagery for Olympic swim races. One Objective of this project was to "capture a single image of each swim-lane at the moment the first athlete finished the race." This would provide photographic evidence of who finish first. A tolerance of less than 0.01 seconds was allowed which proved extremely difficult for the team to achieve.

Eventually, the team questioned "**why** a single image?" The reasoning was to limit the data storage required. The team proposed a continuous video feed from which a single frame would be captured and saved. Doing so easily met the 0.01 second tolerance and required the same data storage as the Sponsor's original plan.

Analysis:

"Capturing" versus "Processing" a single image is a design decision. By the time the team asked "why," they had already purchased an expensive, high-end underwater camera. Although they found a significantly cheaper camera was able to achieve the Objective when the "single image" phrase was eliminated, the expensive camera was not eligible for a refund. Had this aspect been questioned during the INITIATING phase, before they started to build a more complicated solution, they would have saved themselves significant time and resources. If the data storage was a concern, it should have been documented as a Constraint.

- Ask to personally observe the environment.

   Firsthand knowledge of the Sponsor's concerns will help you fully understand their Motivation and Goal. This firsthand knowledge will also put much of the Requirements Document into perspective for you personally. Additionally, direct observations may allow you to take necessary measurements that will be important when developing values for your Specifications.

EXAMPLE 7.11 Requirements Gathering Process #3

A 2020 Senior Design team was tasked to automatically detect the presence of robotic cars passing a certain point along a "typical track." The Sponsor did not define a "typical track" with quantifiable values.

The team requested to personally observe the intended track and independently measure the distances and angles the sensor was required to cover.

Analysis:

This team gained valuable information during their personal observations. Although the Sponsor had appropriately described the track in a qualitative way, the students' interpretation was inaccurate and would have caused major technical issue down the road. These misconceptions were easily resolved during the personal observations.

EXAMPLE 7.12 Requirements Gathering Process #4

A 2017 Senior Design team was tasked to automate a quality control test cycle where a particular hard drive had to be repeatedly inserted/removed from a computer server. They were told that to articulate the drive, they would need to push a release button, pull a lever down, grab the hard drive, and pull or push it into position. This would have been a *complicated* mechatronic system with many moving parts.

Upon visually inspecting an actual test cycle, they realized they could remove the manufacturer's locking mechanism and replace it with one of their own. This would eliminate the need to press the button, move the lever, or grab the drive. The entire process could be simplified to simply pistoning the drive in and out. Limiting the automation to a single movement significantly reduced the design effort.

Analysis:

*This performance team was able to overcome the Sponsor's initial project description that focused too much on **how** the solution should operate. Even after being exposed to preconceived ideas as to how the solution had to operate, they were able to creatively analyze the actual problem, look beyond the preconceived notions, and develop a very simple solution. This problem had previously been addressed by professional engineers who could not look past the preconceived ideas and, therefore, were unable to develop a solution. The students' ability to focus on the real problem and creatively develop their own solution allowed them to produce a functional system for the Sponsor that is still in use today.*

Industry Point of View 7.1 Requirements Define all the Details

Understanding the requirements of a project may seem straightforward but you cannot, I repeat, take for granted that even some of the simplest details can and will trip you up.

I was the Project Manager for a small VLSI (very large-scale integration) design team in the mid-1980s. Each member of the team was experienced but had never worked together before. Our first design was as simple as it came; we were merely responsible for creating a digital interface between two other products. In fact, this was so simple, we never considered developing a formal Requirements Document. However, even with such a small team collaborating on an easy project, the lack of a formal document turned out to be a huge mistake.

Back then there was no standard convention for numbering bits. Today, we always refer to the most-significant-bit (msb, located on the left of a byte) as #7 and the least-significant-bit (lsb, located on the right of the byte) as #0. Although it was never common, back then some designers numbered the msb as #0 so that you counted up as your read left-to-right.

One of the products we were supposed to interface with used this reverse counting system, but we never considered that possibility. Without a formal Requirements Document, we made that same insidious mistake that has derailed so many other engineering projects – we assumed.

Industry Point of View 7.1, continued

We circumnavigated a few of the Design Life-Cycle processes and produced our first prototype in record time. But when we integrated our design into the larger system, nothing worked. After a lengthy and comprehensive review, we identified that our msb was connected to their lsb and our lsb to their msb.

The fix was easy enough, but the time required to rebuild a VLSI prototype back then was at least 6 months. We went from "ahead of schedule" to "dangerously behind schedule" overnight. Worse, the mistake delayed us from releasing the product to market. It was estimated that delay resulted in nearly $54 M of profit loss.

I have developed quite a bit of management experience since the 1980s. Occasionally, my newest engineers want to jump ahead, shortcut the process, and make assumptions. I know the young guys sometimes get frustrated by the process, a process they perceive to be slow, but I continue to insist we "get it right" on the first attempt. We are not perfect; we still make mistakes, but I have lost count of how many times the process has saved us. By taking the time to define *every* technical detail, we save time and money in the long run.

A Final Note about the Requirements Gathering Process

The Requirements Gathering process is always an iterative one, so be flexible and patient. Modifying a Goal Statement will certainly mean your Objectives must be rearranged. Identifying what was thought to be a Constraint as actually a Requirement will mean you need to develop new Specifications. Sometimes recategorizing a Requirement into an Objective will simplify the entire structure of the Requirements Document. As you gain more information and understanding, these modifications will be necessary to increase the quality of the entire document.

Start this process with the expectation that you will make numerous revisions before completing it. Simply knowing this will reduce your frustrations as you attempt "get it right."

Do not rush this process. Get started early, put more time and effort into it than you think is required, and make sure you give yourself time to make those revisions. In the end, you have a document that clearly defines the project and, just as importantly, will be easy to use in the upcoming processes. A well-constructed Requirements Document will make many of the upcoming processes easier to navigate. So, take the time to produce an accurate, organized, and thorough Requirements Document!

Industry Point of View 7.2 Keep Asking "Why"

I am the Chief Sales and Marketing Officer (CSMO) for a renewable energy consulting firm. We perform feasibility studies for distributed generation installation and occasionally design customized microgrids. Half my job consists of developing and monitoring Requirements for our customers.

My favorite question to ask is "why." "Why" is that a requirement, "why" is that necessary, "why" cannot we do this other thing. I never know whether my customer's response is "correct," but if they cannot clearly articulate an answer, I know the issue is not resolved enough to document in a Requirements Document.

Recently, we designed a microgrid for an interactive display at one of our National Park visitor centers. Initially, we were asked to design a PV + Battery microgrid. Usually, I do not include this level of design detail in a Requirements Document, so, of course, I wanted to know "why;" why must we use PV power and why must we use battery storage. During our discussions, I realized that our Sponsor did not have an engineering background and was under the impression that *all* microgrids consisted of a PV + Battery solution.

It turned out that funding for the project was partly based on a large federally funded solar energy incentive. So, using PV was, in fact, a Project Constraint. However, this display taught visitors about solar distillation and could only be used with a direct line of sight to the sun. Of course, that meant that the PV panels also would have direct solar access whenever the display was in use. The batteries were not actually a required component of the system. By eliminating batteries from the design, we cut the cost of the project by 35%.

To design an optimal microgrid, it is my responsibility to completely understand the design space. This includes the load that the microgrid will power, the demographics of the user, both a typical use-case and worst-case scenario of the system, the financial resources available to the project, and any special permitting required for construction, just to name a few.

I remember that developing a complete background for my Senior Design project was a daunting task – there was so much I did not know and so much I had to do. But now that aspect of a design is my favorite part of my job. Our technical solutions become so much more meaningful when we understand how the technology affects real people. I admit that, as a student, I did not comprehend how the interpersonal aspects of engineering are related to the career field. After a few years of experience, I now feel that those aspects are actually the most important aspects of my job. I still enjoy the technical portion of the job too but only when put into the context of how a certain technical solution made life better for somebody that I know personally.

7.6 Concept Check

Examples 7.13 and 7.14 provide a variety of valid statements for two different projects so that you can practice identifying whether each statement is either a Requirement, Specification, or Constraint. Examples 7.15 and 7.16 format those statements, respectively, into the formal Requirements Document structure.

EXAMPLE 7.13 Requirements Document for Mars Rover Exhibit

Identify each statement below as either a Requirement, Specification, or Constraint:

1. The replica must communicate with a "ground station" in a manner similar to how NASA communicates with the real Mars Rover.
2. Communication between the base station and rover must be delayed representing travel time to Mars.
3. Communications must be delayed at least 10 seconds.
4. Instructions must be sent in packets.
5. Instructions must be limited to ten movements per packet.
6. Exhibit must contain two Mars Rovers.

Analysis:

Item 1: The student team identified this as an Objective statement describing a fundamental feature of the desired exhibit. Depending upon the greater context, this statement may also have been characterized as a Requirement. But, for this example, we will assume the statement was correctly identified as an Objective.

Items 2 and 3: Item 2 is a Requirement defining Objective 1. Item 2 is a Specification that defines the delay demanded by Requirement 2.

Items 4 and 5: Item 3 is a Requirement further explaining how scientists really communicate with the rover. Item 5 is a tricky Specification. One of the methods to determine whether a statement is a Specification is to ask, "can this be measured?" We could measure whether ten instructions are sent, but we could also simply check with a yes/no statement. Therefore, this might easily be mischaracterized as a Constraint. However, sending ten instructions/packet is something the solution does. Also, this statement is further defining a Requirement. Therefore, it is a Specification. Contrast this with item 5.

Item 5: This statement is a Constraint. This could also be counted like item 4, but it would be difficult to describe a test where two rovers were measured. Additionally, this statement is not something the exhibit actively does during its operation; the statement describes the final deliverable. Therefore, it is a Constraint.

EXAMPLE 7.14 Requirements Document for a Nameplate Reading System

Identify each statement below as either a Requirement, Specification, or Constraint.

1. The solution will need to locate nameplates on the top of buildings.
2. The solution will need to locate nameplates of a variety of sizes.
3. The height of the buildings can reach 40 ft.
4. The smallest nameplate is 6″ × 4″.
5. A single operator must be able to control the solution from the ground.
6. The solution will image the nameplates using a GoPro camera.

Analysis:

Items 1 and 2: These are Requirement Statements. They further define the Objective Statement "The solution must locate the nameplate."

Items 3 and 4: These are Specifications. They quantify the height of the building and the size of the nameplates. Remember, there should be at least two Specifications for each requirement. Can you think of an additional Specification for both statements 1 and 2?

Items 5 and 6: These are Constraints. They limit the solution to include a single operator and the type of acceptable camera.

EXAMPLE 7.15 Requirements Documents for the Mars Rover Exhibit

After categorizing each statement in Example 7.13, the team organized the statements into the properly structured outline. A portion of a Requirements Document for this project might read as follows:

Obj 1 – The replica must communicate with a "ground station" in a manner similar to how NASA communicates with the real Mars Rover.

Req 1.1 – Communication between the base station and rover must be delayed representing travel time to Mars.
Spec 1.1.1 – Communications must be delayed at least 10 seconds.
Req 1.2 – Instructions must be sent in "packets."
Spec 1.2.1 – Instructions must be limited to ten movements per packet.

Analysis:

The basic outline structure of the Requirements is correctly developed here. The Specs are each assigned with a three-digit number and are indented from the related Requirements. Each Requirement is assigned with a two-digit number and indented from the related Objective. Notice that the second line of Obj 1 and Req 1.1 aligns with the first line of their statements. Follow the formatting instructions provided in Sect. 7.4: to accomplish this.

This Req. Doc. is not yet complete. There should be between three and five Objectives, each including a number of Reqs, and each Req should be defined with more than one Specs, if at all possible. What is shown here is a good start to defining the first Objective but still would need further development before completion.

> ### EXAMPLE 7.16 Structuring the Req. Doc. for the Nameplate Reader
>
> After categorizing each statement in Example 7.14, the team organized the statements into the properly structured outline. A portion of a Requirements Document for this project might read as follows:
>
> Obj 1 – This solution will need to remotely locate nameplate.
>
>   Req 1.1 – The solution will need to locate nameplates on the top of buildings.
>   Spec 1.1.1 – The buildings can be 40 ft. tall.
>   Req 1.2 – The solution will need to locate nameplates of various sizes.
>   Spec 1.2.1 – The smallest nameplate is 6″ × 4″.
>
> ---
>
> Analysis:
>
> *Objective 1 has two requirements, the bare minimum. Each Requirement only has one Specification. This indicates that the possibility that the Req. Doc. is incomplete or poorly organized. You will likely be questioned on the validity of the document if you leave the structure like this. That does not necessarily mean you are wrong, but you should reconsider the document and verify you have not omitted something important. If you are sure you have not, be prepared to defend your work.*

7.7 Chapter Summary

In this chapter, you learned about the Requirements Document. The Requirements Document starts with each Objective included in the Problem Statement. The Objectives are each further defined by a set of Requirements. Each Requirement is quantified by a set of Specifications. Any limitations that are placed on the design are listed as Constraints.

Here are the most important takeaways from the chapter:

- Requirements define (or describe) the functionality of the solution.
- Constraints limit the design options available to the engineer.
- Specifications can be quantifiably tested, typically with a piece of equipment.
- There are times when software Requirements do not need Specifications. This exception to the rule should be used *very* sparingly.

7.8 Case Studies

Let us review portions of our case study teams' Requirements Documents. For each team, a segment of their Requirements Document is presented on the left-hand side of the page with comments from the author on the right-hand side. You may wish to refer back to the Problem Statements (Sect. 6.7) for each team to fully understand the context of the Requirements Document.

Augmented Reality Sandbox

Obj 1: Design the structure of sandbox.
    Req 1.1: Must be sturdy enough to withstand public use.

        Spec 1.1.1: Withstand 90,000 kids per year.

    Req 1.2: Sandbox should be big enough to play with.
        Spec 1.2.1: must be 40″ × 30″
    Req 1.3: Sandbox must fit within existing display space.
        Spec 1.2.1: must be less than 8′ tall.

Obj 2: Must project a clear image.
    Req 2.1: Sensor must be close enough to sand to detect topography.
    Spec 2.1.1: less than 10′ tall
    Req 2.2: Projector must have a small throw ratio.
    Spec 2.2.1: Throw ratio of 0.9

The Objectives listed here do not precisely match those in the Problem Statement.

Spec 1.1.1 would be difficult to quantify. What does it mean to withstand 90,000 kids/year?

This structure does not look like a "tree." Each Requirement only has 1 Spec. When you complete the Verification Process, you must conduct a test for each entry. Having multiple Specs under 1 Requirement will drastically reduce your workload.

Spec 2.1.1 is superseded by Spec 1.2.1. Is it necessary to document this distance when the height of the overall system is already limited to 8 ft?

Smart Flow-rate Valve

Obj 1: Must monitor system parameters.
    Req 1.1: Must monitor flow rate.
        Spec 1.1.1: Accurate to within 3%
        Spec 1.1.2: Range from 2 to 80 gal/min
    Req 1.2: Must monitor valve position.
        Spec 1.2.1: Accurate to within 1%

Obj 2: Must dispense liquid at desired rate
    Req 2.1: System must be responsive.
    Spec 2.1.1: Must reach initial steady state with 1 s.
    Spec 2.1.2: Must have less than 2% error
    Spec 2.1.3: Must respond to disturbance within 0.5 s.

This was a good start to defining Obj 1. However, the Problem Statement does not specify why the system must monitor the valve position. This "requirement" may not have been necessary to include.

The Objectives listed in the Problem Statement included the ability to set the desired rate using a prescription map or a manual override. The ability to input the desired rate from two sources was not specified in the Requirements Document. During the Requirements Gathering process, the ability to read the prescription map was eliminated from the student's scope of work, but the Problem Statement was never updated. As the project develops, you may have to revisit previous process so that they all flow together

Robotic ESD Testing Apparatus

Obj 4: RESDTA (robotic electrostatic discharge testing apparatus) must perform ESD tests

Req 4.1: Must accurately locate test points.

Spec 4.1.1: Operator must be able to easily place DUT in home position within ± 0.005".

Spec 4.1.2: RESDTA must place gun tip on a test point within ± 0.05".

Spec 4.1.3: RESDTA must locate up to 50 points on the DUT.

Spec 4.1.4: RESDTA must be able to find a combination of desired test points up to 500 times.

Req 4.2: Must deliver ESD shocks.

Spec 4.2.1: Must deliver air shocks with gun tip 0.05" + 0.005" from surface of DUT.

Spec 4.2.2: Must deliver contact shocks with less than 1 lb. of force on the DUT.

Req 4.3: Must quickly find test points.

Spec 4.3.1: Operator must be able to place DUT in home position within 2 minutes.

Spec 4.3.2: RESDTA must move in manual mode faster than 6 in/s.

Spec 4.3.3: RESDTA must move in test mode less than 0.5 in/s.

The term "easily" does not belong in the Spec. Consider "how would you quantitatively measure "easily""?

Be careful to consider cascading tolerances. If the DUT was placed + 0.005" off center, is RESDTA still allowed 0.05" of tolerance, meaning a total tolerance of 0.0055", or should this spec have been to place the tip within 0.045" which could result in a total tolerance of 0.05"?

Notice that Req 4.2 refers to delivering shocks, but the Specs relate to tip placement – these Specs should have been included under Req 4.1. There should have been additional Specs listed here defining "shocks."

The team considered all aspects of operation while developing their Req. Doc. See how RESDTA specified both the operational speed and the setup speed. Consider the user's point of view when building the Req. Doc.

Chapter 8: **Final INITIATING Processes**

Although a *potential* project was identified during the STARTING phase of the Design Life-Cycle, that does not necessarily mean Stakeholders are ready to commit large sums of money toward an unvetted concept. The intent of the INITIATING phase is to make sure all Stakeholders understand and agree upon the *details* of the project.

The first step in convincing Stakeholders to move forward with the project, as we have already discussed, is to define the problem carefully and completely. This work is documented in the Problem Statement and Requirements Document. Next, we must determine how risky it is to pursue a solution to that problem. Significantly more information is required before the Stakeholders can assess the risk of going forward. The final initiating processes will provide that information.

Much of the additional information required to assess the project is developed during an investigation called a Literature Review. During this process, you must completely document the context within which the solution will reside, investigate any applicable industry standards, and explore potential solutions. The Literature Review is presented as part of the "background" chapter of the Project Charter (Sect. 10.3).

Another aspect of the Stakeholders' risk assessment will be the project schedule. Stakeholders will want to know whether the project can be completed within a reasonable timeframe. Although the details of the project are still unknown, a *rough* project schedule should be created to verify whether the project is feasible during the timeframe allotted. A rough project schedule can be used to estimate whether the Triple Point Constraint can be balanced given the time allotted to the project. Since a Senior Design project has a very precise allotted timeframe, the Scope and Resource legs of the triangle may need to be modified after the Literature Review.

© The Author(s), under exclusive license to Springer Nature Switzerland AG 2023
C.J. Mettler, *Engineering Design*, https://doi.org/10.1007/978-3-031-23309-8_8

Finally, two important visuals are developed near the end of the INITIATING phase in preparation for the Project Kickoff Meeting. During this meeting, you will be required to explain everything you learned and developed for this project, meaning that you will have to condense many weeks of work into a few minutes. Visual aids are necessary to do so effectively. The Project Schedule is presented using a tool called a Gantt Chart. The Requirements Document is summarized using a tool called a Conceptual Block Diagram.

Depending on the complexity of the project, a number of additional processes may also be enacted. These could include an initial budget analysis, a stakeholder management plan, a communication management plan, and a risk analysis. Due to the complexity of a typical Senior Design project, and the time allotted for the INITIATING phase, this textbook will not address these processes.

In this chapter, you will learn:

- What to include during your Literature Review.
- How to use a Gantt Chart for schedule management purposes.
- How to generate a Block Diagram that will effectively communicate the Requirements Document to an audience.

8.1 Literature Reviews

By now, you have developed an articulate Problem Statement, so you should understand the problem at a high level. However, the Problem Statement is limited to a few sentences. Significant additional background is required. The team must now put the Problem Statement into the appropriate context by gathering more information on the details of the problem including the environment your solution will operate within, existing solutions, possible regulations, and many other details. This process is called a **Literature Review**.

A **Literature Review** provides the background necessary to fully understand the project.

There are many issues which a Literature Review might address. Although no two projects will require exactly the same considerations, here is a list of common issues you should consider when researching your project.

- In what environment will your solution operate?
- What is the existing status of your project (e.g., is this completely new or are you taking over an existing project, was this attempted by someone in the past)?
- Who will use your solution? What characteristics might they exhibit?
- Are there any major competitors producing similar solutions?
- Are there any legal regulations that must be adhered to?
- What technology is available that you can leverage while designing your solution?

There are a variety of sources which you should investigate while conducting your Literature Review, including the following:

- Interviews with your client
- Discussions with experienced engineers or advisors
- Engineering journals and publications
- Critically evaluate and cited (Sect. 10.1) Internet searches

You should carefully document all relevant information in your Engineering Notebook while you perform this review. You will need to report on any pertinent information at the end of the INITIATING phase and will need to cite your references. So, documenting the sources as you initially review them will be a significant time-saver.

The Literature Review is an ongoing process that requires significant effort. It will be formally conducted during the INITIATING phase but continue to grow and develop throughout the Design Life-Cycle.

The importance of this Literature Review cannot be overstated. A thorough review can significantly reduce risks that your project may incur and dramatically improve project efficiency. Ultimately, you will have completed a thorough Literature Review when you know *everything* there is to know about your engineering problem. Knowing "everything" is a high standard which cannot truly be achieved. But the more effort you put into the review now, the fewer surprises you will encounter later.

Components of a Literature Review

There are three primary components of a Literature Review.

- Project Background (i.e., Project Context)
- State-of-the-Art Concerns (i.e., Technology Review)
- Applicable Standards

Although the exact order of these primary components is somewhat arbitrary, it often is best to start with a complete description of your project's background in order to put the rest of the information into the appropriate context.

Project Background Component of the Literature Review:
The details of the **Project Background** will vary drastically between projects. What is important for you to include within this topic is a full description of the problem's context. Recall that the Problem Statement had to be delivered in one short paragraph. This typically includes one Motivation Statement, one Goal statement, and a few Objective Statements. In many cases, the Problem Statement might also include one or two sentences of background or supporting information as well. Those supporting statements are rarely enough to fully explain your project's problem. The Project Background component of the Literature Review is meant to provide a *complete* explanation.

The term Project "Background" may inadvertently infer that the project currently exists and has a history. That may or may not be true. Certainly, if previous iterations of the same project have been attempted, then that history should be included here. Alternatively, if your project is meant to somehow interface with an existing product, that background should be included as well. Even if no one has ever attempted to address your Project Motivation, there is still likely Project "Background" or "Context" to provide. You may need to provide a more detailed description as to *why* the Motivation exists or why the problem should be solved. You may need to discuss an existing solution that your project is intended to replace or a process that your solution will rely upon. Examples 8.1 and 8.2 will provide topics which previous Senior Design teams encounter during their respective Literature Reviews.

EXAMPLE 8.1 Potential Background Topics

The Project Background component of the Literature Review will vary drastically between projects. This example is only meant to give you a starting point for your own considerations.

Consider this Problem Statement for the Industrial Exhaust Fans nameplate reader we have already discussed.

Industrial exhaust fans generate a lot of excess heat and therefore are installed on top of buildings to promote natural cooling. Reading industrial exhaust fan nameplates is necessary for correct maintenance but is often difficult when fans

---

EXAMPLE 8.1, continued

---

are installed in hard-to-access locations. Therefore, a method to remotely read nameplate information of industrial exhaust fans will be developed. This solution will need to remotely locate the fan, locate the nameplate, and record a high-resolution image of the nameplate.

Here is a representative list of possible background topics for this project:

- How tall are these buildings?
- How far away (remote) should an operator be?
- What information is on an Industrial Fan's nameplate?
- How is this information printed?
- How many nameplates will have to be read in a single study?
- How often will the nameplates be read?
- Why are nameplates difficult to read using the current method?

EXAMPLE 8.2 Potential Background Topics #2

Since each project has dramatically different concerns, the topics of the Literature Review may be significantly different from project to project. Consider the Mars Rover Exhibit team's Problem Statement and how it drove a different list of Literature Review topics than the Nameplate Reader project.

Problem Statement:

The Kirby Science and Discovery Center has a Mars Rover exhibit which they would like to modernize. The existing exhibit is out of date and a number of features are no longer operational. The new Mars Rover exhibit must allow children to interact with a Mars Rover replica. The Mars Rover replica must communicate with a "ground station" in a way similar to how scientists actually communicate with the real Mars Rover.

Literature Review Topics:

- How large of a space will the exhibit encompass?
- What is meant by "ground station" in terms of the exhibit?
- How many children will interact with the exhibit at a time? In a day? Over the course of a year?
- What is the mission of the Kirby Science and Discovery Center, and what do they want children to take away from a standard exhibit?
- What is wrong, or not operational, with the original exhibit?

The specific set of questions an audience might be concerned about for the Nameplate Reader project is significantly different than the set of questions related to the Mars Rover project. However, the two sets do have some similarities. The topics that should be covered in the Project Background component of the Literature Review develop the understanding of the *problem* in greater details. These topics should include issues like the following:

- The physical size of problem space:
  - How tall are the buildings?
  - How large is the exhibit?
- How a user might engage the solution:
  - How far away should the user be?
  - What should children learn from the exhibit?
- What is the concern with the existing situation?
  - Why is reading nameplates difficult using the current method?
  - What is wrong with the current Mars Rover exhibit?

This is only a short list of possible topics. You will need to investigate your project in enough detail so that you completely understand the Sponsor's concerns. The Background section of the Literature Review should convey your understanding to a broad audience.

This component of the Literature Review should almost certainly contain figures or pictures.

The use of figures and/or pictures in this section will dramatically improve the reader's understanding of the problem while also reducing the number of details you are required to write.

State-of-the-Art Component of the Literature Review

The **State-of-the-Art** component of the Literature Review is a fancy term for a Technology Review. This section should focus on technologies specifically related to your problem.

Again, your project will include a unique set of topics, so it is difficult to define precisely what should be included here. You should use your judgment when investigating this material. This section will provide a few representative examples of issues you may want to consider.

It is possible that a technical term was used in the Problem Statement that must be explained to a general audience. Often, technology already exists that could be used to reduce the engineering efforts on your project; this could be a product, method, or a technique. Sometimes a complete solution may exist for certain aspects of your project. Other times intellectual property exists that prevent you from solving your problem using a particular method. If a solution does exist that completely solves your problem, you should explain why the Sponsor should continue with *your* project rather than simply purchase this preexisting solution.

This section of the Literature Review should be well documented with numerous references.

The State-of-the-Art discussion should contain many reference citations.

Performing Internet searches on keywords is a good place to start investigating this component. Another source of many State-of-the-Art Literature Reviews is patent searches. Example 8.3 provides a few examples of State-of-the-Art topics.

The State-of-the-Art topics presented in Example 8.3 focus on potential solutions and related technology. At this point, you should not be deciding upon any solution; this will cause problems later in the design process. However, one aspect of the Literature Review is to compile as large of a list of *possible* solutions as you can. The purpose is to develop as much knowledge as you can so that you have options later.

Sometimes, determining whether a topic is "Project Background" versus "State-of-the-Art" is difficult. For example, the Mars Rover Exhibit design team included a discussion on how scientists communicate with the real Mars Rover in their State-of-the-Art discussion.

EXAMPLE 8.3 Potential State-of-the-Art Topics

This example presents some of the state-of-the-art topics the two projects discussed in Examples 8.1 and 8.2 included in their Literature Reviews. Only a few representative topics are presented here for each project. The final state-of-the-art component of their respective Literature Reviews actually each discussed *many* additional topics besides those presented here.

Nameplate Reader Literature Review:

- Review of commercially available drones including size, payload capacity, imaging capability, and cost.
- Discussion of image processing techniques specifically focused on text recognition, resolution, and recording.
- Review of customizable carrying cases for a drone. *(This topic was not important enough to directly mention during the Problem Statement but was a concern of the client.)*

Mars Rover Exhibit Literature Review:

- Diagrams and images of a "modern" Mars Rover
- Research and interview responses related to how scientists communicate with the real Mars Rover
- A review of existing Mars Rover exhibits at other museums

However, that topic *might* have been considered background information about the existing situation; it would not have been wrong to do so.

Here are few general guidelines for when you are attempting to categorize your topics:

- If the topic is relating to the *problem*   → Project Background
- If the topic is relating to the *solution*   → State-of-the-Art
- If the topic is relating to something that currently   → Project Background
  exists *for your client*
- If the topic is related to something that could   → State-of-the-Art
  be used *by your client*

Applicable Standards Component of the Literature Review

The **Applicable Standards** component of the Literature Review documents any industry regulations, standards, or guidelines to which your project must adhere. Governmental regulatory bodies provide many *regulations*. For example, the Federal Communications

Commission (FCC) regulates the frequency, bandwidth, and transmission power that certain communication devices are allowed to use. Failure to adhere to such regulations is a legal issue and one that should be carefully avoided. Professional societies such as IEEE (Institute of Electrical and Electronics Engineers) provide industry *standards* to ensure that devices work well together. For example, a renewable energy developer must document compliance to IEEE 1547–2018 before receiving permission to connect their design to the US power grid. Failure to adhere to standards is not usually a legal issue, but your design may not be allowed to interface with existing products if you fail to meet these standards. Additionally, there are many other sources of *guidelines* that should be followed as a matter of best practice but may not actually be mandatory. An example of a guideline that may not be mandatory is Ingress Protection (IP) Code 67 or 68. These two codes define how "waterproof" a device is; IP67 means a device can be submerged in water to 1 meter, and IP68 means that device can be submerged to 1.5 meters. Some smart phones meet IP67 ratings; others do not. Meeting a particular IP rating is not mandatory for a product to be sold. However, documenting that your product *does* meet the guideline allows you to claim that your product is better, and therefore more valuable, than your competitors. Any standards that apply will need to be properly sited in your Project Charter.

Not all Senior Design projects are subject to Applicable Standards. After a thorough investigation of possible concerns, if you believe that your project is not subject to any regulations, standards, or guidelines, document what you have investigated and claim in the Literature Review that "no standards apply."

Make sure to consult with your project advisor regarding this topic, particularly if you believe there are no Applicable Standards that apply to your project. If your client has a technical background, they should also be consulted.

8.2 Using a Gantt Chart

During the INITIATING phase, a rough Project Schedule should be developed. This schedule will define the major milestones which the project team must meet. At the very least, the Project Schedule should, at this point, include milestones such as the following:

- Dates of any reviews (PDR, CDR, FDR)
- Dates of the final deliverable
- Time periods during which each of the Design Life-Cycle phases will be addressed

The rough schedule is occasionally developed by the design team itself and other times dictated to the design team by an administrator. Typically, in Senior Design, course instructors will have created the rough project schedule for the class at this stage. For now, you should be aware of this schedule. During the INITIATING phase, you will become familiar with the basics of schedule management. Then, during the PLANNING stage, you will learn more schedule management skills. Eventually, you will be asked to modify the rough schedule with your project's unique details. All of this scheduling is accomplished using a tool called the **Gantt Chart**.

A Gantt Chart is used to manage a project schedule over an extended period of time. This visual representation provides detailed information in such a way that a calendar or list of due dates cannot. On simple projects, or projects which are allotted with an abundance of time, Gantt

Charts may not be necessary. However, when there are many resources to manage, the project timeline is particularly long, or the project schedule is particularly tight; it is critical to use the time and resources allotted as effectively as possible. The Gantt Chart will aid in this effort.

A Gantt Chart can be used to track due dates and task assignments; however, it is much more useful than that. This tool will allow a Project Manager to easily determine whether all members of a team are being tasked effectively and that the order in which assignments are scheduled is as optimal as possible.

It is extremely difficult to schedule every task accurately at the start of a project. Of course, you will want to develop a realistic schedule, but it is not a reasonable expectation to determine every detail up front. Many tasks will require more time than you originally anticipate, some tasks will finish earlier than expected, and you will likely encounter a few tasks which you did not initially account for at all in the initial schedule. Fortunately, attempting to determine all the details up front is not necessary.

> The purpose of the Gantt Chart is to **manage the unexpected changes** to the project schedule that you will encounter in a year-long project.

This section will explain the purpose of a Gantt Chart and instruct you on how to use one that has already been created. You will learn to develop your own Gantt Charts during PLANNING phase (Chap. 11).

Standard Components of a Gantt Chart

Gantt Charts may look slightly different depending on the software package used to create it. Microsoft PROJECT is an industry standard software package that is often used. PROJECT is powerful but expensive. Fortunately, there are many cheaper or open-source options that will suffice for most projects. This textbook will make use of an application called *Smartsheet*. Regardless, all Gantt Charts have the same basic features.

Figure 8.1 displays the most important features of a Gantt Chart. The top of a Gantt Chart displays a calendar (e.g., we can see the weeks of July 4th and 11th). A "dateline" (e.g., the vertical dotted line) progresses along the calendar to indicate the current status of the project. The dateline on this figure is the vertical dotted line approximately in the center of the figure; this figure was created early Wednesday the week of July 4th.

The first vertical column provides a list of tasks to be completed; in this case, there are three tasks. Tasks are represented by colored bars on the graphical portion of the Gantt Chart. Tasks can be grouped for convenience. The other item listed in this column is a "milestone" or due date. Here we can see that three tasks must be completed before Milestone 1.

We can also see that Task 2 and Task 3 are dependent upon Task 1; we know this because of the arrows leading from Task 1 to the other tasks. The arrow leading from the end of Task 1 to the beginning of Task 2 is called a Finish-to-Start dependency; Task 1 must finish before Task 2 can start. This is the default dependency. The arrow leading from the beginning of Task 1 to the beginning of Task 3 is called a Start-to-Start dependency. This means that the start of Task 3 should occur simultaneously with Task 1.

This Gantt Chart also displays who is responsible for each task. This visual representation will make it easier to arrange assignments. Here we see Clint is assigned to Task 1, whereas Abodh is assigned to Task 2, and Maria is assigned to Task 3.

The lightly shaded portion of the taskbar represents the completion status of the respective tasks. Here we see that Clint has started his task but is behind schedule; this is indicated by the shaded portion of the task bar ending prior to the dateline. Maria, on the other hand, has made progress on her task and is actually ahead of schedule; this is indicated by the shaded portion of her taskbar extending past the dateline. Abodh has not started his task; this is acceptable because his task was not scheduled to start until the following week.

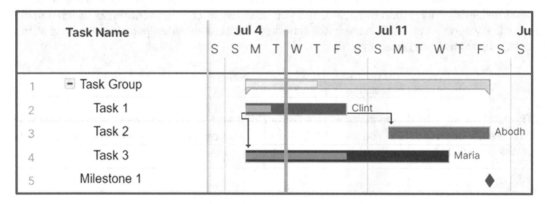

Figure 8.1 Standard components of a Gantt Chart

The Purpose of a Gantt Chart

The purpose of a Gantt Chart is to help you manage a complex schedule. This includes scheduling due dates, called milestones, and allocating engineering resources appropriately to meet these milestones. More importantly, the Gantt Chart is used to monitor your progress and manage any necessary modifications to the schedule.

It is *expected* that your project activities will not precisely match the planned schedule. This is normal and unavoidable. However, deviations from the planned schedule *can* cause major problems, even project failure, if left unchecked. So while it is not important for the project activities to *exactly* match the schedule, it is critical that any deviations are detected early and accounted for in the management plan.

> The Gantt Chart is used to detect deviations from the planned schedule and develop acceptable modifications to avoid project delays.

Understanding that you do not have to perfectly predict reality months in advance will make developing the Project Schedule much less daunting. Examples 8.4 and 8.5 provide more detail on the two primary purposes of a Gantt Chart: initial scheduling and adapting to schedule changes.

EXAMPLE 8.4 Using a Gantt Chart #1

A Gantt Chart is used to *efficiently* schedule tasks for your team members. Let us use the situation portrayed in Fig. 8.1 to discuss this further.

In order to complete your project on time, each individual team member will need to consistently contribute the expected hours to the project every week.

On the original Gantt Chart, we clearly see that Maria has been fully tasked, meaning she has an assignment for both weeks. However, Clint and Abodh have each been assigned only 1 week worth of work. This needs to change! But how?

First, we need to establish whether it is necessary that Abodh is assigned to Task 2 or not. If not, we could assign the sequence of Tasks 1–2 to Clint and free up a team member to address another technical issue as shown in Fig. 8.2. This solution allows Clint and Maria to complete the tasks necessary for Milestone 1 and frees up Abodh to make progress on another task group.

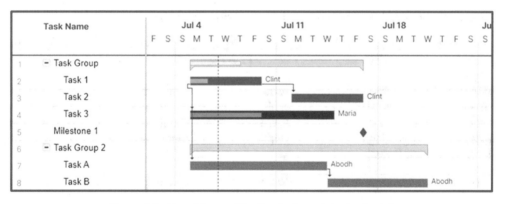

Figure 8.2 Possible modification to the original schedule

It is equally conceivable that it *was* necessary to assign Task 2 to Abodh, and the modification presented in Fig. 8.2 is not feasible. In this case, it is still necessary to assign Abodh and Clint 2 weeks' worth of work during this time period. There are often many extra tasks to consider that can be used to fill these gaps. Some examples include preparing for upcoming presentations, writing reports, or researching ideas for upcoming issues. Figure 8.3 provides one alternate option.

EXAMPLE 8.4, continued

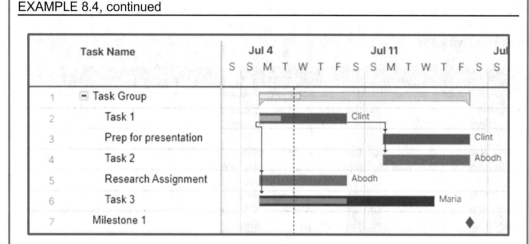

Task Name		Jul 4							Jul 11						Jul
		S	S	M	T	W	T	F	S	S	M	T	W	T	F	S	S

1   ⊟ Task Group
2   Task 1   Clint
3   Prep for presentation   Clint
4   Task 2   Abodh
5   Research Assignment   Abodh
6   Task 3   Maria
7   Milestone 1   ◆

Figure 8.3 Another possible modification to the original schedule

EXAMPLE 8.5 Using a Gantt Chart #2

Although scheduling tasks with a Gantt Chart is one use of the tool,

> The primary use of a Gantt Chart is to detect deviations from the planned schedule and develop modifications to avoid significant delays.

Let us use the modified schedule presented in Fig. 8.3 and assume this is the schedule presented during a team meeting (recall all team meetings should start with a review of the Project Schedule). Upon a quick glance, we see that the Progress Line for Task 1 is behind the date line, suggesting Clint is behind schedule. Considering he is only 1 day behind; this will be easy to address.

It is possible that Clint was unable to dedicate much time to the project over the past few days. This is understandable given students' busy schedules as long as Clint plans to make up the time. A short discussion of Clint's plans might alleviate any concerns and the meeting can continue.

However, another possibility is that Clint has worked very hard on this task and is stuck. If this is the case, it is likely, the completion of this task will be delayed. The Gantt Chart makes it easy to observe the ramifications of this delay. Figure 8.4 shows the new expected timeline for Task 1 and how it affects Task 2 – Abodh will not be able to start Task 2 on time and, therefore, is not even scheduled to complete Task 2 until *after* Milestone 1.

It is very easy to portray these results in Smartsheet. You can simply drag the end of the taskbar to the right in order to extend the duration. Then, all tasks dependent upon Task 1 will automatically update for you.

EXAMPLE 8.5, continued

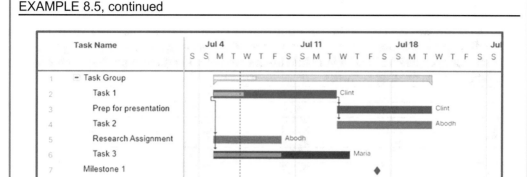

Figure 8.4 Ramifications of a delayed task

Somehow this schedule must be adjusted so that Task 2 is completed before Milestone 1. There are many ways to do this, including the following:

- Abodh could commit to completing Task 2 in 3 days rather than five. This is the easiest method to make the graphic look nice but has a large risk of failure. This is not the best option.
- Maria is ahead of schedule and her task should end early. She could pause the work on Task 3 and support Task 1. This plan has less risk than the first option, but what happens when Maria becomes focused on Task 1 and loses the momentum on Task 3? She might jeopardize Task 3.
- Abodh's Research Assignment was a time filler; the priority is completing Milestone 1 on time. He could team up with Clint to speed up Task 1. This is the best option.

Using a Gantt Chart during the INITIATING phase

Developing an initial project schedule can be a time-consuming and difficult chore, but your ability to complete this chore will improve with experience. Fortunately, the Project Schedule during the first many weeks of Senior Design will look similar for most student teams. Therefore, rather than having teams create their own Project Schedule, one may be provided to you by the course instructor. At a minimum, this schedule should include the major due dates (milestones) of the course and the expectations of the first few weeks.

This initial schedule will provide a strong framework for you to eventually build a more detailed schedule for your specific project once you have a better idea what the Requirements are for your unique project. In the meantime, you can use the provided initial schedule to become accustomed to using a Gantt Chart and effectively guide your team through the first critical weeks of the project.

Specifically, use this tool in the following ways:

- Refer to the Gantt Chart at the start of every team and advisor meeting. Verify you are working on the correct tasks for a given week. Keep an eye on what is coming up.
- Track your progress. When a task has been complete, update the shaded portion of the taskbar to indicate that progress.

"Programming" a Gantt Chart

In order to create the Gantt Chart, a number of fields must be programmed. These can be uncovered by sliding the dividing line between the programming fields on the left and the graphic on the right back and forth to reveal as much/little text as you desire. This dividing line is highlighted orange in Fig. 8.1; this is the default presentation position. Sliding the dividing line far to the right produces the view shown in Fig. 8.5.

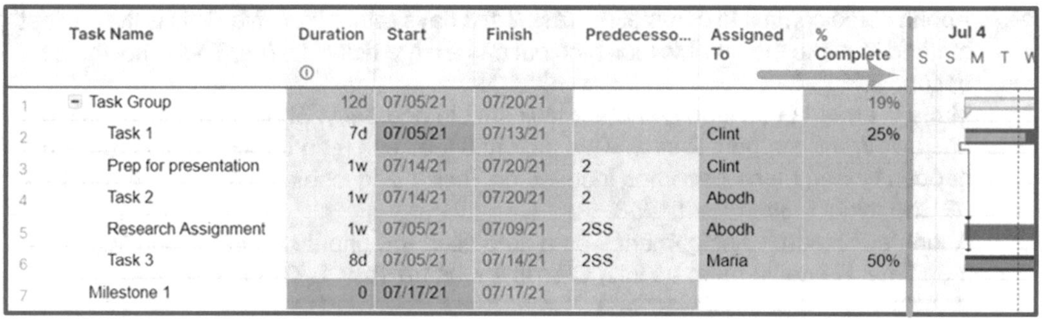

	Task Name	Duration	Start	Finish	Predecesso...	Assigned To	% Complete	Jul 4
1	— Task Group	12d	07/05/21	07/20/21			19%	
2	Task 1	7d	07/05/21	07/13/21		Clint	25%	
3	Prep for presentation	1w	07/14/21	07/20/21	2	Clint		
4	Task 2	1w	07/14/21	07/20/21	2	Abodh		
5	Research Assignment	1w	07/05/21	07/09/21	2SS	Abodh		
6	Task 3	8d	07/05/21	07/14/21	2SS	Maria	50%	
7	Milestone 1	0	07/17/21	07/17/21				

Figure 8.5 Programming fields of the Gantt Chart

When a new **task** is created, you should enter the expected **duration** and the number of that task's **predecessor** (columns highlighted in yellow). The **Start** and **Finish** dates will automatically be calculated. *Do not* program these directly for "normal" tasks. Doing so will lock the task into place so that it will not move when you want to adjust tasks later.

There are a few exceptions when you will need to directly program the Start and Finish dates (highlighted in blue). First, the start date of the *first* task must be programmed directly since there is no possible predecessor before the first task. Also, all **Milestones'** start dates should be directly entered. This is because milestones are *due dates* and, typically, they are out of your control. This means

You do not want milestones to move as you adjust the tasks schedule.

Another exception occurs when creating a milestone. The "duration" of a milestone is 0 (zero); it is a moment in time. Entering 0 into the duration column will automatically change the symbol in the graphic from a taskbar to a milestone diamond.

There are a number of fields you should not program directly (gray cells with red text); these will update automatically based on the other fields you do program. In particular, you should not program any dates (other than those already mentioned). Nor should you program any field related to a task group other than the group's "name."

Finally, you should assign each task to a team member (highlighted green). In most cases, you should assign each task to exactly one member. This does not mean the other members are not responsible for helping with that task. Rather, a task assignment means one person has the responsibility of tracking and reporting on that task. Typically, they complete the majority of the work but not necessarily all of the work.

Also highlighted in green is the **Percent Complete** column. Use this column to create the lightly shaded portions of the taskbars to indicate your progress.

Displaying Gantt Charts

Notice that the name of this tool is a Gantt *CHART*, not a Gantt Table. Gantt Charts are useful because they display a lot of materials in an easy to comprehend *graphic* or chart. The entire point of the Gantt Chart is to use this graphic to effectively display the scheduling details. All presentations and meeting discussions should focus on the *graphic*. The programming fields as displayed in Fig. 8.6 to the left of the orange line include the same information as the graphic to the right of the line. Presenting the same information twice reduces the space available to clearly display the graphic and causes the visual to be difficult to read or use. Simply slide the dividing line back to the left after you have completed programming new information to create a useful graphic.

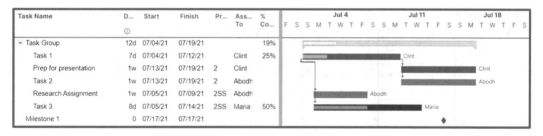

Figure 8.6 *Ineffective display of a Gantt Chart*

Even with the dividing line properly placed, displaying a Gantt Chart can still be challenging at times. Consider collapsing some of the task groups or breaking the chart into multiple images so that any necessary details are clearly visible on your slides or documents. Before presenting the Gantt Chart, ask yourself "would I find this graphic useful if I was in the audience?"

Industry Point of View 8.1: Schedule Management by Everyone

I manage a department of engineers for a communication technology company in Sioux Falls, SD. At any given time, I have between three and five teams working on different projects. Each team is led by a Technical Lead and a Project Manager. I no longer directly manage the Project Schedules; that task is left to my project managers, but I do meet with the leaders in monthly project reviews. During those reviews, we assess the state of each project and determine whether our resources are allocated correctly. To accomplish this, my Project Managers typically present their project's Gantt Charts.

Industry Point of View 8.1, continued

In a recent project review, one of my teams was working on a high-speed switch for a new communication system we were installing. They were falling behind an important schedule, an issue that would have cascaded into the work of a number of other departments company wide and put the entire project in jeopardy. They had encountered an unexpected technical obstacle that they could not seem to fix. Shari worked on a different team in my department. She had experience with this specific issue but was fully tasked on her primary assignment. Her team was exactly on schedule, and I was hesitant to reassign her – doing so may have jeopardized her primary project without guaranteeing she could get the troubled team back on track.

Gantt Charts make it relatively easy to see how all of a department's resources are being allocated. A third team had two engineers, Shrihasha and Jason, who were not fully tasked at this time (their project was wrapping up). Since Shari's work was important, but not technically difficult, we tasked both Shrihasha and Jason to back fill Shari's workload. Assigning two resources to this task probably was overkill, but doing so provided a safety margin in case we lost time during the transition. That freed up Shari to be reassigned to the high-speed switch design team and help solve their technical issues.

I need my employees to understand Gantt Charts. Obviously, this includes the Project Managers as Schedule Management is half of their job description. But many of my entry-level engineers are surprised when they are expected to use their Gantt Charts as well. Shari, Shrihasha, and Jason are not Project Mangers, but had they failed to use the Gantt Charts correctly, my leadership team would not have been able to identify this creative solution. The Project Managers create and manage the schedule, but my leadership team cannot use the tool effectively unless everyone involved with the project keeps their responsibilities up-to-date.

8.3 Block Diagrams

Recall that one of the primary concerns of the INITIATING phase is to convince all Stakeholders that the engineering team has the knowledge and ability to solve the Problem Statement. Part of this effort will include presenting the entire Requirements Document for review and critique.

This presentation is part of an event called the Project Kickoff Meeting (Chap. 9). It will be necessary for the audience to fully understand, and agree with, the Problem Statement and Requirements Document. If the Decision Makers do not understand the details of your presentation, they will reject the project. Therefore, it is critical that you convey this material effectively and succinctly.

For large projects, the Requirements Document may consume many, even hundreds, of pages, and the Project Kickoff Meeting may require hours, or even days, to complete. It is often difficult for Stakeholders to retain that much information and capture how all the details relate to each other. One useful tool for summarizing a Requirements Document is called the **Conceptual Block Diagram**. Like in most communication, the old adage is true: "a picture is worth 1000 words."

After the Project Kickoff Meeting, the Requirements Document is regularly referred to but rarely reviewed in its entirety. Instead, the introduction of all future project presentations will include the Block Diagram as a tool to quickly convey the important concepts of the Requirements Document.

> The Conceptual Block Diagram is a tool which graphically depicts the most important concepts within the Requirements Document and demonstrates how they are related to each other.

Recall that the Objectives should have been organized in some logical manner when discussing them during the Problem Statement; this organization should be maintained in the block diagram. Often, the Objectives were listed from input to output, so the Block Diagram should display the inputs on the left, the processing in the middle, and the outputs on the right.

It is likely that the Requirements Document contains too much information to display on a single figure; so, it is important to select the most important pieces that will communicate a clear understanding of the project to any audience. Typically, this can most easily be accomplished by focusing on the Requirements level of the Req. Doc.

There is no singularly correct diagram that perfectly represents any given project. Effort to exhibit your project properly will require a little creativity, and often a few iterations, to perfect. As the project evolves through different stages, it is common to modify the diagram to accurately represent the project in its current stage.

When building the Block Diagram, consider effective use of at least some of the following:

- Color – make the diagram interesting and informative
  (e.g., power blocks might be colored red).
- Arrow styles – differentiate different types of signals
  (e.g., wireless transmission might use a dotted line).
- Shading – distinguish the most important topics
  (e.g., a gray box might represent an existing feature which you are required to interface with but are not responsible for designing).

Examples 8.6 and 8.7 exemplify student projects that made effective use of their block diagrams.

EXAMPLE 8.6 Block Diagram #1

Polly, a nursing professor at South Dakota State University, desired a device to assist her in a study on cortisol, a hormone in the body. This study required participants to collect saliva in a test tube and record the exact time the sample was collected. Unfortunately, it was suspected that participants were not recording the times accurately enough. A solution was required which accurately time stamped the instant a particular test tube was removed from a kit.

A list of the project Requirements included the following:

1. The system must be enclosed in a shippable container.
2. The system must contain test tubes.
3. The system must sense when a test tube is removed from the kit.

EXAMPLE 8.6, continued

4. The system must time stamp the event.
5. The system must indicate which test tubes have been used.
6. The system must save the time stamp data for the duration of the study.
7. The system must communicate data to a researcher at the conclusion of the study.

The block diagram produced by the design team looked like this (Fig. 8.7):

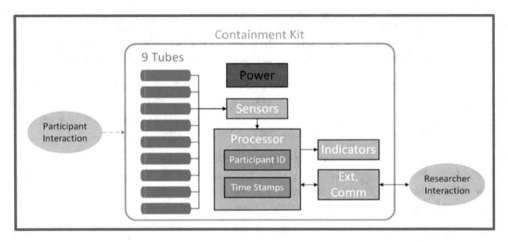

Figure 8.7 Block Diagram for Example #1

Consider how each Requirement is represented:

1. The shippable container is represented in the "Containment Kit" block.
2. Nine test tubes are represented inside the kit. Notice this feature captures the Requirements of containing the tubes and the specification that nine tubes are used.
3. Sensing the removal of tubes is represented by the "sensors" block.
4. Time stamping is represented by a block embedded in the "processor" block. Although the need for a processor is not specified in the Requirements, it is certainly useful to consider when communicating about the project.
5. An indication feature is represented by the "Indicator" block. Notice this probably does not represent the physical design. The physical placement of the indicators will likely be near the tubes. Remember, representing the conceptual flow of the project in a block diagram is more important than showing the physical layout of the solution.
6. The Requirement to save the data for the duration of the study is not explicitly represented. However, it is addressed by the inclusion of the "Power" block, assuming that the Power block is designed to power the system for the duration of the study.
7. The communication feature is represented by the "Ext. Comm" block.

Additionally, notice that the participant and researcher interaction are represented in gray ovals. The team was not responsible to design anything for these ovals but needed to represent where the humans would interface with the product.

EXAMPLE 8.7 Block Diagram #2

Tyler, the owner of Trigger Engineering, wanted a game to enhance a target-shooter's experience at a gun range. The Objectives of the project included:

1. Design a smartphone app to randomly activate a particular target and report shooting success ratios.
2. Design a set of targets which the controller could individually activate, and which could report when the target was hit.
3. Design a round counter that could communicate with the controller and report how many times the shooter fired his weapon.
4. Design a communication hub to handle both short range and long-range communications.

These four Objectives generated 43 Requirements and over 120 specifications! The Trigger Engineering project team's conceptual block diagram would likely not be able to describe all the Requirements in as much detail as the previous example. This team decided to ensure the high-level concepts of the project were represented well in the conceptual block diagram, and then create what they called "detailed block diagrams" to further describe some of the details.

This is the team's Conceptual Block Diagram (Fig. 8.8):

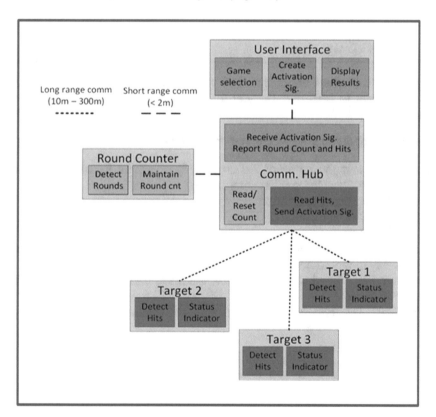

Figure 8.8 Block Diagram for Example #2

EXAMPLE 8.7, continued

The team clearly depicted the four Objectives and further defined what was meant by each Objective by indicating the most important Requirements.

One of the most significant challenges of the project was the drastically different communicating range specifications. The User Interface, Communication Hub, and Round Counter were all positioned at a shooting bench. The Targets were positioned anywhere between 10 and 300 m from the shooting bench. Because this was one of the most critical aspects of the project, the team decided to include these specifications on the conceptual block diagram even though the point of the block diagram was primarily to communicate details about the Objectives. The use of different line styles effectively communicates these details.

The team also made effective use of color to group Requirements within the Round Counter, User Interface, and Target Objectives. This was useful because the Communication Hub managed facets of each of the other three Objectives.

Although the so-called detailed block diagrams were necessary to complete the description of the Requirements document, the team did an excellent job developing a Conceptual Block Diagram to effectively communicate the most significant aspects of their project before discussing the details.

8.4 Chapter Summary

In this chapter, you learned about the final processes in the INITIATING phase. These processes are meant to help prepare you for the upcoming Project Kickoff Meeting and Project Charter.

Here are the most important takeaways from the chapter:

- Literature Review is the name of the process in which the Problem Statement is fully investigated. There are three main components:
  - Project Background or Context
  - State-of-the-Art or Technology Review
  - Applicable Standards
- The Gant Chart is used to visualize and adapt to deviations in the Project Schedule. It is not necessarily meant to be an exact prediction of reality.
- When presenting a Gantt Chart, only show the tasks and the chart; do not show all the programming fields.
- Use the Block Diagram to display the Project Requirements in an organized and appealing way.

8.5 Case Studies

Let us look at Block Diagrams each case study team produced during the INITIATING phase.

Augmented Reality Sandbox

Figure 8.9 displays the AR Sandbox team's Block Diagram.

- What went well: This colorful diagram is easy to read and clearly depicts the important aspects of the project. The team was able to use this diagram effectively throughout the project.
- What could have been improved: Typical Block Diagrams show the inputs on the left and outputs on the right. Here we see inputs such as the "Water Drain Button" coming from the right. We also see the output of the "AR Software" going left before arriving at "Projector." This structure makes it difficult for the reader to identify exactly where to start looking at the diagram.

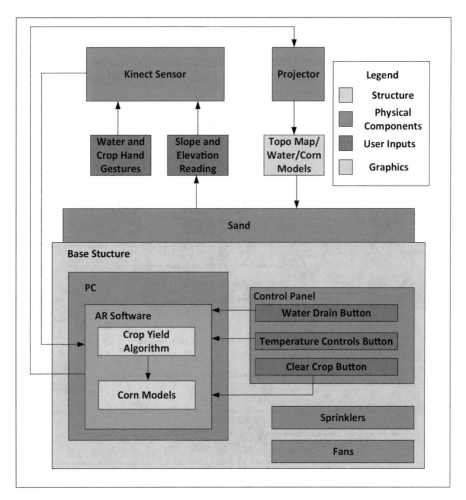

Figure 8.9 AR Sandbox Block Diagram

Figure 8.10 displays the Block Diagram for the Smart Flow Rate Valve project.

- What went well: This diagram shows that all the components necessary for the Flow Rate Valve will be included in a single housing. Within the HOUSING block, we see the typical closed-loop feedback topology commonly used in control algorithms. The inputs are on the left and outputs on the right. There are two feedback loops, one from the valve's encoder and one that measures the output flow with the flow meter. Using a standard configuration makes it easy for anyone with a controls background to understand this figure.
- What could have been improved #1: Although the team correctly oriented the inputs and outputs of the automated control system within the housing, they chose to orient the *USER INPUT FLOW RATE* block on the right. Whenever possible, *all* inputs should be on the left. The *USER INPUT* and *9–33 V Power* blocks could easily have been swapped. This would also have allowed the *9–33 V Power* block to be directly connected with the *POWER CONVERTER* block.
- What could have been improved #2: This Block Diagram could also have been improved by a better color scheme. There is no clear reason why two boxes were red. Red often denotes a power block, and at first glance, it appears that the team used this standard for the *9–33 V Power* block. But the *Output flow* block is not a power box and is also red. The red boxes could potentially have denoted the interface points between the student project and the existing system. However, in that case, the user input (which came from an existing field computer) should have also been red. The team could have improved this Block Diagram by using a consistent color scheme for the blocks and included a legend with the diagram.

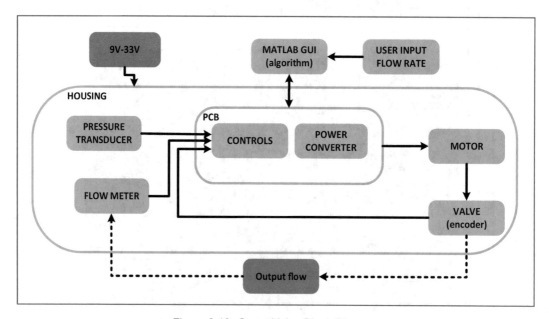

Figure 8.10 Smart Valve Block Diagram

Figure 8.11 displays the Block Diagram for the RESDTA project.

- What went well: This Block Diagram includes a clear legend. The color gradient on the "MOTORS" block suggesting that the motors addressed both mechanical and electrical Specifications was a nice touch. Using gray blocks to indicate features outside the scope of the project which the RESDTA product must interface with was also well done.

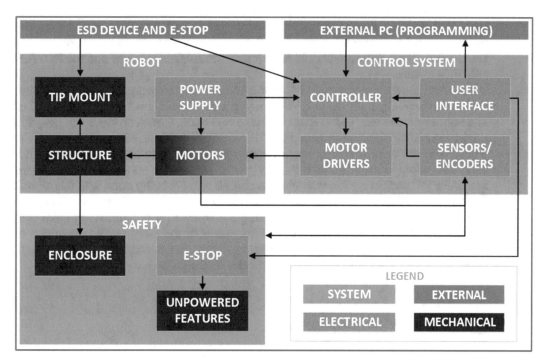

Figure 8.11 RESDTA Block Diagram

- What could have been improved #1: This Block Diagram is extremely "high level." Perhaps this was necessary in RESDTA's case due to the size of their Requirements Document. The team effectively used this Block Diagram to introduce the high-level Objectives but also proceeded to discuss the low-level details (which are not shown here) while presenting this diagram. Secondary Block Diagrams could have been used to discuss Requirements of each "system" block (e.g., ROBOT, CONTROL SYSTEM, and SAFETY). Doing so may have been useful to guide the reader's transition from the high-level Objectives, through the Requirements, down to the low-level details.
- What could have been improved #2: There are two "E-STOP" blocks here (one gray, one green). It is not clear what the difference between the two blocks are. The gray E-STOP block was intended to represent a mechanical "kill switch" connected directly to the power source. The green E-STOP block represented a digital signal that

interfaced with the controller. Therefore, the names of the blocks should be different. Furthermore, it is unclear why the gray E-STOP device shares a block with the "ESD DEVICE." These are two distinct features and should have been two distinct blocks.

- What could have been improved #3: Considering that the gray E-STOP block represented a mechanical switch connected directly to the power source, the diagonal arrow pointed from the gray E-STOP block to the "CONTROLLER" should have been pointed directly to the "POWER SUPPLY." The team drew this incorrectly because the physical wiring of the E-STOP feature did go through the "CONTROLLER" PCB but not because the switch was designed to interface with the controller; using the PCB for connection purposes just happened to be a convenient layout. Remember, the Block Diagram should be a conceptual representation of the Requirements, not a physical representation of the wiring schematic. The Requirement that dictated the need for this mechanical E-STOP feature stated that "the system must be completely powered off in case of an emergency." Therefore, the Conceptual Block Diagram should have linked the E-STOP feature and the POWER SUPPLY feature.

- What could have been improved #4: Using two arrows connected to the "EXTERNAL PC" block to clearly indicate that PC received signals from the "USER INTERFACE" and sent signals to the "CONTROLLER" was well done. However, the Block Diagram uses a double-sided arrow between the "SENSORS/ENCODERS" and the "SAFETY" blocks. It is not clear what signals are sent from the sensors to the safety block nor what the safety block does with those signals. This arrow should terminate inside the SAFETY block in such a way that we can tell what is being done with this signal. Notice also that this is the only arrow pointed to a blue SYSTEM block rather than one of the feature blocks. Perhaps labeling the arrows (or differentiating them with dashes or colors) could have improved the clarity of the diagram. It is common to differentiate power versus signal arrows, types of communication protocols (e.g., HDMI vs. UART), and electrical versus mechanical signals.

Chapter 9: **Project Kickoff Meetings**

The first major milestone of any project is the **Project Kickoff Meeting**. Throughout the STARTING and INITIATING phases, the project team has been studying, learning, listening, and digesting. At the Project Kickoff Meeting (PKM), the team begins to evolve into the authority on the project. By this point, you should know more about the project than anyone else in the room! This can be a strange feeling for many students considering that their *professors* are a part of the audience we are talking about. However, you are becoming *colleagues* with the experienced engineers in attendance and should be equally engaged in the discussion, as opposed to students' learning from those professors. Your opinion is valid, and it matters. You are still students and still learning, but it is also time for you to start taking a leadership role in your project. Practice approaching this meeting with confidence.

> The purpose of the **Project Kickoff Meeting** is to convince all Stakeholders that continuing to support the project is an acceptable and worthwhile risk.

The PKM provides a chance for the engineers (i.e., you) to summarize all the information they have gathered thus far and present one cohesive explanation of the project to the Decision Makers (Sponsors, Instructors, and Advisors). Your priority is to convince those Decision Makers that this is a project worth pursuing and that the risk of applying additional resources to the effort is acceptable and worthwhile.

In this chapter you will learn:

- How to prepare and present a Project Kickoff Meeting
- How to use a template and assessment rubric to increase the probability of success of any review or report

C.J. Mettler, *Engineering Design*, https://doi.org/10.1007/978-3-031-23309-8_9

9.1 Purpose of a Project Kickoff Meeting

Before learning how to prepare and present a Project Kickoff Meeting, it is useful to understand what is at stake. This meeting is a "risk mitigation" strategy. All projects have risks from the moment they are conceived until the moment they are concluded. One of the primary purposes of Project Management is to reduce and manage that risk. In particular, one way of ensuring that investors do not lose their entire investment on a project is to release control over that investment in stages.

The STARTING and INITIATING phases of a project essentially require no resources to complete, other than a small portion of an engineer's salary. All that is required are discussions, Internet searches, and maybe a small amount of travel. However, the PLANNING phase is going to require some more significant expenditures. A significant portion of engineering salary will be required, and at least a few components will be purchased. Investors do not want to spend that extra time and money if the project has not been initiated properly; doing so will almost certainly result in a failed project and complete loss of expenditures.

Therefore, in order to complete the INITIATING phase, a PKM is held to discuss the Problem Statement and Requirements Document or, in other words, to verify that the engineering team understands the problem and is capable of solving it.

There are four possible outcomes of a Project Kickoff Meeting:

1. The Stakeholders are convinced that the project team is suited to solve the problem. The team is directed to complete the INITIATING phase by writing a Project Charter (Chap. 10) which documents the details of what was presented at the PKM. Everyone is confident that the project will be formally accepted once the Charter has been produced. Obviously, this is the desired outcome.
2. Minor issues are raised, and the project team is directed to address those issues in the Project Charter before final acceptance is granted. When the issues are relatively minor, the modifications can usually be summarized in the Project Charter in such a way that everyone is satisfied, and no further review is necessary. The project is likely initiated once the Charter is reviewed. This is the most likely and common outcome.
3. Significant issues are raised, and the project team is directed to revisit most/all of the INITIATING phase processes. The Project Kickoff Meeting will need to be repeated. Unfortunately, this is not an uncommon outcome but one which is best to avoid. In industry, this will be an expensive proposition and will likely make Stakeholders unhappy; it might even call into question the competency of the design team. In academia, when projects are very tightly bound by the academic calendar, repeating any significant part of the INITIATING phase puts the project at great risk. This outcome periodically occurs in Senior Design and will require your team to work overtime for the rest of the semester to pass the course. The Project Management techniques taught here should mitigate this scenario.
4. Major issues are raised, and the project is outright cancelled. This is an extremely rare occurrence in Senior Design. If this occurs, it is because the project team simply did not put forth the proper effort toward the course. This almost certainly results in a failed project and an opportunity for the students to come back next year and try again. A failed PKM is easy to avoid. In fact, this textbook was written, in part, to avoid this outcome at all costs. A cancelled/failed project will not happen to you if you are serious about your project and address the feedback your Instructor and/or Advisor gives you.

9.2 Formatting a Project Kickoff Meeting

There is no catch-all formatting rule for Project Kickoff Meetings (PKM). A PKM for a Small Business Innovation and Research (SBIR) government-funded grant will not look identical to a PKM for an internally supported project at a Fortune 500 company like IBM. The PKM which NASA conducted for the Landsat-8 satellite development team required a full week and dozens of engineers, whereas a PKM conducted by the Northern Plains Power Technology for a standard consulting project required approximately 1 hour.

However, all Project Kickoff Meetings address the same basic concerns:

* Do we understand the problem?
* Is a solution achievable?
* Are we prepared to accept the risks and proceed with the project?

During the meeting, the team members will summarize, in their own words, the Problem Statement. This may include a small portion of the Literature Review as background if necessary for the audience to understand the Problem Statement. Although everyone in attendance should know what the problem is, specifically articulating *your* version of Problem Statement assures the Sponsor that the team is addressing the problem exactly as the Sponsor intended. Repeating the Sponsor's wishes in your own words is a communication technique called "parroting." "Parroting" your understanding of the problem back to the Sponsor is an excellent technique used to ensure that everyone is on the same page. While this may seem trivial, it is not uncommon that misunderstandings are uncovered by this process. The PKM provides an opportunity to correct any of those misunderstandings.

Next, the team will present the complete Requirements Document in detail. Each Objective of the Problem Statement will be described using Requirements and refined in detail with the full list of Specifications. The Stakeholders will be monitoring this part of the discussion for contradictory or omitted Requirements or Specifications. Also, this discussion is the chance for the "experts" to identify any details which may not be achievable within a reasonable amount time or money; these will need to be modified on the Project Charter.

The team will then summarize the technical discussion by presenting a Conceptual Block Diagram (0). The focus of this part of the meeting is to show that all the Requirements are integrated into the project in an understandable way.

Finally, the team will discuss any concerns they may have regarding the project's likelihood of success. While this may seem counterproductive since you are trying to convince the stakeholder to support you, showing that you have a thoughtful understanding of the risks in the proposed project actually provides Decision Makers with a certain level of confidence. If a team claims that "there are no concerns related to this project," an experienced businessman will immediately doubt the competency of the engineering team; they know that *all* projects contain risks. When the engineering team openly discusses project concerns, they make it clear that they have studied the topic in depth and are being honest about their findings. The purpose of the PKM is not to say, "there are no risks," but rather the message should be "we understand the risks and we are able to mitigate them."

Of course, it is possible that the team cannot claim to understand all of the risks they uncovered. It is also possible that the team is uncertain whether they are capable of addressing them. The PKM is also an opportunity for the engineering team to walk away from the project to avoid a loss of investment. Better to move on to a new project which can be successfully completed than to claim you are able to complete a project and fail to do so. This situation is not typically applicable to Senior Design since the course instructors have pre-emptively vetted project and assigned appropriate students so that a failed outcome is extremely unlikely.

9.3 Effective Use of Templates

In the next section, we will study one possible version of a Project Kickoff Meeting. It would be a mistake to assume this is "the" template that everyone uses; in fact, you probably will never perform PKMs *exactly* this way again. Every project is different, every reviewer will expect different material to be covered, and every presenter has their own style. There is not a "correct" way to present a PKM. This means that a "perfect" PKM template that is applicable to *every* situation is not possible to create. However, all PKMs are similar in nature, and this template can be used, with slight modifications, to present all future PKMs you may experience.

Remember, when a template is provided for an event, your audience is expecting information to be presented using that template. Minor project-specific modifications can improve the overall delivery of that information; however, too many modifications will leave the audience confused. Remember, the template was provided to the engineering team to make *the reviewers'* lives easier, not to reduce the workload of the engineering team. Templates are provided to enhance the *reviewer's* ability to complete a thorough assessment of many dissimilar projects. If each team presents in drastically different styles, the reviewer will have a difficult time making accurate assessments and/or comparisons.

To use a template effectively, you must first understand what the audience expects to hear. Deviating from these expectations is not "wrong" if you have a justifiable reason, but you must make it clear to the audience that you are doing something unexpected. This will prepare the audience and enhance their ability to follow you.

> Ensure that any justification for making changes to a provided template is not based on individual preferences, but an obvious improvement to the delivery of the project-specific material.

If you are going to make any modifications to the template, verify that your reviewers (e.g., Instructors) accept your modifications before the formal event.

Once the template is understood, attempt to organize material in the exact order the template requests. Review the entire presentation and verify that your document cohesively and completely explains your project. Consider whether there is anything unique to your project which was not directly requested in the template; if there is something, include that information. Also consider whether the audience will understand the material as it has been organized. If not, consider how the presentation might be improved. Decide whether the risk of confusing your audience with a change to the template is worth the improvement. If the benefits of the change supersede the risk of confusion, make the change, but include a comment about that change during the presentations. Example 9.1 contrasts two projects, one in which a change was not necessary and one where a change benefited the presentation and was successfully implemented.

EXAMPLE 9.1 Modifying the Order of Template Slides

In Sect. 9.4: a Project Kickoff Meeting template will be presented. In that section, you will learn that the first three slides of a PKM are the following:

1. Introduction/Title Slide
2. Problem Statement
3. Necessary Background

Consider these two Problem Statements and determine how one might benefit from modifying the template:

Problem Statement #1: Hard drive temperature chamber

HGST performs many temperature tests on hard drives. During these tests, the drives are exposed to extreme temperatures for specified periods of time. HGST would like to automate some of these tests. A temperature chamber will be developed for this purpose.

Problem Statement #2: Floating Island International

Floating Island International produces man-made floating islands for the purpose of passively cleaning waterways. They would like to incorporate a NanoBubbler in their floating island design. A photovoltaic system will be designed for this purpose.

Analysis:

Problem Statement #1 is completely understandable as it is presented. Most reviewers can be expected to understand the concepts of temperature, time, and automation. The average person is familiar with hard drives. At the end of the Problem Statement slide, every person in the room should have a reasonably accurate understanding of what this temperature chamber design will entail.

Some background is still necessary. The terms "extreme" temperature and "specified" periods of time should be quantified. But those quantifications are not necessary to understand the Problem Statement.

Unlike Problem Statement #1, Problem Statement #2 is less understandable to the average reviewer as stated. The most concerning issue here is the term "photovoltaic." This is a solution-specific technology and typically would not belong in the Problem Statement, so what is it doing here? Additionally, the average reviewer probably does not know what a "floating island" is or how they "passively clean waterways." They probably do not know what a "NanoBubbler" is either.

*This team aptly presented their background slide **before** presenting their Problem Statement so that Problem Statement was completely understood by everyone when they eventually presented it.*

EXAMPLE 9.1, continued

On their adapted background slide (Slide 2), they explained that

> "A Floating Island is a lightweight plastic matrix developed from repurposed material. This matrix encourages the growth of naturally occurring biofilms which filter many toxins from the water. Islands are often installed in remote areas where grid-power is not accessible."

They also explained that

> "A Nanobubbler was a type of aerator with a power pump."

By the time the team presented their Problem Statement on Slide 3, reviewers had a mental image of Floating Islands, understood that a NanoBubbler would require a significant power source and that photovoltaics were the only realistic power source due to the remote location of the islands to be installed.

Proper Implementation of Changing the Template

Deciding to modify the template and present the Necessary Background before the Problem Statement was a good choice. However, this team did not simply switch the order; they also prepared the audience for the switch. As they transitioned from the Introduction slide to the Background slide, the speaker stated:

> *Floating Islands are an exciting product that produces many positive impacts to the environment. Before discussing our team's Problem Statement, we would like to explain what they are and define some important terminology.*

This short transition statement got everybody's attention by claiming that this was an "exciting" product with "positive impacts." At the same time, the statement prepared the audience for what was coming, that is, that they changed the template order and were about to present Necessary Background rather than the Problem Statement.

Notice that the team did not refer directly the modification. They did not say

> *We modified the template and are going to present the Necessary Background slide before the Problem Statement slide.*

Doing so often sounds rigid, as if you struggle to present fluidly. By simply explaining what they were doing, rather than directly address the modification, the presenter sounded extremely professional, and the PKM was a huge success.

9.4 A Project Kickoff Meeting Template

This section describes one method of presenting a Project Kickoff Meeting (PKM) effectively. The primary topics that every PKM must cover include the following:

- Introduction of the Team
- Problem Statement
- Necessary Background

- Block Diagram
- Requirements Document
- Project Concerns

Keep in mind that the entire point of performing the PKM is to convince the reviewers that you completely and accurately understand the problem. Those reviewers (i.e., your primary audience) will include your Instructor and Advisor and potentially your Sponsor. Other classmates may also be in attendance.

Introduction

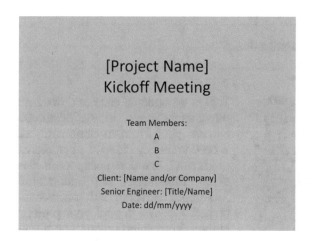

This is *your* meeting, take charge of the room at the scheduled time. Ask if everyone is ready, and when you have the audience's attention, introduce the project and the primary stakeholders (including each team member). If this is a multidisciplinary team, it is appropriate to provide a comment on each member's background or specialty.

It is respectful to acknowledge the Sponsor and thank that person for attending.

You might want to embellish this slide with a picture that is directly related to what you will be discussing. Consider adjusting the text to make room for the picture.

Problem Statement

The most effective communication during any presentation typically starts with the big picture and then drills down to the specific topic of interest. The Problem Statement is an excellent starting point for *most* PKMs.

Presenting this slide with confidence will convey to the audience that you are knowledgeable and prepared. Conversely, reading this slide, or referring to notes, will convey to the audience that, after many weeks of work, you do not even know the very basics of the project. The use of notecards is a perfectly valid presentation technique, *but not here* – this slide needs to be presented in a polished and confident style.

It is tempting to read the slide directly ("Our Problem Statement consists of a

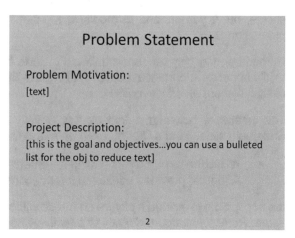

Motivation Statement and Project Description. . . ."). Presenting that way will come across stiff and rehearsed. Instead, present this slide as if you were telling a story. Review the project-related examples previously presented (Examples 6.4, 8.6, and 8.7). Notice the Problem Statements never actually use the terms Problem Statement, Goal, and Objective.

This is the first numbered page. It is important for the reviewers to have page numbers on the slides. They will likely be taking notes and will want to refer back to specific content from the presentation during the Q&A portion of the meeting. Without the page numbers, this becomes cumbersome. Ensure that your edits to the template have not removed page numbers.

Background

Necessary Background

- Focus ONLY on what is necessary to understand THIS presentation.
- MUST include figures!

3

This is an optional slide and does not necessarily have to be slide #3. Review your Problem Statement, will it be completely clear to *all* audience members? Are you using terms that *every* stakeholder will understand? Are you interfacing with an existing project that will impact your project? Is there something unique about your sponsor's environment?

If you believe some background information will provide clarity to the Problem Statement, it should be included here. Rather, this is a chance to present a small subset of information with the specific intent of clarifying the Problem Statement.

Now consider whether the Problem Statement should be stated first followed by background to provide further clarity, or is the background necessary to understand the Problem Statement? This is probably the most common deviation from the standard template. Make your choice, but if you choose to modify the order, refer to Example 9.1 to ensure you do so professionally.

Remember, the only reason you would include a Background slide is to provide additional clarity about an important aspect of the Problem Statement. Demonstrating additional clarity is difficult without an effective figure.

The difference between a "picture," potentially used on the Introduction and/or Problem Statement slides, and a "figure, used on this Background slide, is that

- A "picture" embellishes a slide for entertainment purposes.
- A "figure" provides additional informational value to the slide.

It is acceptable to include a *picture* on a slide without directly discussing it. That is not true for a *figure*. Refer to the figure directly to effectively communication your message.

Conceptual Block Diagram

After the design challenge has been clearly explained, provide a high-level description of what you are going to do about it using the Conceptual Block Diagram.

Remember, the point of the meeting is to discuss the Requirements Document in minute detail. This slide is simply intended to prepare the audience for the technical details of what is coming, so it is not necessary to provide those details here.

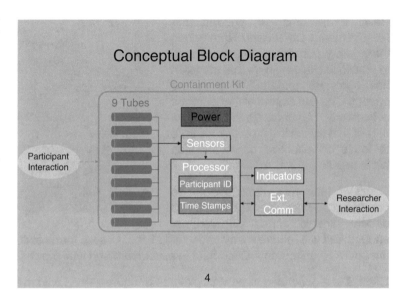

Using the project presented in Example 8.6, the presenter might present Slide 4 by saying "The team will be developing a kit that contains 9 test tubes. A method to detect when a specific tube has been removed from the kit will be implemented. When this happens, the system will timestamp the event and save the data until a researcher is able to download it. The processor will be configurable to contain a specific participant identification. The processor will also provide an indication for each tube which was previously used in order to avoid confusing the samples. A robust power system and containment kit will also be significant design challenges." The presentation would then continue to the next slide to explain what each of those statements meant using the Requirements Document.

The Conceptual Block Diagram is only useful if it is readable. Consider this figure from the audience's point of view. It should be large, easy to read, and succinct. It should fill the slide. One team member should physically point to each item as it is discussed to ensure the audience follows the discussion. If the PKM is being conducted virtually, you will need and still need a method to highlight detail as you speak about. This can be accomplished with a digital laser pointer or through creative animations in Power Point.

Now, it is time to present the Requirements Document. Remember, the primary purpose of this meeting is to convince Stakeholders that the team has complete understanding of the project. One key aspect of that purpose is to clearly communicate the technical details that define the problem. Be sure to focus on effective communication while presenting this material!

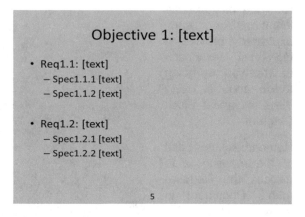

Consider how much text should be presented on one slide. Too much text and the slide will appear cluttered; not enough text will require many additional slides. Use as many slides as necessary to present the entire Requirements Document in a readable and well-formatted style.

Often, it is a good idea to provide pictures or figures along with the text. That is true for these slides too, but this is one time where you should not allow the picture to dominate the text. Consider presenting a representative portion of the Block Diagram on each slide to clarify what part of the conceptual solution this particular objective is referring to.

Another effective communication technique is to provide the audience with a set of handouts that includes the Conceptual Block Diagram and Requirements Document. This way the Stakeholders can refer back to the big picture as you present new details.

Remember that Constraints are an important part of the Requirements Document even though they are not presented within the standard outline format. Typically, Constraints are presented in a bulleted list on an additional slide immediately after the Requirements outline.

The remainder of the PKM can vary depending upon the specific audience and project being presented. Issues such as an initial budget analysis, high-level project schedules, and risk analysis might be desired at some PKMs. However, while these topics are definitely useful in convincing Stakeholders to support the project, it is usually too much to ask students to develop much of this material so early in the semester schedule. The template recommended in this textbook contains just one slide in this category, Other Project Requirements. This textbook will cover budget analysis, schedules, and risks in later chapters.

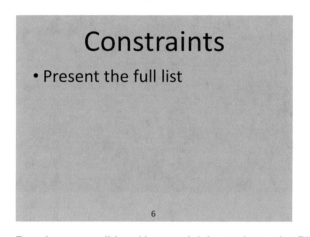

The "Other Requirements Slide" is included here because it affects a large subset of Senior Design projects. However, many projects will not have anything to specifically include here.

"Other Requirements" may include a research or outreach component of your senior project. These concerns must be addressed to successfully satisfy your Sponsor and/or Advisor but are not directly part of the Design Life Cycle's Requirements Document.

It is a good idea to discuss the Other Requirements slide with your Advisor prior to the PKM. Commonly the Other Requirements slide can be omitted. However, if you do have any project requirements that have not previously been addressed, this is the place to include them.

Project Concerns, Conclusions, and Open Discussion

Project Kickoff Meetings typically end with an open discussion session. This starts with a "Project Concerns Slide" to voice any significant issues *you* have uncovered thus far. Then, wrap up the discussion with your "Conclusions." Finally, transition the meeting into a Question/Answer session for the audience to address their concerns.

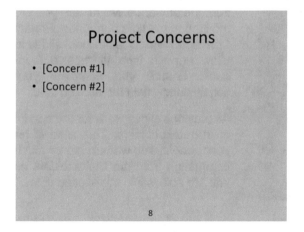

The "Project Concerns" Slide provides the team a chance to discuss any issues they have uncovered during their Literature Review or Requirements Gathering research. Issues you may want to consider include the following: Whether the solution to this is possible based on what other people have accomplished, whether the budget will be reasonable, any perceived technical challenges, or anything else which keeps you up at night thinking about completing your senior project.

By disclosing your concerns about the project, you demonstrate that you have thought deeply about the project. This will instill confidence in the stakeholder and make it easier to support your efforts. Every project has risks, but the worst risks are the unknown risks. Unknown risks are impossible to prevent or mitigate. Known risks can be analyzed and dealt with. This is also a chance for you to request additional help from the other Stakeholders.

Presenting the "Conclusion" slide requires some quick thinking and advanced presentation skills. You should prepare this slide with your conclusions, but you should also consider any discussion that occurred during the meeting. You may have to adapt your message in real time.

For example, if one of your project concerns is that you have been provided a potentially insufficient budget and that was prepared on the conclusion slide, but your Sponsor said they would increase the budget during the "Project Concerns" slide, do not conclude with *another* (obsolete) concern about the budget even if a statement on this slide refers to that concern. Instead, modify your message by saying "I believe, as we have already discussed this, insufficient funding is no longer an issue."

Conclusions

- Discuss the impact of this project if it should be successful

- Discuss your ability to complete this project successfully.

9

Remember, Stakeholders will have three primary concerns at a PKM. A strong conclusion directly addresses those concerns:

1. Probability of success: Will this specific team be able to achieve a successful outcome?
2. Impact: Assuming the problem is solved, who will care? Who will be affected? Is there profit to be made or can a valuable service be provided?
3. Risk: There is always risk even if the probability of success is reasonable, but, will the impact be worth the risk?

Questions or Concerns from the stakeholders?

Finally, allot time to receive feedback from the audience. Make sure you are monitoring the time during the "Project Concerns" and "Conclusions" slides to allow enough time to receive feedback. Ensure you complete your conclusions within the allotted time.

All previous slides were for the benefit of the stakeholders. This slide is for *your* benefit. You want to get as much information from the Stakeholders as you are able within the allotted time.

A reviewer's job at a PKM is to ensure this project has been thoroughly initiated. They will push and challenge you, sometimes aggressively.

Reviewers' questions can come across as challenges, and sometimes they feel like personal attacks; you have worked hard preparing for this event after all – there *is* personal aspect to your work. But the questions and challenge are not personal! Reviewers are attempting to uncover weaknesses in your project now so that you are not surprised by them later. They will push and probe and challenge. They are trying to help you succeed. Recall, the sooner an issue is addressed in the project life cycle, the easier and less expensive it is to fix. As uncomfortable as it is, you *want* this challenge.

It is acceptable, and expected, that you stand your ground. But avoid being defensive. Listen carefully to the reviewers' concerns. Take time to consider what they are really trying to say. Then, provide thoughtful answers. If you do not know how to respond, or have not considered the issue, the responses "I am not sure about that" and "we will look into that and get back to you" are perfectly valid.

9.5 Advanced Presentation Techniques

The Project Kickoff Meeting is critically important. Therefore, presenting well is also critical. Consider for a moment the consequence of a poor presentation: Stakeholders will not formally initiate a project which they do not understand. If there is a gap in understanding at the PKM, who do you think that will most severely impact? You, of course! It is your responsibility to explain this project. If the Stakeholders do not understand something, you will be asked to do more work. There are a number of techniques you can employ to reduce the chance of these types of misunderstandings.

- Use the template. We have already discussed this.
- Practice your presentation as a team. Critique each other's presentation. Are your partners using effective language? How is the enthusiasm, tone of voice, and confidence?
- Consider practicing presenting to another team in your class. If you can communicate well to another group of students, you likely will be able to communicate effectively to faculty.
- Use a rubric to assess yourself if one is provided. Rubrics are an excellent tool for a reviewer to communicate what they expect from the presentation. If you are practicing with your team or reviewing another team, use the same rubric your reviewers will be using when you perform the formal PKM.
  Caution: this is a complete waste of time if everyone is "nice" to each other. The point here is to tear the presentation apart *in order to improve the final results*.
- Use the slides to your advantage. Physically point to features on a figure. The speaker can perform this task themselves, or a team member could assist the speaker if necessary.
- Record all feedback provided. Ensure a team member is assigned to record details of the conversation and action items in the team's Engineering Notebook during the open discussion.

Avoiding Common Pitfalls 9.1: Avoid Cumbersome Slide Transitions

When transitioning between slides, novice presenters will often read the title of the slide, or sometimes turn the title into a question which they then proceed to answer.

Doing this will come across to the audience as if you were unprepared, although, in reality, this has less to do with your preparation and more to do with your experience presenting.

Consider the "Problem Statement" slide from the team discussed in Example 6.4.

Problem Statement

Problem Motivation:
Raven Industries plans to introduce an auto-leveling feature to a spray-implement but must develop a method to detect the height of a crop's canopy and the distance between the implement and that canopy. Simultaneously detecting the crop canopy and the ground directly underneath is a difficult task for most sensors, the optimal method of doing this is unclear.

Project Description:
A test system will be designed to analyze the performance of different sensors. For the solution to be effective, this system must deploy a variety of sensors over a crop canopy. The distance between the sensors and the canopy and the actual height of the canopy must be independently adjustable. Finally, the output of the sensors must be clearly displayed and recorded for analysis.

A novice presenter might state the following:

What is our problem statement?
Problem motivation. Raven plans to introduce an auto-leveling feature... (Presenter proceeds to read the slide verbatim.)
Project description. A test system will be designed to analyze the performance of different sensors... (followed by more direct reading)

Conversely, a prepared and experienced presenter knows that the slide is simply a visual representation of the verbal presentation. This presenter knows it is not effective to directly read this slide. Instead, an experienced presenter will deliver the content on this slide in a confident and entertaining voice. They will probably not even use the terms "Problem Statement," "Problem Motivation," and "Project Description." These categories simply break up and organize the visual; it is not necessary to use them in the verbal delivery.

This issue applies to all slide transitions. Delivering the Problem Statement is used here as a representative example. For another example of this issue, reconsider the discussion a few pages back referring to the Block Diagram. Notice, the term Block Diagram is never used in the provided narrative.

Avoiding Common Pitfalls 9.2: Avoid Using Pronouns

Pronouns make our literary writing more interesting to read, but they are imprecise. Technical communication authors must excel at being precise. Consider this statement from a 2017 student status report:

Leah identified a mistake in her battery-capacity calculations when testing the current on the supply cabling. She has corrected the calculations and determined new ones will have to be purchased. Yi (her sponsor) has reviewed the corrections and approved

Analysis:

This is not an unreasonable status report. Anyone carefully reading the report should be able to understand what is meant. Right? Wrong – there are two ambiguous pronouns in this statement. You must have made an assumption about these statements, perhaps unconsciously, and possibly inaccurately while initially reading the report. Leaving interpretation to the chance of a reviewer's assumption will eventually cause problems. Consider the following:

- *What does the term **ones** stand for on the second line? Did you assume new batteries or new cables had to be purchased? This is a big difference. The cost of replacing batteries will amount to hundreds of dollars; cables are much cheaper.*
- *Who does the term **she** refer to in the last sentence? Is Leah, the poor college student, or Yi, the sponsor, paying for this mistake?*

Correction:

Leah identified a mistake in her battery-capacity calculations when testing the current on the supply cabling. She has corrected the calculations and determined that new CABLES will have to be purchased. Yi has reviewed the corrections and approved them. LEAH has agreed to pay for the additional costs WHICH AMOUNT TO $5.00.

After reading the corrected version (which is what Leah had meant in her 2017 entry), consider whether you correctly interpreted the original entry.

Industry Point of View 9.1: Kickoff Meetings

I am the Project Manager for a Department of Defense subcontractor. Our contracts are typically divided into three phases. A Phase I contract allows us up to 6 months to complete the INITIATING phase of a project. A Phase II contract allots 2 years to complete the PLANNING and EXECUTING phases, although this time frame is negotiable depending on the project. A Phase III contract is a manufacturing and service contract and, for the most part, do not require Project Management.

During Phase I, we analyze the problem space by performing a thorough Literature Review and then develop our Project Requirements. After approximately 5 months, we hold a Kickoff Meeting with all the primary Stakeholders. For a typical project, the Stakeholders include a Project Sponsor from the DoD, the Primary Subcontractor's

Project Manager, and representatives from all Secondary Subcontractors. The Secondary Subcontractors include business partners who have been included on the contract to support a particular technical development and university partners who have been included to support any necessary research component.

Our Project Kickoff Meetings are scheduled for 3 days. The first 2 days have a defined agenda and the third day provides time for an open discussion. During the first morning, we present the Problem Statement. Every Stakeholder is asked for conformation that the Problem Statement correctly defines the desired project. Any concerns are addressed and the wording of the Problem Statement is usually modified, at least slightly before we all agree.

Also on the agenda for the first morning is a reading of the Project Objectives and Requirements. We typically have between 8 and 12 Objectives. We start by reviewing the list of Objectives to determine whether the modifications to the Problem Statement necessitate a modification to the Objectives. Once satisfied that the Objectives are correct and complete, we start over and review each Objective and its related Requirements.

The agenda for the first afternoon is a review of Project Specifications. We start back at the beginning of the Requirements Document and review each Requirement's Specifications. Our team meets that evening to incorporate all changes to a new draft of the Requirement s Document. The agenda for the second morning is to start over and review every detail of the new draft.

The agenda for the second afternoon is an acceptance discussion. Each Stakeholder, starting with the Secondary Subcontractors, asked to confirm whether they support the project as defined by the newly drafted Requirements Document. Assuming that at least the Primary Subcontractor supports the Requirements, the DoD Sponsor makes a final decision.

Approximately 50% of our projects move past Project Kickoff Meetings into a Phase II project. I am proud to say our company regularly surpasses the national average of Phase II awards, which is 30%. The DoD Sponsor and Primary Subcontractor are about equally responsible for canceling projects.

The Sponsor may cancel a project due to a budget limitation. The DoD supports approximately three times as many Phase I contracts as Phase II contracts, meaning they are obligated to cancel 66% of Phase I projects. The Primary Subcontractor may cancel a project if a particular Specification is so restrictive that they do not believe the project can be successful and are unable to negotiation a change to the Specification. Secondary Subcontractors occasionally back out of the project after the PKM, but they typically can be replaced if the Primary Subcontractor and Sponsor wish to proceed with the project.

9.6 Chapter Summary

In this chapter, you learned about the Project Kickoff Meeting. The PKM is an important milestone that occurs near the end of the INITIATING phase. One of the purposes of the INITIATING phase is to investigate a project's potential. By the time of the PKM, the entire design team (including the engineers and the Decision Makers) must come to an agreement whether the project is worth pursuing or not. Here are the most important takeaways from the chapter:

- The greatest risk to any project at the end of the INITIATING phase is that the design team does not fully understand the Project Motivation, Problem Statement, and/or Requirements.
- Purpose of the PKM is to determine whether all Stakeholders are in agreement that the design team does, in fact, fully understand the project.
- The four outcomes of a PKM are as follows:
  - Complete acceptance of the project (rare)
  - Conditional acceptance of the project with a few minor modifications which will be addressed in the upcoming Project Charter (goal)
  - Temporary rejection of the project requiring the team to revisit part/all of the INITIATING phase (not uncommon, but hopefully avoided in Senior Design)
  - Outright rejection of the project (avoid)
- The design team has a responsibility to present information effectively to the Stakeholders at a PKM. Effective communication will enhance the Stakeholders' ability to assess the project's risks and determine whether to move forward or not.

Chapter 10: **Project Charters**

The INITIATING phase culminates with a document called the **Project Charter**. A "charter" is a "written instrument or contract...that guarantees rights, franchises (responsibilities), or privileges" [23]. Projects need approval before they can be executed. A Project Charter is a contract that defines the rights (payment) and responsibilities (workload) of the project [24].

A lot of discussion, brainstorming, and ideating occurs during the INITIATING phase. Stakeholders tend to lose track of details, miscommunicate with each other, and misinterpret intentions. Hopefully, all the good ideas have been incorporated, all the poor ideas omitted, and any miscommunication is eliminated by the end of the Project Kickoff Meeting, but we need to verify and document the details so there is no confusion.

The Project Charter is a *legal* document, a contractual agreement between the sponsor/client and the engineers/consultants. Meaning, once signed, the Project Charter constitutes a binding and enforceable contract. As such, it may be scrutinized in a court of law for any little miswording or ambiguity.

Accurately conveying the project definition within the Project Charter is critical!

Although much of what you have learned in previous English/Literature courses still applies, writing a Project Charter requires a special skillset called "technical writing." Technical writing is the skill of conveying accurate technical information in the most concise method possible. Nobody reads technical documents "for fun." The reader wants the information as clearly and quickly (i.e., as easily) as possible. To accomplish this, authors are required to present the information in an agreed-upon, standardized format. When the author correctly applies the format, the information flows more easily to the reader.

Admittedly, the idea of a "correct" style of writing is completely foreign to *literary* authors. Few people claim that reading Shakespeare, Tolkien, or Dostoevsky is *easy*, but to assert that their style was "incorrect" would be nonsense. These authors continue to enrich reader's lives *because* of their unique and challenging styles, but it does take effort to immerse yourself in their works. Readers of *technical* documents are attempting to gain information quickly so that they can make accurate assessments of technical concepts. They expect the information to be presented in an easy-to-understand manner. Writing "standards" exists to facilitate that content delivery. Authors who conform to these standards are more likely to communicate effectively with their reviewers.

In this chapter you will learn:

- Basic technical writing strategies
- Tip for reducing your formatting efforts in long documents
- How to format a Project Charter

C.J. Mettler, *Engineering Design*, https://doi.org/10.1007/978-3-031-23309-8_10

10.1 Basic Technical Writing Strategies

There are many rules and strategies that should be applied to technical writing in order to be the most effective author possible. This section will cover a few of those strategies which will make the most difference in your ability to communicate your project in the Charter. These include the following:

- Effective use of figures
- Antecedent method
- Appropriate citation referencing
- Use of third person – most of the time

Effective Use of Figures

Including figures in your documentation is beneficial to both the author and the reader. But in order to be effective, it is important to understand *why* figures are useful. For the most part, figures are *not* included as *evidence* that the statements made are true. Rather, they are included to increase understanding about the statements in the first place. This is an important distinction!

When starting to write about a new idea, novice authors will often make the mistake of focusing on what they should *say* rather than what they should *show*. This leads to a long narrative that can be difficult to understand. Then, as an afterthought, the author includes a figure as evidence about what was just explained. However, an experienced author first considers "what image would best demonstrate the idea" before they start writing, and they will tend to introduce the figure much earlier in the narrative. Once a figure is introduced, the narrative can utilize the figure. This pattern will significantly reduce the amount of writing required by the author, as well as make the topic easier for the reader to understand.

Example 10.1 illustrates this point.

In order to ensure that authors *effectively* use figures in their writing, there is an important formatting rule you must abide.

All figures must be introduced before displaying the figure or discussing the topic.

When this rule is taken out of context, the rule can seem like an unnecessary or nitpicky hindrance. However, the intent of this rule is easy to understand when considered in the context of *why* figures should be used in technical writing. Following this rule will ensure that you focus on the figure when you write and therefore clarify your intent.

Again, the rules states: First, introduce the figure. Then, present/discuss the figure. Typically, it is best to show the figure before discussing it. However, depending on the page layout, that might not be convenient. Notice Example 10.1 was introduced above this sentence, but there was not room on the page to present it – it is displayed on the next page.

This rule also applies to the following:

- Tables
- Equations
- Lists

EXAMPLE 10.1: Effective Use of Figures

Consider the following two narratives explaining how a DC motor functions. Which is easier to understand?

Option #1:

Permanent magnet direct current (PMDC) motors operate under Lenz's principle that current flowing perpendicular to a magnetic field induces a force orthogonal to the direction of current. As current flows around a coil of wire in the armature, it crosses a magnetic field created by permanent magnets located in the stator which flows perpendicular to the coil. As current flows past the magnetic field's north pole, it creates a force in one direction (perhaps pointed upward). As the current returns past the south pole, it will create a force in the opposite direction (in this case downward). The opposing forces create a torque on the motor which in turn causes the motor to spin. This is explained in Fig. 10.1 below.

Option #2:

Figure 10.1 depicts the operation of a permanent magnet direct current (PMDC) motor. The motor has two primary parts: the Stator and the Armature. The Stator (#1) creates a magnetic field (represented by the blue dashed lines). The Armature consists of a DC battery (#2) and a coil of wire (#3).

The operation of a DC motor is as follows:

- As current (left red arrow) flows out of the battery, it passes the north pole.
- This creates an upward force (left green arrows), dictated by Lenz's law.
- As current flows around the coil, it passes the south pole.
- This creates a downward force based on the same law.
- The opposing forces create a torque and force the motor to spin.

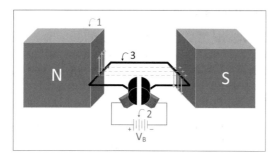

Figure 10.1 PMDC motor's operation

Analysis: Option #1 required 124 words whereas Option #2 only required 116 words. More importantly, Option #2 naturally forced the reader to periodically refer to the figure while reading the narrative. This dramatically improves the reader's understanding of the topic.

> ### *Avoiding Common Pitfalls 10.1: Introduce All Lists*
>
> Forgetting that "sections" under a "chapter" is a list is common mistake. Consider this chapter's outline:
>
> Chapter 10: Project Charters
>
> > Section 10.1: Basic Technical Writing Strategies
> > Section 10.2: Helpful Formatting Tips
> > Section 10.3: A Template for a Project Charter
>
> You will notice that at the start of this chapter, there exists a few paragraphs introducing the topics of the chapter before Sect. 10.1 starts. This pattern is repeated at the beginning of Part III: THE INITIATING PHASE where the upcoming chapter topics are introduced and again in Sect. 10.1 where the technical writing strategies discussed in this section are introduced. The introductions to Part III and this chapter are technically necessary; a set of numbered chapters and/or sections is definitely a "list." The introduction of writing strategies in Sect. 10.1 was the author's choice. The topics are not necessarily organized into a numbered list, so the choice is more ambiguous. If you are faced with a similar choice, consider whether the introduction will make the upcoming information easier to read and understand.

The Antecedent Method

The next technical writing strategy we will discuss is called the "**Antecedent Method**." This is an advanced concept but one of critical importance. Before defining the method, we should put the concept of an *antecedent* into context.

In the opening scenes of *Wizard of Oz*, we see a tornado whisk young Judy Garland off to Oz. For the next hour and 52 minutes, audiences are entranced by her escapades of trying to get home. Of course, when she clicks her ruby slippers, she wakes up in her own bed – it was just a dream, whew! Now consider, would the story have been as captivating had we known at the *beginning* that she was under the watchful care of Aunt Em the entire time?

Thespians called this plotline "reverse chronology." Reverse chronology creates suspense or intrigue through the use of misdirection. It makes the *Wizard of* Oz more entertaining. But try this in a courtroom and a lawyer is likely to jump up and shout "Objection! Leading the witness, your honor." We have all heard that line in some TV courtroom drama, but do you know what it actually means? Understanding this familiar objection will help clarify the "antecedent method." Example 10.2 explains this concept in detail, but in essence, lawyers are required to establish antecedents in order to avoid ambiguity.

Similarly, the "antecedent method" is a technical writing term that is used to ensure you communicate effectively to your reader. Since *clarity* is critically important in technical communication, as it is in a courtroom, understanding the "antecedent method" is important. This is a relatively simple concept that will make a dramatic difference in your communication skills.

EXAMPLE 10.2: Leading the Witness

In a court of law, clarity is crucial so that everyone understands "the truth." Leading questions either create a biased or confusing response [27] and, therefore, are not allowed. Here is an example:

Lawyer: Did the defendant strike you in the face with a book?
Witness: No.

What does this negative response truly mean?

- No, the witness was not struck in the face?
- No, the defendant did not use his/her fist?
- No, it was someone else who struck the witness?

The problem here is that a logical chronology (or antecedent) has not yet been established. This leads to confusion. The lawyer might rephrase this question:

Lawyer: Did the defendant strike you?
Witness: Yes.
Lawyer: Where did the defendant strike you?
Witness: In the shoulder.
Lawyer: With what did the defendant strike you?
Witness: Her fist.

Failure to build a "logical" argument is one example of, in legal terms, "leading the witness"; in technical writing terms, it is called "failure to establish antecedent basis."

In order to understand the Antecedent Method, we must define three terms:

- An antecedent is a word or phrase referred to by another word.
- An indefinite article is a word (such as *a* or *an*) that *generally* introduces a noun.
- A definite article is a word (such as *the* or *said*) that *specifically* introduces a noun.

For example, consider this sentence:

"Get certified as a PMI Project Manager; it will help your career in the years to come."

- Pronouns such as "it" can be confusing to an audience if the author has not been clear about what the word "it" refers to. In this example sentence, the meaning of the word "it" is clear – "it" refers to "getting certified." The phrase "Getting certified" is the *antecedent* of the pronoun "it."
- The word "a" in the phrase "a PMI Project Manager" is an *indirect article*. An article introduces a noun; an *indirect* article introduces that noun in the general sense. The example sentence is not referring to a specific Project Manager; it is suggesting that you become one of many Project Managers.
- The word "the" in the phrase "the years to come" is a definite article. The word "the" specifically introduces "years," not just any random years but specifically "*three* years to come."

Consider the importance of the proper use of indefinite versus definite articles. If a course assignment instructed you to "download *the* Charter template," you would assume there is one specific (definite) template you should download. Conversely, if the instruction was "download *a* Charter template," you may reasonably assume that *any* Charter template would be acceptable. Using the wrong article may cause confusion as to what the instructions actually meant. Using the proper article, along with establishing an "antecedent basis" will dramatically increase the clarity of this (and any) discussion.

An "antecedent basis" is a clear establishment of a phrase so that the phrase may be used without ambiguity in future sentences. The "antecedent method" of writing is to establish an antecedent basis for a phrase before assuming that the reader will understand that phrase. In other words, the antecedent method states as follows:

> All elements [of a discussion] must be properly introduced before the element can be modified or qualified [or discussed] [25].

Basically, this ensures that complicated concepts are laid out in such a way that avoids confusion or that requires the reader's interpretation. In fact, Title 35 Section 112(b) of the US Code (i.e., law) requires "that patent applications point out and **distinctly** claim what the inventor regards as the invention." Not surprisingly, "patent claim lacks an antecedent basis" is one of the most common ways to show a claim does not follow patent law, specifically §112(b), and therefore must be rejected.

Of course, a Senior Design Project Charter is not subject to the same scrutiny as a US patent, but industry-based Charters *are* a legal document that describes complicated and technical information. The result of ambiguous terminology in a Charter can very realistically result in the engineering consultant not being fully compensated for the years of work dedicated to the project. This could be devastating for an engineering firm. Using the antecedent method will improve the clarity of your writing and avoid these ambiguities.

Following the Antecedent Method is actually quite easy. Most of the time, the method simply boils down to referring to a noun using an indefinite article ("a" or "an") *before* referring to the same noun using a definite article ("the" or "said") [26]. Example 10.3 discusses using the proper article further.

EXAMPLE 10.3: Applying Antecedent Method #1

The Antecedent Method can, and should, be applied at all levels of technical writing. Consider, for example, how many times this method was used in the previous paragraph alone:

> Following the Antecedent Method is actually quite easy. It simply boils down to referring to a noun using an indefinite article ("a" or "an") before referring to the same noun using a definite article ("the" or "said").

How many times was the method applied?

The Antecedent Method was applied three times in this example paragraph alone.

1. The author did not assume you (the reader) remembered what an "indefinite article" was. So, the general concept "***an** indefinite article*" was first introduced with a clarifying

EXAMPLE 10.3, continued

    definition (**"a" or "an"**). From now on, the author can refer to indefinite articles assured you (the reader) know what is meant.

2.   The method was applied again when referring to the word "noun." The first time "noun" was used was in the indefinite sense. The phrase "referring to **a** noun" applies to any noun (an indefinite number of nouns). But the next time "noun" was referred to, the author specifically meant "**the** same noun" (definite/specifically) that was just correctly introduced.

3.   The point of this paragraph is that following the antecedent method "boils down" to the proper use of an article. The word "It" in the second sentence needed to either be removed or defined with an antecedent. With the sentence "Following the antecedent method is actually quite easy." immediately preceding the word "It," there is no ambiguity about "It" refers to.

Analysis:

The antecedent method most often refers to the introduction of nouns as in examples 1 and 2. The method also can refer to phrases or ideas as in example 3. Simply by introducing "a" new design component before discussing it in detail will improve the clarity of your discussion of "that" new component.

In patent law, a rejection caused by "a lack of antecedent basis" is most often caused by improper use of articles [25]. Using the proper article will eliminate most causes of confusion in your writing. However, there are a few other, slightly more, complicated antecedent issues to be aware of. It is possible to create an antecedent issue by the *lack* of an article, as opposed to the *proper* use of an article. Example 10.4 provides an example that contains both of these common issues.

EXAMPLE 10.4: Applying Antecedent Method #2

Another common antecedent issue is the *lack* of a proper article, as opposed to the improper choice of article. Consider this Project Description from a Project Charter submitted in 2020 that includes both types of antecedent issues.

> *Logging details about the power fluctuation of the residential systems will help to understand the changing requirements of those systems. The data logger collects information that could be used as statistical data to create accurate simulations of the residential grid system. These simulations could help engineers to downsize the infrastructure of the power grid.*

All of the statements in this paragraph are true, but when taken as a whole, it is still unclear what the Project Description is. Assumptions can be made about what the author *meant*, but we should not leave the Project Description open to interpretation.

EXAMPLE 10.4, continued

The first sentence alone has three phrases which lack antecedent basis, two obvious and one more subtle, including the following:

- **"The** residential systems" – There is no basis to understand which residential systems are being discussed. The authors did not mean any "specific" residential system, but rather they meant "all" (generic) residential systems.
- **"The** power fluctuations" – There is no basis to understand which power fluctuations are being discussed.
- "Logging details" – There is no basis to understand *details* of a system which has not yet been introduced. This error is more ambiguous because this antecedent flaw is due to the *lack* of an article, rather than an *improper* article.

Let us fix the "lack of basis" by using the antecedent method to rewrite this paragraph.

> *A small residential system, such as a home, experiences more drastic power fluctuations than most other types of electric loads. These power fluctuations can be caused by natural changes in a residence's lighting profile between day and night or a heating profile between summer and winter.*

- The antecedent basis of "residential system" now has been established with a clear introduction (starting with the indefinite article "**A**") and followed with a concise explanation.
- Grammatically, "power fluctuations" in the first sentences no longer needs an article. So it is slightly unclear whether a basis has been established for this term. When there is doubt, add an explanatory sentence such as sentence number 2, starting with "These power fluctuations. . . ."
- Notice the new sentence does not cause another antecedent basis issue by avoiding the use of definite articles.

In the original text, the term "**the** data logger" in the second sentence also lacks an antecedent basis. The existence of the data logger needs to be established. By using the *definite* article "**the**," we must assume there is no need to introduce the data logger because it is known to exist. However, in actuality, the data logger is the product this team is trying to get funding support to design! Do you see why there might be a problem if the reader assumes one already exists? Fortunately, this is an easy fix.

> *A data logger will be designed to collect information on these power fluctuations.*

- Establishing the indefinite (or thus far "abstract") idea of a "data logger" will allow the author to clearly and uniquely refer to this term in the remaining discussion

Another antecedent issue that commonly creeps into technical documents is an assumption that the reader will understand a particular term without a formal definition of that term. This issue is a little more difficult to address because the author cannot possibly define *every* term in the entire document. Example 10.5 presents an excerpt from a 2021 Senior Design team's documentation that illustrates this type of antecedent error.

EXAMPLE 10.5: Applying Antecedent Method #3

Consider the following statement made by a 2021 Senior Design team:

The charge controller will be placed in series between the PV panels (source) and the batteries (storage) in order to measure the batteries' voltage.

Let us assume that the team appropriately introduced "**the** PV panels" and "**the** batteries" at some previous point in the discussion. So the use of the definite "**the**" is correct here. However, the first use of the term "charge controller" lacks an antecedent basis. Let us rewrite the sentence using the appropriate indefinite article:

__A__ charge controller will be placed in series between the PV panels (source) and the batteries (storage) in order to measure the batteries' voltage.

Does this modification fully eliminate the lack of an antecedent basis concern?

Analysis:

There is still a "hidden" antecedent issue here that stems from the author's assumption that the reader will correctly understand the term "charge controller."

Perhaps this assumption would be more reasonable if the statement was as follows:

*A charge controller will be placed in series between the PV panels (source) and the batteries (storages) in order to **Control** the batteries' charge.*

It is reasonable to *assume* the charge "controller" would "control" the batteries' charge. If that was what the author meant, this statement might not need further explanation. But that is not what the charge controller is being used for in this case. The actual claim above was that the charge controller would *"**measure** the batteries' voltage."*

Should the majority of engineers understand that a controller is used to measure?

Analysis:

Since the intended use of the charge controller is more advanced or less obvious that a standard implementation, this sentence needs further correction to eliminate the antecedent concerns.

One such fix might look like this:

*A charge controller **with voltage monitoring capabilities** will be placed in series between the PV panels (source) and the batteries (storages) in order to measure the batteries' voltage.*

This sentence not only appropriately introduces the new concept of "**A** charge controller," it also defines what that device is by providing additional context regarding the "**monitoring capabilities**."

Appropriate Citation of References

All technical documents include references. Doing so generates credibility to your statements and can often reduce the effort of explaining many thoughts. Ensuring that references are appropriately sited is important, so it is clear which thoughts are your personal contributions and when you are relying upon someone else's intellectual property.

When researching a topic, authors must consider *which* source (or sources) to cite. The first links on a Google search often are excellent starting points for your research but are not necessarily best or most reliable resources. Under most circumstances, you should only cite *reviewed* sources in your final document.

Consider a set of references used in this textbook:

- For most topics, reviewed sources are used.
  - When introducing important Project Management terms, the Project Management Institute *Handbook* [22] or papers published on PMI's website [1] was referenced.
  - Information about historical characters such as James Cook [17] or Ernest Shackleton [20] were obtained through published *books*.
  - When discussing patent law, statements were quoted directly from the *US Code* (and the exact title and section were provided), and interpretations were used from *established legal sites* [27].
- There are a few cases where non-reviewed sources are used. But these sources were used sparingly and judiciously. In particular, when discussing Gene Kranz:
  - A *Wikiquote* [12] was cited. A Wikiquote is certainly the most questionable citation used in this textbook and should be considered a *secondary* source, at best:
    The reference used was a well-worded quote about Kranz's management style. Quotes are interesting and useful but not particularly important to your overall education. Therefore, a secondary source is acceptable here.
    Also, the quote matched what the author personally learned about Kranz while researching this topic. The quote was supported by numerous other sources of information.
  - Information from a blog was used to explain Kranz's team leadership. Blogs are also a somewhat questionable resource that required additional investigation. This particular blog was posted by Virginie Briand. The author of this textbook was not familiar with Ms. Briand at the time of writing the passage on team leadership. So an additional search on this potential source of information was necessary. Ultimately, Ms. Briand's publication was used for the following reasons:
    Turns out, Ms. Briand is a prolific writer on business teams and leadership. So when speaking about Kranz's teams, she definitely has credibility.
    The use of Ms. Briand material was limited in this textbook to Kranz's *leadership* – a topic which corresponds to her expertise. Although she mentions factual events that occurred at Mission Control on April 13, 1970, other sources were found regarding *those facts*. In particular, archival data on the Apollo 13 mission [10] (published directly by NASA) was used for the factual timeline.
    Most importantly, Ms. Briand's narrative matched a number of other sources that were reviewed but not ultimately cited here. Ms. Briand's authoritative discussion was particularly well worded and inciteful. So this reference was used over the other options.

Avoiding Common Pitfalls 10.2: Choosing Your References

Choose your references wisely.

When initially investigating a topic, feel free to use secondary sources like Wikipedia or personal opinions posted on blogs. But do not consider your investigation complete at that stage. Dig a little deeper. Follow the secondary reference's sources to primary sources. For example, use the supporting links in a Wikipedia article as the references rather than the Wikipedia site itself.

Often, this is very easy to do:

- Wikipedia is often one of the first links that appear in a Google search.
- Wikipedia will quickly provide you with some initial thoughts and facts about your topics.
- Wikipedia articles usually provide references in hyperlinks, follow them, and verify they support your initial thoughts.

Developing these primary sources should not be difficult or time consuming if the initial information is valid. Consider the following:

- If the Wikipedia site does not provide easy-to-follow references, the information may not be valid in the first place – move on to other sources!
- If the Wikipedia site's references are obscure or they do not clearly support your interpretation of the initial information, the Wikipedia article may be misleading you – definitely move on to other sources!

10.2 Helpful Formatting Tips

Proper formatting of technical information in a *Word* document can be time consuming. Fortunately, there are a number of shortcuts that will reduce the overhead effort so that you can focus on the writing itself. This section will introduce three such shortcuts.

Inserting Citations (aka References)

Most technical documents will rely on external sources. To avoid plagiarism, these sources must be appropriately credited. *Word* refers to the sources as "citations." This textbook will be using IEEE formatting and, therefore, refer to sources as "references."

Word provides assistance in formatting and tracking references. This means that you do not have to reorder the reference should you choose to insert a reference in the middle of the document; *Word* will do that automatically. Also, *Word* will assist you in formatting the bibliographical information.

To illustrate how to do this, let us assume you are discussing Newton's laws of motion and need to reference the statement in Fig. 10.2. To do so, type the statement as shown. Then navigate to "References" and verify that the "style" is set to IEEE.[1]

Figure 10.2 A claim that requires a reference

Next navigate to "Insert Citation" and select "Add New Source as shown in Fig. 10.3. Choose the "Type of Source" and fill in the requested information.

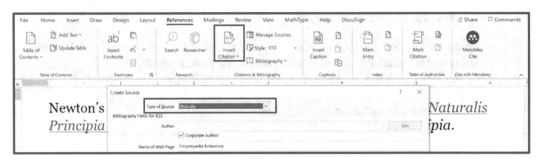

Figure 10.3 Accessing Word's automated citation feature

Later, when you are finished typing, you can

- Highlight all the texts.
- Right-click.
- Select "Update Field."

This is shown in Fig. 10.4, and all the references will be sorted in chronological order. If there are references inserted before this statement, the reference shown in Fig. 10.4 will no longer be numbered as [1].

---

[1] You may use styles other than IEEE with permission of the Project Charter's reviewer.

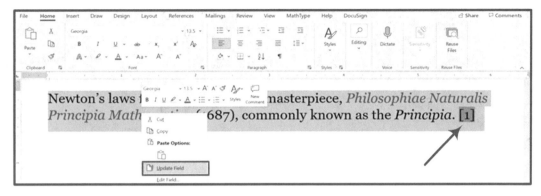

Figure 10.4 Updating a document's citation list

Moreover, *Word* will format your Bibliography (or "Reference") page automatically. To insert this at the end of your document, place the cursor where desired. Then, navigate to "References" and select "Bibliography." Then, since you should be formatting using IEEE standards, select "References," Fig. 10.5. The complete bibliography, formatted in proper order, will be inserted. If you make a change to the document, you may simply click on the title and select "Update Field" here as well.

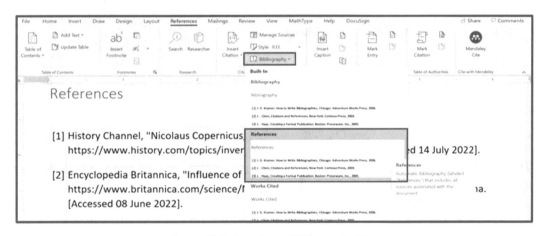

Figure 10.5 Inserting a Bibliography Page

Inserting Captions

Technical documents often include Figures, Tables, and Equations. *Word* refers to all of these additions as "captions." Captions can be inserted and managed just like citations. For example, *Word* can automatically manage equations as shown in Fig. 10.6. To do this,

- Navigate to "Resources."
- Select "Insert Caption."
- Choose the "Label" (in this case "Equation").
- Click OK.

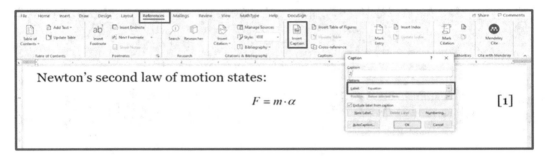

Figure 10.6 Managing captions

Later, you may use the "Cross-reference" option, Fig. 10.7, to refer back to a caption. *Note: the caption must be **inserted** before it can be **cross-referenced**.* Choose the appropriate "Reference type" and how you want to "Insert reference." Use the "Insert Caption" option to initial title the object you are displaying (e.g., equations or figures). Use the "Cross-reference" option in the body of the text to refer to a previously captions object.

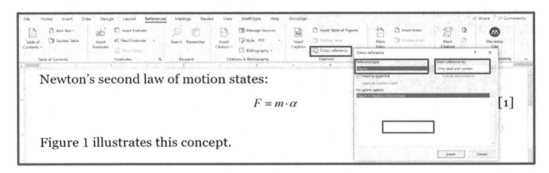

Figure 10.7 Cross referencing captions

Use the "Insert Caption" option directly on figures and tables, as well as equations. Simply choose the appropriate label in the "Insert Caption" dialog box.

Inserting Equations

Typing equations directly into *Word* is often cumbersome. *Word* does include an embedded equation editor, but it is not much more user friendly than *Word* itself. Fortunately, there are numerous equation editor packages that interface well with *Word* that make this take much easier.

One such app is called MathType. Currently, at the time of the textbook's publication, there were annual licenses available for students for $1.00/student/year. At this price, the app is well worth the cost.

Inserting Variables

Another cumbersome aspect of technical writing is typing variables. In particular, Greek symbol variables with sub- (or super-) scripts are not necessarily easy to type on a standard keyboard. However, with a little preparation, *Word* can be configured to easily handle these characters.

In order to easily access your most commonly used symbols, navigate to "Insert" >> "Symbols" and select "More Symbols." In the popup window, set the "Subset" to be "Greek and Coptic," and select the symbol you want (e.g., Δ, Fig. 10.8), and click "Shortcut Key."

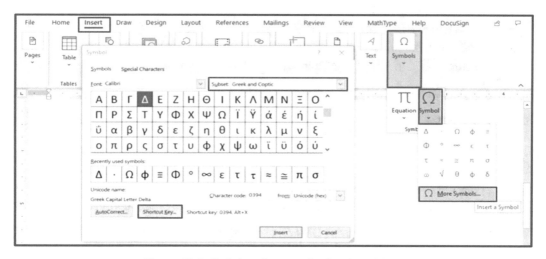

Figure 10.8 Defining shortcuts for Greek variables

In the next popup window (not shown here), you can define your own shortcut for this symbol. The following list provides some suggestions of easy-to-remember shortcuts for commonly used symbols. Of course, there are many other possibilities – define shortcuts that are easy for *you* to remember.

Under the Subset = Greek and Coptic:

- α: alt + a
- β: alt + b
- δ: alt + d
- ω: alt + w
- τ: alt + t
- π: alt + p
- Δ: shift-alt + d
- Ω: shift-alt + w

Under the Subset = Mathematical Operators:

- ≠: alt + =
- ±: shift-alt + =
- ≈: alt + ~
- ≡: shift-alt + −
- ∞: alt +8
- °: shift-alt +8

- $\geq$:   shift-alt + $>$
- $\leq$:   shift-alt + $<$

Many variables are specified using a subscript or a superscript. For example, the first resistor is often labeled R_1, or the "effect of voltage-source-1 on the output-voltage" might be labeled as V_{out}^{V1}. *Word* has a built-in shortcut to assist with this type of formatting.

To type a subscript:

- First type the variable
- Press Ctrl + $=$
- Type the subscript variable
- Example: R [Ctrl + $=$] 1 results in R_1

To return to normal font, simply press Ctrl + $=$ again.

To type a superscript:

- First type the variable
- Press Shift-Ctrl + $=$
- Type the superscript variable
- For example, if you type: V [Shift-Ctrl + $=$] x WORD will produce in V^x.

To return to normal font, simply press Shift-Ctrl + $=$ again.

10.3 A Template for a Project Charter

Keep in mind that the specific format of an organization's Project Charter may differ and/or be more inclusive than what is presented here. This section covers the most important and most common topics reviewers would expect to see in a "typical" Charter. Topics include the following:

- Project Motivation
- Project Description
- Project Background
- Project Requirements
- Technical Description

Project Motivation

The Project Motivation is typically found early in the Project Charter's narrative. This section provides a concise explanation about the motivation behind the project. It essentially answers the question:

Why should this project be supported?

The Motivation Statement was previously described in Sect. 6.1. At that point in the Design Life-Cycle, the focus was creating a concise and impactful Problem Statement. The Motivation Statement was condensed into one or two sentences, but the Project Motivation in the Project Charter does not have to be *as* condensed. This section should be written in a short paragraph format. This allows you to expand *slightly* on the motivation. Be careful, however, to avoid providing too much background here. Provide a complete picture of the motivation, but save most of the background materials for that upcoming section.

Your primary audience is the Decision Maker Stakeholders (e.g., high-level executive or CEO). Consider how much information is critically necessary so that this audience will understand the need, and be motivated to pursue this project.

Project Description

The Project Description immediately follows the Project Motivation in most Project Charters. The motivation established the problem, the description explains as follows:

What should be done about that problem.

This section should also be a short paragraph written to the primary Decision Maker. It presents the Goal and Objective Statements. If necessary, you can expand on the limited sentences developed for the Problem Statement. However, because the audience is the primary Decision Maker (likely a nontechnical person), this is not the time to present any technical details. The Project Description provide a concise starting point that puts the remainder of the document into the proper perspective.

> ### Avoiding Common Pitfalls 10.3: Distinct Motivation and Description
>
> These two sections are meant to grab the attention of the primary decision-maker. This person will often only read the introduction of a long document and then assign the responsibility of digging into the details to a subordinate. This means you have to grab the decision-maker's attention, educate this person, and recruit them as a project support in a very few lines of text.
>
> First drafts of the Motivation and Description sections often have repetitive statements. Repetitive statements make the information less concise and perhaps less impactful. This flaw is created when the author has not categorized their thoughts in a precise manner. This issue is compounded when more than one author contributes to these two sections.
>
> Fortunately, the first draft is still valuable and solving this issue is relatively easy. After writing the first draft of both statements, consider what information directly defines the problem and what information explains the solution. Then, simply reorganize your statements so the two paragraphs flow together effectively.

Project Background

The Project Background provides the details required to *fully* understand the Project Motivation and Description. It also provides details that will assist the reader to understand the upcoming technical discussion. This section should be written to a general engineering audience. You may assume the reader to have a technical background, but not necessarily on material specific to *your* project. In other words, the intended audience is no longer the primary Decision Maker.

The Project Background is often split into three topics, similar to the Literature Review:

- Project Background
- Technology Review
- Applicable Standards

Depending on the project's details, these subsections may be more or less applicable. In fact, in some cases, one of these sections can be omitted when it does not apply to the specific project. Occasionally, the exact titles of these subsection may be modified for some projects to provide a more applicable description of what will be presented. Also, in rare cases, there might even be an additional topic that must be covered for a particular project which does not conveniently fit under one of these three titles. In that case, it is permissible to add a fourth subsection.

Notice these topics are essentially categories of the Literature Review you have already performed. The Literature Review is a process which you performed to collect important information; the Project Charter's Background section is where the important aspects of the Review are published. Note: The entire Literature Review should be documented in your Project Notebook, whereas the material published in the Project Background section is a condensed version of the most important and applicable topics.

The Project History subsection provides a detailed description of the project's history. These may include the following:

- A description of a previous project or predecessor
- The operating environment that causes the Project Motivation
- An explanation of a system with which the proposed project must interface

The Technology Review subsection provides a summary of important technical aspects related to the project. These may include the following:

- Explanation of a technology or process that will likely be implemented
- Data on potential components that will likely be incorporated
- Review of potential competing (existing) solutions

Most engineering products are subject to certain standards. Understanding which standards *your* project must adhere too is important. The Applicable Standards subsection explains these standards and provides citations to authoritative references.

Most of the topics in the Project History and Technology Review should include a figure. Recall that the writing should first introduce the figure at (or near) the beginning of the topic and use the figure to effectively describe the concept.

Nearly every topic in all three subsections should include at least one reference.

Each of the topics presented within these three subsections should be structured in a cohesive manner. Often, it is useful to create numbered sub-subsections to organize these topics. Alternatively, clear and distant headers might be preferred. For example, this paragraph is located in Sect. 10.3 under heading "Project Background." Remember, a list of subsections and/or headings should be introduced (Sect. 10.2).

Project Requirements

After providing the primary Decision Maker a stimulating (but not technical) Project Motivation and Description, and documenting and explaining all the important issues related to the project for the engineers, the Project Requirements are provided.

Project Requirements are the technical definition of your project.

This section should be written to an audience with specific and detailed knowledge of the engineering required to complete the project. The Project Requirements topics are usually broken into a few important categories:

- Objectives
- Project Requirements Outline
- Design Constraints
- Additional Project Concerns or Deliverables
- Team Member (or Stakeholder) Responsibilities

The *Objective* section is *essentially* a repeat of the Project Description section, with one notable difference. The Project Description was written for a nontechnical audience, whereas the Objective section under Project Requirements is written for an advanced technical audience. The more complicated and sophisticated the project, the more difference there will be between these sections, but for most Senior Design projects, the two sections are typically pretty similar. This should be written in paragraph format.

The *Project Requirements Outline* section is a formal publication of the Requirements Document that has already been produced. This information can be introduced in a single sentence and then presented in outline format.

The *Design Constraints* section is a formal publication of the Constraints that have already been identified. This information can also be introduced in a single sentence and then provided as a list.

The *Additional Project Concerns or Deliverables* section documents any requirements placed upon the *team* rather than the *product*. Recall that a *Requirement* is a statement that defines what the *product* must do. A *Constraint* is limitation on the designer's ability to complete the solution. Sometimes, there are additional concerns that must be addressed but do not fall into either category. Examples include the following:

- Additional documentation such as a user manual
- Required research that will be completed using the final product
- Installation of the final product
- Travel required by the Sponsor

The Additional Project Concerns or Deliverables section may also include a discussion of priorities. Perhaps it is unclear whether a project can accomplish certain things due to Triple Point Constraint limitations. This section can be used to define priorities or conditional changes. Typically, these discussions should refer to the Triple Point Constraint and explain how the conditions will be balanced.

The *Team Member's Responsibilities* section is meant to define who will be responsible for what project tasks. Typically, this can be broken into Critical Subsystems + a few additional (project specific) categories. This is particularly important when the Sponsor/sponsor is responsible to deliver or design a portion of the project. For a Senior Design project, you should discuss with your Instructor whether it is appropriate to "assign" your Sponsor a task in this context.

Technical Description

The amount of detail provided in the Technical Description can vary drastically between organizations. Most Project Charters provide at least a *high-level* description so as to be clear what the ultimate deliverable should look like. But different organizations may want more (or less) detail at this stage. You should follow your organization's (or Instructor's) policies on this section.

EXAMPLE 10.6: *Additional Project Concerns*

A Sponsor of a 2021 Senior Design project stated they:

Desired the volume of their project housing to be limited to $1.0m^3$ but would accept a housing under $1.5m^3$.

The appropriate way to handle this ambiguity is to define the least restrictive Specification in the Requirements Document (in this case, volume <1.5 m^3). Then, discuss this issue in the Additional Project Concerns subsection.

The Sponsor would like you to surpass the Specification (volume <1.0 m^3). But under which conditions? Are they willing to pay more than the allotted budget? Are they willing to give up some functionality to accomplish the size reduction? Are you willing to extend the time-constraint and run a produce a condensed prototype?

If a concession like this is included in the Additional Project Concerns subsections, you should define two important parameters:

1. When will the decision be made to pursue the upgrade?
2. Who will decide to purse the upgrade?

The Project Requirements Document should include the minimum viable definition that the Sponsor will accept as "successful." The "Additional Concerns" include any desired but unnecessary" upgrades.

10.4 Chapter Summary

In this chapter, you learned how to write a Project Charter. The Project Charter is the final output of the INITIATING phase. The reality of conducting business in today's society means that engineers must consider the legal ramifications of their work. The Project Charter is a legal contract that defines the project, identifies responsibilities, and specifies deliverables. Here are the most important takeaways from the chapter:

- The Project Charter is a legal contract that precisely defines the project. This document will be used to assess the quality of the project at the end of the Design Life-Cycle.
- Because the Project Charter must be written in precise language, proper technical writing techniques must be used.
- Start each new section or topic by considering what figure will best represent the idea. Properly introduce the figure and make the figure the subject of your discussion.
- Apply the Antecedent Method continuously throughout your writing. Establish each new idea or item using indefinite articles before assuming a reader knows what you are talking about. After an idea or item has been established, refer specifically to that topic using definite articles to prevent the reader from assuming you are intending to establish a different idea or item.
- Rely predominately on "primary" resources for your Literature Review. Cite those sources appropriately. Only cite "secondary" resources which you have vetted.

PART IV: THE PLANNING PHASE

"To the optimist, the glass is half full.
To the pessimist, the glass is half empty.
To the engineer, the glass is twice as big as it should have been."
— Yves Behar, Swiss Entrepreneur

"The consequences [of poor planning], *in my world, at the extreme, is death. But, also failed objectives, a needless waste of resources, a negative impact on the indigenous population and local government – we always have to consider how our actions affect others."*
— Conrad Anker, American Mountaineer

Conrad Anker has been pushing the limits of mountaineering for the last 30 years, evolving into one of America's top alpinists [28]. Anker is a visionary athlete who has established new routes on mountains from Antarctica to Patagonia, from the Karakoram to the Alaska Mountain Ranges. Between 1992 and 2018, he served as the captain of the North Face Climbing team and continues to support climbing endeavors across the globe.

Part of what makes Anker so successful is the attention to detail he puts into planning each of his expeditions. Even if he is only headed up to a local ice floe in Hyalite Canyon thirty minutes from home, he packs his gear the evening before the climb so that he has a night to digest his packing list before setting off the next morning. Of course, the more extreme the expedition, the more detail he puts into the planning.

Ed Viesturs, another exceptional mountaineer, reminds adventurers that "Reaching the summit is optional, getting down is mandatory." Conrad Anker takes that sentiment a step further, "a successful expedition starts and ends at your front door. Even though we may have expended 80% of our emotional energy reaching the summit, we have only traveled 50% of the required distance. You must plan for the *entire* adventure."

When Conrad speaks about the "entire adventure," he uses much of the same terminology as a Project Manager. Because, he is, in fact, managing a very complex solution to a difficult challenge (i.e., "a project"). He talks about the ideation of a project (STARTING phase), defining the specific mission objectives (INITIATING phase), preparing for the mission (PLANNING phase), and executing the mission (EXECUTING phase). By the time he publishes articles documenting the mission, he has, in some sense, completed a formal CLOSING phase.

Successful planning for an expedition to the immense mountain ranges of the world, like the Himalaya, is a significant undertaking. For every day required to execute the expedition, Anker's teams spend one-and-a-half days preparing. A typical climbing team dedicates at least 2 months to an excursion on Mount Everest. This means if Anker wants to climb in the Himalaya in April and May, he must start planning the expedition late the previous year.

Permits must be procured from multiple local, regional, and national governments. Teams must be formed and organized. Routes must be researched and vetted. Gear must be assessed and shipped. Supplies must be carefully measured and metered out for each stage of the journey. Contingency plans must be made for any number of adverse situations. Further, a thousand other details must be prepared. If any one of those details is improperly accounted for, Anker puts the success of the entire expedition at risk.

By planning every detail of a round-trip expedition, Anker is, in essence, considering the "system-level" point of view. As with all major projects, a holistic plan dramatically increases the potential for success. Imagine, if you will, the consequences of merging the planning and execution of a mission. Anker *could* book tickets to Kathmandu, Nepal, and *then* begin navigating the permitting process. He *might* be successful and procure permission to trek to one of the mountain's basecamps. Perhaps the indigenous people he meets along the way could provide information about which specific route up the mountain he should attempt. Once at camp, he *possibly* could scavenge, borrow, and otherwise collect enough climbing gear to attempt a summit bid. Of course, if you have ever attempted even a simple camping trip, you are probably laughing right now at such a disorderly course of action. No adventurer could possibly hope to be safe and successful navigating a mission in such a haphazard manner.

Fortunately, Conrad finds enjoyment in the planning. If the planning phase requires more than half of time dedicated to the mission, "you might as well enjoy it," he says. He finds beauty in the PLANNING process, knowing each moment planning is time spent ensuring the success of the mission, each successfully organized detail generates a little higher chance of success.

A similar mindset should be used while planning a solution to an engineering problem. A carefully planned solution has a much greater chance of success than a haphazardly developed solution. Each detail carefully considered during the PLANNING phase reduces the risk of encountering an issue during the expensive EXECUTING phase.

During the PLANNING phase, the entire system design is produced "on paper." The first step is to lay out a Project Schedule so that all major due dates and milestones are known. Then, each subsystem is designed. Often this includes some basic prototyping efforts and may require that a small amount of the Project Budget be expended. Each subsystem is carefully incorporated into the overall system design. Detailed plans for the fabrication and verification of the product are also developed. Two risk-reduction reviews are conducted, the Preliminary Design Review approximately midway through the PLANNING phase and the Critical Design Review at the end of the phase.

There are six primary processes which must be accomplished during the PLANNING phase of your project. These are covered in Part IV of this textbook.

Processes of the PLANNING Phase	Presentation Structure
A project schedule must be produced and managed during the Scheduling process.	**Gantt Chart**
Options for any crucial components must be analyzed and the optimal item selected.	**Alternatives Analysis & Selection**
The most important subsystems must be designed (or simulated) "on paper"	**Subsystem Designs**
Prototypes are developed for key subsystems to increase confidence in the initial design.	**Preliminary Prototype Demos**
A plan must be prepared for the complete system integration.	**System Design**
The system-level design must be accepted by stakeholders.	**Preliminary Design Review & Critical Design Review**

Chapter 11: **Developing Gantt Charts**

In Sect. 8.2, you learned a few basic concepts about *using* a Gantt Chart. This section will provide instruction on *developing* your own Gantt Chart or at least modifying the rough schedule you may have already been provided. Typically, a team will produce a high-level, rough Project Schedule that includes primary milestones and generated workflows during the INITIATING phase and present that at the Project Kickoff Meeting. Then, one of the first processes during the PLANNING phase is to produce a more detailed Project Schedule. The detailed Project Schedule should define intermediate milestones and specify when certain tasks of the PLANNING phase will be accomplished. With experience, a Project Manager will also schedule many of the details of the EXECUTING phase many months in advance; however, this is much to ask of a Senior Design team until the team has developed experience in conducting an EXECUTING phase. Project Schedules are documented using a tool called a Gantt Chart.

Most Senior Design courses have a fairly structured schedule during the first many weeks of the first semester, and so there is often little need for students to create their own Gantt Chart during those weeks. However, by the time the PLANNING phase starts, each project will proceed on a unique path. The course will likely still have a general framework to structure your Project Schedule, but you will need to define and manage the specifics for your project. By the time you reach the EXECUTING phase, project details will have diverged so much that it will not be possible for your Instructor to predefine even a framework for your Project Schedule. After this point, students must be ready to take control of their own schedules. This chapter will guide you through developing and managing your project schedules.

In this chapter you will learn:

- How to define and properly scale project tasks
- How to enter those tasks into a Gantt Chart
- How to manage those tasks and keep the Gantt Chart up to date

© The Author(s), under exclusive license to Springer Nature Switzerland AG 2023
C.J. Mettler, *Engineering Design*, https://doi.org/10.1007/978-3-031-23309-8_11

11.1 Use a General Framework as a Guide

The first step in the Project Scheduling process is to define the tasks that need to be accomplished. Then, each of those tasks must be arranged in a sequential manner that ensures all important milestones are met. As you format those tasks into a sequential schedule, you will likely identify additional tasks that need to be included. This should be expected and is not a difficult issue to deal with. However, developing the schedule is most efficient when you previously know the majority of the tasks you will be entering before attempting to sequence them. It is a good idea to brainstorm a list of tasks before attempting to arrange them in the Gantt Chart.

To begin this process, start by identifying any major milestones that you will have to meet. Pay particular attention to when each phase of the Design Life-Cycle is scheduled to end and to any major deliverables you must submit at those points in time. Remember to consider the entire project, not just the PLANNING phase. Incorporate these milestones on the Gantt Chart to create a framework that you must adhere to for the remainder of the Project Scheduling processes.

Figure 11.1 displays the framework schedule used throughout the version of the Design Life-Cycle presented in this textbook. This framework starts with the PLANNING phase under the assumption that you have previously been provided a rough schedule for the STARTING and INITIATING phases. The framework continues through the EXECUTING and CLOSING phases. Milestones are designated by diamonds and indicate the dates when a major deliverable must be completed or an event (e.g., PDR) is scheduled. The blue, purple, and yellow bars represent categories of tasks within the PLANNING, EXECUTING, and CLOSING phases, respectively. Currently, these bars *appear* to be "tasks" in Fig. 11.1. However, once details are included and correctly formatted under each category, the bars which you currently see will auto-update to become brackets similar to the gray brackets corresponding to the Part 4, Part 5, and Part 6 parent categories.

Once a framework has been developed, you can schedule the necessary task details.

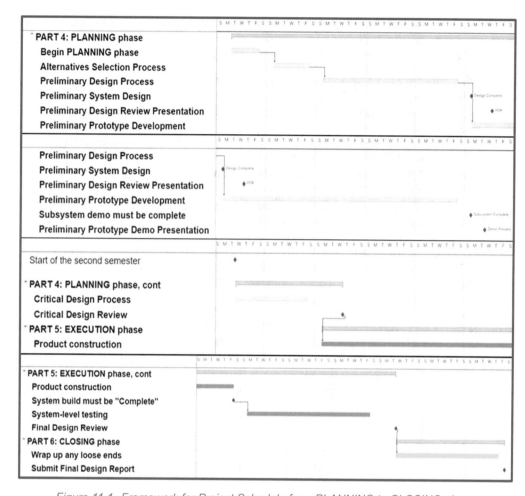

Figure 11.1 Framework for Project Schedule from PLANNING to CLOSING phase

11.2 Scheduling Detailed Tasks

Once a general framework has been developed, specific project tasks must be scheduled for each category of work. Each task should be assigned to a specific team member. There should be multiple tasks scheduled under each category.

These tasks should require *approximately* one week each to complete. Work assignments which require longer than a week become difficult to track and manage. It is better to break these assignments up into subtasks of shorter duration. This will provide your team the means to update your individual progress each week. On the other hand, if you attempt to schedule the details down to a day, rather than a week, you spend too much effort tracking and managing minute details. Your expected weekly time commitment to your Senior Design project corresponds to approximately one or two days of effort by a full-time employee. Thus, in the context of the Senior Design class, scheduling tasks approximately one week in duration has been found to work well. Scheduling a reasonable set of tasks over the course of a week will provide you flexibility to manage and rearrange your entire weekly schedule so that all your responsibilities are accomplished. It is acceptable if a *few* tasks are slightly longer or shorter than a week, but in general one-week tasks make for the best schedule.

Each team member should be assigned at least one task every week. It is probable that each team member will sporadically be assigned more than one task per week. This may be necessary to complete the project on time but should be avoided if possible. Exceptions to this rule may include the following:

- Sometimes one team member has a significantly lighter workload in a given week and will have extra time to dedicate to a new aspect of the project. Assigning this member an extra task may help the team to get ahead of schedule.
- Sometimes there is extra work that must be accomplished in order to meet a milestone. Assigning multiple tasks to a particular team member may be the only way to stay on schedule.
- Unfortunately, all team members will fall behind the schedule occasionally. Falling behind is not a problem so long as that member dedicates additional time to the project in the near future to get caught up. This should not be intentionally scheduled, but the Gantt Chart may be adapted if necessary.

The remainder of this section will provide a brief description of the type of tasks that should be scheduled under each of the categories presented in Fig. 11.1. The discussion in this section is meant to provide you a general understanding of the type of tasks to consider for each category, as opposed to defining the exact tasks you should schedule for your own project. The details of what is required within those categories will be covered in detailed in upcoming chapters.

Scheduling the PLANNING phase

Under the PLANNING phase parent category shown Fig. 11.1, there are at least five subcategories of tasks which must be scheduled. The first four of which are typically completed during the first semester of the Senior Design sequence; the fifth is usually accomplished during the start of second semester.

Beginning the Planning Phase
This category is an exception to the rules discussed above. This task only requires about one week, and all subtasks are probably handled best when the entire team works together rather than allocating individual assignments. During this week, you should do the following:

- Verify that any and all feedbacks, comments, or concerns raised during the INITIATING phase have been addressed.
- Review the Block Diagram, verify if it has been organized into clearly defined subsystems, and modify it if necessary.
- Identify the system's Critical Components (0).
- Complete the Gantt Chart (as instructed in this section) in as much detail as possible through the PLANNING phase, and provide at least an outline of the expected schedule through the EXECUTION phase.

Alternatives Selection Process

Although this category is also only scheduled for one week, there is enough work within the category to divide the entry into individual tasks. The Alternatives Selection taskbar shown in Fig. 11.1 will automatically be promoted to a "category" when detailed tasks are inserted below the existing entry. During this process, multiple options for each of the system's Critical Components are carefully analyzed. Eventually, a specific solution is selected for each Critical Component. The additional tasks you should include during the Alternatives Selection process are as follows:

- Engineering Alternatives Analysis of component per team member (Sect. 12.2)
- Finalization of Alternatives Selection process at the system level (Sect. 12.3)

Preliminary Design Process

Ideally, each Critical Component analyzed in the Alternatives Selection process was located within a unique subsystem. During the Critical Subsystem Design process (Sect. 13.2), individual team members should complete the design of the subsystem related to the Critical Component which they were responsible for selecting. To accomplish this subsystem design, each individual should be assigned with tasks related to their subsystem that will guide them through the entire subsystem design process. A representative list of these tasks might include the following:

- Selecting any additional components required to operate the Critical Component
- Calculating frequency responses, power requirements, or current limits
- Conceptualizing the processing flow
- Verifying each subsystem interfaces well with all other Critical Subsystems

Additionally, the ideation of the system-level design (Sect. 13.3) should be considered within this category. That does not mean the system has to be designed just yet, but you should verify that what you have designed will not hinder the design of the system's remaining features. Some specific tasks to consider include the following:

- Consideration of any subsystems which were not related to a Critical Component
- Consideration of the mounting, housing, and packaging of all subsystems

The Preliminary Design process ends with a Preliminary System Design. It would not be possible to have a truly finalized design at this point of the PLANNING phase. The Preliminary System Design is a framework for the Complete System Design. This includes the selection of all Critical Components through careful Alternatives Analysis, primary development of all Critical Subsystems, and consideration of all interconnections between subsystems. However, it is highly likely that some of the details are not finalized yet. For example, a particular signal might need amplification, but the exact gain may not be known; a generic design of an OpAmp-based amplifier without the exact details would be acceptable at this time. Another reasonable Preliminary System Design would be a conceptual drawing of the product's packaging, even though the exact dimensions may not be finalized. The purpose of this process is to produce a preliminary design that is as complete as possible, to understand all the outstanding issues, and to have a plan to complete the design before the end of the PLANNING phase.

Example 11.1 provides an example of a well-planned scheduled for the Preliminary System Design Process.

EXAMPLE 11.1: Alternative Selection and Subsystem Design Gantt Chart

A 2010 Senior Design team was tasked to monitor the "hoop strain" in a pressurized tube. Hoop strain is caused by a force radiating out of a cylindrical tube causing the diameter of the tube to momentarily expand. The device needed to be installed in a remote location within a nuclear plant. So the data needed to be recorded locally and then transferred to a reading device upon request.

During the "Begin PLANNING phase" task, the three-person team identified three Critical Components: Strain Sensor, Processor, and Data Transfer Method. Then, they developed the following detailed Project Schedule. The yellow, green, purple, and orange highlights were later added for purposes of this example.

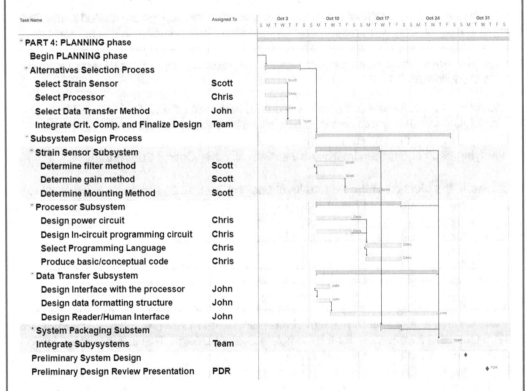

During the Alternative Selection Process (yellow), each of the Critical Components was analyzed and selected by individual team members (Chap. 12). Then, the team jointly integrated the analysis into final selections.

During the Subsystem Design (i.e., "Preliminary Design" Sect. 13.2) process, each team member was assigned to complete the subsystem design related to their Critical Component. The Strain Sensor Subsystem (green) required three considerations: Filtering, Gain, and Mounting. Detailed designs for these considerations would have been difficult until the sensor was procured. So the tasks were to simply determine methods or plans. The *details* will need to be addressed later. This allowed Scott time to design an extra subsystem, the System Packaging Subsystem (purple).

EXAMPLE 11.1, continued

The team had more experience with the Processor and Data Transfer Subsystems (not highlighted) at the beginning of the project. So it was possible to complete detailed designs for these subsystems during this early time period.

Most of the tasks were appropriately scheduled to be accomplished within five days. A few simple tasks were schedule for two days. This is acceptable in some cases but typically should be avoided. In this particular case, John had experience developing this type of subsystem and did not feel he needed much time for these particular tasks. Conversely, the "Design Reader/Human Interface" task was scheduled for 11 days. Occasionally, it is necessary to schedule a task for longer than a week. An 11-day task is not necessarily "inappropriate," but doing so will make managing the project slightly more difficult. While John's schedule is not terrible here, it could have been improved by reorganizing the work so that there was a clear focus for each week of his Preliminary Design process.

This team insightfully scheduled all tasks to be accomplished prior to the Preliminary System Design milestone with enough time to include one additional task. This team included time in the schedule to "Integrate Subsystem Design."

After the team develops their Preliminary System Design, a Preliminary Design Review, PDR (Sect. 13.4), is held so that the project Stakeholders are able provide feedback to the team. At the PDR, each Critical Subsystem's design is reviewed in detail and the complete system's ideation is discussed.

Preliminary Prototype Development

It is very common that a complete design of a subsystem is difficult to complete without physically interacting with a subset of the necessary components. There are many potential reasons for this. Datasheets can be challenging to interpret, specific operational parameters for a particular scenario might not be available, or the designer may simply be unfamiliar with how a component functions. In any case, developing Preliminary Prototypes and analyzing a set of challenging design aspects are often useful.

In Senior Design, you will be asked to produce a Preliminary Prototype of one aspect of the Critical Subsystem for which you are responsible. This does not mean that you should necessarily build the entire subsystem. Section 14.4 will guide you in selecting which aspect to develop. Tasks you may want to consider during this process include the following:

- Procuring hardware
- Procuring software license
- Building test circuits (often on breadboards)
- Producing drafts of three-dimensional models
- Writing test code

Example 11.2 continues from Example 11.1 by displaying the 2010 team's Preliminary Prototype Development Gantt Chart.

EXAMPLE 11.2: Preliminary Prototype Development Gantt Chart

The team discussed in Example 11.1 included details for their Preliminary Prototype Development in their Gantt Chart as well.

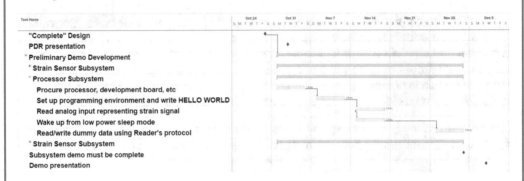

After the Preliminary System Design milestone was achieved, the Preliminary Prototype Development starts. Often, the first step in this development process is to simply procure parts. Placing an order during the week of the PDR will overlap shipping time while you are focused on other responsibilities.

In the Gantt Chart presented above, the Strain Sensor, Processor, and Data Transfer subsystem prototypes were scheduled to be accomplished between the completion of the Preliminary design and the Demo presentation milestones.

Only the details of the Processor Subsystem development are displayed here, but all three subsystems would need to be scheduled in detail at this time. The + symbol on the other two subsystem's rows indicates that drop-down lists are available with these details.

Each task under the Processor Subsystem is appropriately scheduled for one week. No work is scheduled on the days highlighted in pink. These represent vacation days. If your schedule starts to slip, one potential mitigation strategy would be to make up some hours during this unscheduled time period.

This Gantt Chart does violate the standard rules in one aspect. Chris is assigned two one-week-long tasks during the week of Nov 14. In general, scheduling multiple tasks in the same week should be avoided, although occasionally it is necessary. In this case, the decision to accomplish two tasks in a week makes sense; let us look at the details to understand why.

Notice that the first two tasks are logically sequenced. Obviously, the very first task for any hardware development is to procure the hardware! Setting up the programming environment and writing a simple code would be a logical next step. The following three tasks are not necessarily sequenced "logically." This term is not meant to imply the schedule is

---

EXAMPLE 11.2, continued
_____

illogical but rather that the tasks did not directly depend upon each other and probably could have been scheduled in a different order with an equal chance of a successful outcome.

"Reading the analog input," "waking up from low power mode," and the "Read/write data" were all necessary, but are relatively unrelated, tasks; so the order in which they were sequenced was somewhat arbitrary.

"Reading the analog input" and "waking up from sleep" were both relatively easy tasks; neither was expected to require a full week. These tasks could have been independently scheduled for two sequential days each but that would imply a precise order. Since the order did not matter, they were scheduled as simultaneous tasks that both had to be accomplished within the given week.

The "Read/write data" task was anticipated to be a more difficult task and would require a dedicated week worth of effort. It was scheduled to be completed in the subsequent week.

One improvement that could have been made would have been to schedule the more difficult task (Read/write data) first. If the order of a set of tasks do not matter, then addressing the higher-risk task first is typically the best strategy.

Critical Design Process

The Critical Design process is scheduled after the Preliminary Prototype Demonstration and is conducted between the end of the first semester and the start of the second semester. Often design teams learn a great deal about their design from the Preliminary Prototype Development process. Previously unconsidered issues are uncovered and shortcomings in the Preliminary Design are discovered. The Critical Design process is scheduled to ensure all necessary design considerations have been accounted for in the final plan and that a complete system design is produced.

Furthermore, it would be challenging to schedule every detail of the EXECUTING phase at the start of the PLANNING phase. However, by the time you are wrapping up the PLANNING phase, you should have enough information to complete the project schedule through the end of the EXECUTING phase. So part of the PLANNING wrap-up process is to complete the schedule for remainder of the project.

At this point, it is acceptable to leave the wrap-up as a stand-alone task on the schedule until you understand your project more fully. So including subtasks under this category is not required just yet. However, you may *wish* to add subtasks in the near future as the project develops.

Scheduling the EXECUTING and CLOSING phases

The design developed during the PLANNING phase is implemented during the EXECUTING phase, and final documentation is prepared during the CLOSING phase of the project. Most of the scheduling details for these phases cannot be determined yet. However, it is still a good practice to plan out a broad framework of the EXECUTING phase during the PLANNING phase. This includes incorporating any known milestones and blocking out large sections of the schedule for the major task categories within each phase. The EXECUTING phase typically starts shortly after the beginning of the second semester of the Senior Design sequence and ends approximately one month before the end of that semester. The schedule for these two phases provided in Fig. 11.1 is probably as detailed as possible at this early stage of the Design Life-Cycle. As the last two phases approach, additional scheduling details will be added. However, each project is different, and if you are able to add detail to these phases at this time, you should do so.

11.3 Schedule Management Process

One of a Project Managers' responsibilities is to manage the Project Schedule. This section will discuss a number of strategies and techniques to help you manage your schedule.

Keep the Gantt Chart Up-to-Date

The Schedule Management process begins with keeping the Gantt Chart up-to-date. The best way to do this is for each team member to update the Gantt Chart immediately after completing *any* project-related work. Just like you should always open your Engineering Notebook and start a new entry when beginning any work session, you should also open the Gantt Chart, and quickly verify that the work you are about to begin is focused on the appropriate task. Then, at the completion of the work session, you should quickly update the Gantt Chart to show your progress.

Another method to ensure the Gantt Chart is up-to-date is to start every team meeting with a quick review of the schedule. If any work has been completed on a task, but not recorded in the Gantt Chart, use this time to update the chart. This should not be the primary method to update the chart; doing so regularly would be an inefficient use of your meeting time. This method should only be used by the team as a backup to ensure no individual neglected this important responsibility. Example 11.3 illustrates how to do this.

EXAMPLE 11.3: Progress Updates to a Gantt Chart

To update the Gantt Chart with your progress, open the Smartsheet document and expose the programming fields as shown in Fig. 11.2. Notice that the dateline (the orange vertical dashed line) is located on Friday afternoon the week of Jul 3. We can also observe that Thien successfully completed Task 1; 100% was entered in the %Complete column and the light-blue status bar completely covers the dark-blue task bar for Task 1. However, Thien is only 10% complete with Task 2, a task that should be complete by now. We can see that the light-blue status bar only covers a small portion of the Task 2 bar.

EXAMPLE 11.3, continued

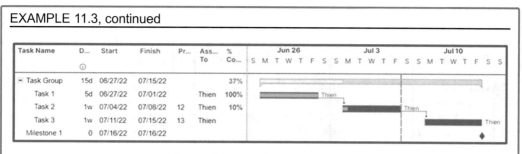

Figure 11.2 Schedule Management

Finally, notice that the %Complete status of "Task Group" is currently at 37%. This value was not entered by Thien; the Gantt Chart will automatically update that for you.

Consider what it means to be a certain "percent complete" on a project task. You might say a week-long task is 20% complete after the first day but that would be pretty misleading if no work was accomplished on that day. Perhaps you could define percent complete related to the percentage of the budget spent; however, on a software task, you might spend 90% of the effort before 1% of the budget is expended. So, again, the budget is not necessarily a useful indicator of progress.

Some Project Managers must incorporate complex statistical analysis while monitoring the progress of a project. Metrics such as dollar-per-functionality, expected-budget-versus-actual-expenditures, or resource-allocation-per-completed-feature might be used. Doing so will produce highly predictive metrics that can be used to assess progress on large projects. However, an introduction to Schedule Management does not need to be this complicated.

A less rigorous, but often just as useful, method of tracking progress is for an organization to define agreed-upon benchmarks. These benchmarks are very generalized and meant to be easy to use. For Senior Design, an appropriate set of benchmarks are as follows:

- 10% complete   means you have started some sort of work.
- 50% complete   means you have made some reasonable progress.
- 70% complete   means you are complete with the bulk of the work and are just putting the finishing touches on the task.
- 100% complete   means you are ready to move to the next task.

So looking back at Example 11.3, we can see that Thien was complete with Task 1 (100%) and had started some amount of work on Task 2 but had not made any significant progress (10%).

Managing the Project Schedule

Recall that the primary purpose of the Gantt Chart is not to perfectly predict the future allocation of project resources throughout the entire project. Rather, the Gantt Chart is a tool that provides an estimate of resource allocation that can be used to deliberately manage deviations to those expectations.

To make use of the Gantt Chart, the scheduled events must be appropriately scheduled. This means that the milestones are hard-programmed to occur on particular due dates but that the tasks are relatively programmed based on prior *predecessors* (Sect. 8.2) which allows the chart to update the entire schedule when a single deviation is made.

When the reality of completing a particular task does not match the original expectations, the Project Manager must modify the schedule to account for this deviation. When a task is scheduled based on a predecessor and the predecessor's schedule is modified, the task's timeframe will automatically update relative to how the predecessor updated.

EXAMPLE 11.4: Modifying a Task's Schedule

In this example, we see that Thien is behind on Task 2 and will obviously need more time to complete the task than what was originally allotted to him. The team identified this issue during their regularly scheduled Friday afternoon team meeting and developed a mitigation plan.

Thien, and his partner Jacob, had prioritized a test in a different course this week. Jacob was ahead of schedule on his project tasks but needed some extra help on the test. Thien agreed to help Jacob on the test with the condition that Jacob help Thien get back on track over the upcoming weekend. They estimated that by Monday afternoon, Task 2 would be completed.

This update is documented in Fig. 11.3 by modifying the duration of Task 2 from one week (five days) to six days. Now we can see the real purpose of the Gantt Chart. When the duration of Task 2 was extended, the schedule for Task 3 was pushed back. In this case, if the original schedule is maintained going forward, Thien will only be 50% complete with Task 3 at Milestone 1.

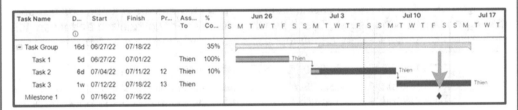

Figure 11.3 Modifying a Task's Schedule on a Gantt Chart

Mitigating Project Slippage

Large companies with a wealth of experience repeating the same type of projects get very good at predicting the necessary schedule for the "next" project. Knowledgeable Project Managers use statistics, data, and experience to define a reasonable Project Schedule. But, even then, project slippage is unavoidable because the original Project Schedule is impossible to perfectly predict – the Project Manager uses a Gantt Chart, not a crystal ball, to schedule the project. Advanced Project Management techniques have been devolved to improve the accuracy of schedule predictions, but the real point of Project Management, and Schedule Management in particular, is to identify and mitigate the inevitable slippage.

The term "mitigation" in this context means to reduce the effect of a negative event. In other words, a plan must be developed and enacted so that the effect of slippage on one aspect of the Project Schedule does not endanger the overall success of the project.

Example 11.5 will illustrate this point further.

EXAMPLE 11.5: Using a Gantt Chart to Mitigate Slippage

In Example 11.4, we saw the Thien's Project Schedule slipped. The team has a plan to finish Task 2, but they also need to modify Task 3's schedule in order to complete the Task Group before the milestone as shown in Fig. 11.4.

Figure 11.4 Mitigating Schedule Slippage

Here we see that the duration of Task 4 was reduced from one week to four days and that the graphic now shows Task 3 completing before Milestone 1. We also see a "comment" documenting where the mitigation plan can be found.

The "Comments" programming field is found far to the right of the programming options. Many of the programming columns are hidden in Fig. 11.4 for purposes of this example.

The Gantt Chart should be updated to account for the project slippage (like Task 2 above) and the mitigation plan (like Task 3 above). Also, the mitigation plan should be documented in the comments. Typically, this means simply recording the Engineering Notebook entry where the mitigation plan was discussed.

By detecting a small schedule slippage early, you will have many more options to mitigate that slippage. It would be better to reduce four tasks each by one day than to reduce one task by four days. Of course, this is no longer an option when you detect the slippage immediately before a milestone. In that case, your mitigation can only affect a few upcoming tasks (e.g., ones scheduled immediately before the milestone).

One further point must be made here. The purpose of a mitigation strategy is not to force the Gantt Chart's graphic to look "pretty" or adhere to some predefined format.

A mitigation strategy explains HOW the schedule will be adjusted

In other words, it is not enough to simply modify the duration of a task so that the task now is shown to complete "on time." If a task was originally scheduled to take five days, you cannot force it to only require four days just because you need to complete it faster. The mitigation strategy must explain *how* that time reduction will be possible. Here are a few mitigation plans that might be applied to your project when you experience slippage:

- Slightly modify many tasks rather than make a major modification to one task
- Work during nonscheduled hours such as weekends or holidays
- Receive assistance from another team member who is ahead of schedule
- Adjust your project priorities and rearrange your schedule accordingly
- Request additional assistance from your Advisor and/or Instructor,
- Request that the milestone be modified (used in only exceptionally rare cases)

11.4 Chapter Summary

The first process of the PLANNING phase is to schedule the details of the remainder of the Design Life-Cycle. The general framework for the entire project must be developed by this point in time. However, it is not necessary to develop every *detail* throughout the Design Life-Cycle. The focus should be a detailed Schedule for the PLANNING phase and framework for the EXECUTING phase. Here are the most important takeaways from the chapter:

- The details of the PLANNING phase should be scheduled now. This includes the following:
  - Alternatives Selection process
  - Preliminary Design process
  - Preliminary Prototyping process
  - Critical Design process
- In order to effectively schedule the PLANNING phase details, you will need to continue reading so that you understand what each process entails.
- The purpose of a Gantt Chart is to identify and mitigate project slippage. The earlier slippage is identified, the more options you will have to mitigate the situation.
- Each team member should regularly update their progress on the Gantt Chart. The team should regularly review those updates.
- When project slippage occurs, you should take the following steps:
  - Adjust the duration of the task in question.
  - Observe how that change affects the remainder of the schedule.
  - Develop a mitigation strategy to reduce or rearrange the schedule so that the deliverable for the upcoming milestone remains unaffected.
  - Document that strategy in your Gantt Chart and Engineering Notebook.
  - Work with your team to develop the mitigation strategy for the best results.

11.5 Case Studies

Let us look at our case study team's Gantt Charts.

Augmented Reality Sandbox

Figure 11.5 displays a segment of the AR Sandbox's PLANNING phase schedule as it was produced in mid-December 2016.

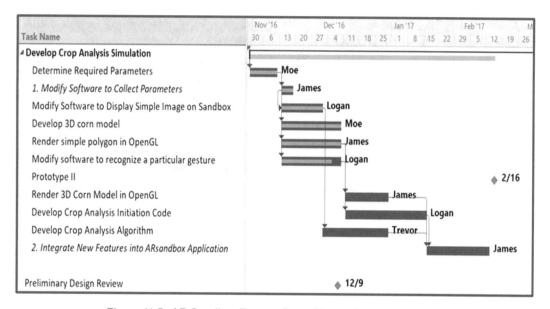

Figure 11.5 *AR Sandbox Team's Gantt Chart – PLANNING phase*

- What went well: The AR Sandbox team did a good job of dividing the workload equally among the team members for the most part. James led the software development, but Moe and Logan provided valuable support. As the software leader, James was regularly tasked with the integration tasks, such as task number 2 (*Integrate New Features...*). Remembering to schedule integration tasks and assigning these tasks to the best qualified resource is an important consideration to make when developing a Project Schedule.
- What could be improved #1: The tasks scheduled here are, for the most part, too vague – notice that they all are scheduled for more than 1–2 weeks. This caused the team scheduling troubles throughout most of first semester. Without the ability to track whether small tasks were on track each week, each team member regularly found themselves far behind schedule at the end of a month-long task and had to scramble to get caught up. The team used Winter Break to reorganize their Project Schedule with more precise tasks and operated much more efficiently in second semester.
- What could be improved #2: Another issue the AR Sandbox team should have addressed relates to task number 1 (*Modify software*). Notice that no task depends upon task #1 finishing (i.e., there is no arrow leading from the end of the task bar to another task). Let us use Fig. 11.6 to further illustrate this concern. For purposes of

this illustration, let us assume that James was unable to complete the task in question for many weeks (as shown by the extended orange line). Yet, the Gantt Chart shows that *all* subsequent tasks could still be finished before the 2/16 milestone. Although this scenario is not impossible, it is highly improbable, and as such, the team should have at least reconsidered the impact of this task. Remember, the point of the Gantt Chart is to observe the effect of a schedule change. If tasks are not appropriately linked together, this observation is not possible.

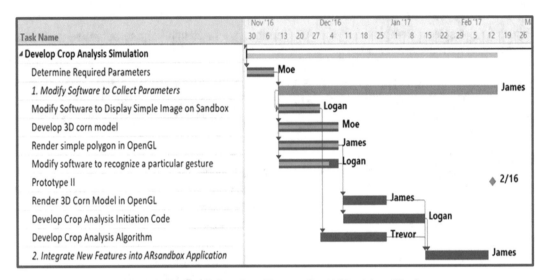

Figure 11.6 AR Sandbox Team's Gantt Chart (modified)

Smart Flow-rate Valve

Figure 11.7 displays a portion of the Smart Flow Rate Valve team's Gantt Chart. There are a number of Schedule Management aspects implemented here that only high-performing teams remember to consider.

- What went well #1: The red taskbar indicates that a modification to the "Layout" task was necessary. In this case, Keith was not behind schedule when the modification was made. Rather, he was re-tasked elsewhere in mid-January to assist on another aspect (not shown) of the project. Of course, this meant Keith could not complete the PCB layout on schedule. So to account for the reassignment, Keith creatively split the taskbar when the modification was made. Splitting the taskbar, instead of simply extending it, indicates that no work was scheduled on the task during the interim weeks. (Note: not all Gantt Chart software tools allow this feature.)
- What went well #2: As team leader, Tyler was assigned to submit the PCB to a PCB house (i.e., PCB manufacturer). Most teams will account for the shipping delay with an extra task. Instead, Tyler used an advanced scheduling technique wherein the following task was scheduled with a Finish-to-Start-with-Delay precedence. In this case, the "Continuity Test" was programmed with the "Design sent to PCB house" task as a predecessor but with a ten-day delay. Scheduling tasks like this and

eliminating "placeholder" tasks (e.g., a shipping delay) reduce the size of the Gantt Chart.

- **What went well #3:** The team used another advanced Schedule Management Technique when scheduling the "User Interface" tasks. Here we see a Finish-to-Start relationship with Tyler's "Solder Components" task. However, this time, the related task was scheduled with an advance (i.e., a negative delay). The scheduling thought process went like this:

  1. The PCB was a bottleneck in the project and needed to be functional as soon as possible, so soldering the components had to be scheduled as soon as the PCB was manufactured.
  2. The shipping delay provided Tyler some buffer time in the schedule.
  3. To fill that time, he was tasked with the unrelated, but important, task of starting the user interface development.
  4. Tyler estimated the amount of time required to finish the user interface and then counted backward from the *Soldering Components* task to determine when the user interface work had to begin in order to be complete before the soldering had to start. This was an advanced Schedule Management thought process to be sure!

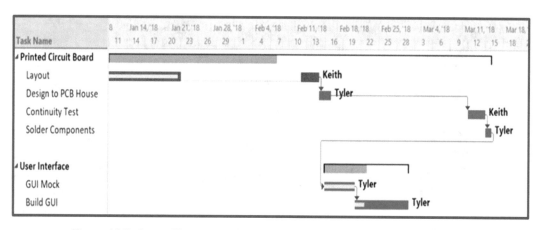

Figure 11.7 Smart Flow Rate Valve Team's Gantt Chart – EXECUTING phase

- **What could have been improved #1:** As we already stated, Keith was re-tasked to another aspect of the project while in the midst of the PCB *Layout* task. He took the first step in documenting the modification correctly by splitting the taskbar and coloring it red but did not complete the documentation. There are no comments in the Gantt Chart nor in the Engineering Notebook to explain this change. Remember, modifying the taskbar and turning it red is the first step of the Schedule Management process. The second, and more important, step is to document the reason for the change and the mitigation strategy.
- **What could have been improved #2:** Tyler's implementation of a Finish-to-Start-with-Advance predecessor demonstrated impressive forethought. However, the "User Interface" tasks had to be completed before the *start* of the "Soldering" task. The relationship was not a Finish-to-Start but rather a Start-to-Start-with-Advance. This minor mistake made no difference in Tyler's schedule because the "Soldering" task

was only scheduled for one day. Had the predecessor been scheduled many days, the Finish-to-Start relationship may have thrown off the intended start date of the successor task.

Robotic ESD Testing Apparatus

Figure 11.8 displays a portion of the RESDTA team's Gantt Chart during the final stages of the Verification process (see Chap. 19). The category "Individual System Tests" has a slightly misleading title when taken out of context. The team was performing system-level testing throughout the entire schedule shown here. The *Individual System Tests* referred to testing Requirements of particular subsystems rather than the Objectives. However, the completed system was used to complete these tests as well as the "Full System Tests." This category should probably have been renamed "Requirements Tests" and the next category renamed as "Objective" Tests.

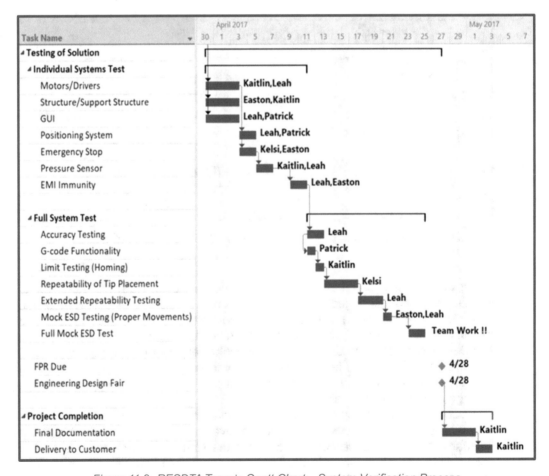

Figure 11.8 RESDTA Team's Gantt Chart – System Verification Process

- What went well #1: The team technically violated a Gantt Chart-building guideline by assigning multiple resources to many of the tasks shown here. However, doing so was intentional and turned out to be a great choice. By this point in the project, many of the subsystems had been integrated. Rather than assigning the testing of an integrated system to one individual, the team chose to assign the final testing to the two resources who were primarily responsible for the individual features of that integrated system. Remember that most Project Management "rules" or "guidelines" should be understood "in context." Occasionally, allowing yourself the freedom to violate these rules can provide more effective Project Management of a unique project. However, there is a difference between "intentionally deviating from the standard with a clear understanding of what you are doing and why" compared to "just winging it" without a clear reason. Discuss any modification you decide to make to the standard with the other Stakeholders in order to prevent confusion later on.
- What went well #2: Notice that there are five different combinations of shared tasks (e.g., Leah coordinated with Kaitlin, Patrick, and Easton on different tasks). This illustrates an exceptional level of team collaboration. Although not explicitly shown here, this team also maintained close collaboration with the Sponsor when scheduling the "Delivery to Customer" task.
- What could have been improved: This testing schedule left no room to adjust if anything went wrong before the final deliverables. Remember to include some buffer into your schedule in case something does not go quite as planned.

Chapter 12: **Alternatives Selection Process**

One of the most important processes of the PLANNING phase is called the Alternatives Selection process. During this process, the Critical Components of a design are identified, analyzed, and finally selected. Selecting Critical Components is relatively easy to do component-by-component; however, integrating all the final selections into an optimized system design is the priority. Integrating the selections can become challenging when the ideal selections for two different components are incompatible.

> During the Alternatives Selection process, a set of components is chosen to optimize the *system* design.

When a single component is analyzed, a ranked priority list of the best options should be developed. Then, the best options for all Critical Components must be analyzed together to ensure compatibility. This analysis is an important step in the Alternatives Selection process because the *priority* of this process is to determine the best *system* of components.

For example, maybe options A, B, and C have been identified as potential options for the first Critical Component of the system, and it is believed that option A is the ideal choice. At the same time, a second Critical Component was independently analyzed, and options X, Y, and Z were identified with X being the preferred choice. It is not uncommon that options A and X are not compatible. Perhaps the combination of B and X will produce a better system.

The first step to the Alternatives Selection process is to develop a prioritized list for each Critical Component, the second, and more important step is to select the *set* of options which will produce the best system design.

This chapter will guide you in the Alternatives Selection process to identify solutions for the Critical Components of your system.

In this chapter you will learn:

- How to identify critical design components
- How to perform an engineering alternatives selection
- How to merge individual component selections into system design

12.1 Identifying Critical Components

The Critical Components of a system must be identified during the beginning of the PLANNING phase. Although the Alternative Selection process is not actually conducted until after the Project Schedule has been developed in the Gantt Chart, the team must know how the project will be divided among team members before the Gantt Chart can be developed. So the Critical Components are identified during the first week of the PLANNING phase while the Gantt Chart is being developed.

The Critical Component Identification process is yet another process where there is no "right" answer, although there are poor choices that should be avoided. This section will help you identify the components of your system that must be analyzed during this process.

Before you learn how to identify your system's Critical Components, let us review the purpose of Project Management and the Design Life-Cycle. It was previously stated that Project Management is as follows:

> An effort to balance the Triple Point Constraint such that a project is completed effectively and efficiently.

Another way of stating this is that "Project Management is all about reducing and managing risk." This interpretation is particularly true during the INITIATING and PLANNING phases. Consider why we spent so much time defining the Problem Statement during the INITIATING phase. The simple answer is *we did not want to produce a "great" design for the "wrong" problem.* By knowing exactly what we were supposed to produce, we drastically decrease the risk of producing an inappropriate solution.

Similarly, the ultimate goal of the PLANNING phase is to produce a *system* design "on paper" without spending much of our finite resources. Only after we are relatively sure the design will produce the desired results, we will risk those resources. Discovering that the calculations do not pencil out, or that the design crashes in simulation, or that the analysis does not meet regulations is significantly more cost effective than *manufacturing* the design only to have it fail. The output of the PLANNING phase is often called the "paper" design, although it is rarely delivered *on paper* by modern engineers. Instead, computer drawings, simulations, and plans comprise the deliverable.

> A "paper design" is inexpensive to produce, but when produced properly, will significantly increase the probability that the physical product will work when finally manufactured.

The fact that we are ultimately trying to reduce risk by conducting every process of the PLANNING is an important aspect to the Critical Component Identification process. When choosing the set of components that will be considered the "Critical Components" for a project, you must be able to articulate exactly how analyzing *this* set of components will reduce your project risk. There are at least three common ways a particular set of components might reduce risk.

- Reducing technical uncertainty.
  By first identifying a set of the most challenging components to analyze, you can reduce your project risk by ensuring there IS a technical solution for the most difficult aspects of your project.

- Reducing the risk of complete project failure.
  By identifying all the components required to complete the highest-priority function, you ensure that at the end of the project, you will have produced something of value. Maybe the Project Goal was not completely accomplished, but at least significant progress was made and can be built upon by future project teams.
- Reducing the risk from lack of knowledge or information.
  Sometimes the engineering team may not have the technical expertise to design certain aspects of the system without first designing and analyzing a different part. By identifying the components of a particular subsystem, you ensure that one subsystem has been completed and can be used as a basis for future design decisions. This option should only be used when one or more aspects of the system are so technically complicated; you have no other choice. Do not select this option without discussing your choice with your Instructor.

Avoiding Common Pitfalls 12.1: Avoid "Poor" Selection Choices

Ultimately, the selection of which components will be considered "critical" should be determined directly by the design team; there is no "right" or "wrong" here. However, there are some "poor" choices that you should avoid:

- Remember, the point of the Alternative Selection process is to reduce project risk. That means that this is the time to tackle the "hard part." Choose components that have the most impact on the end result of the project.
- Be reasonable about the size and difficulty of your set of Critical Components. This is only the starting point of the PLANNING process. Your set should substantially *reduce* the risk of project failure, but you cannot *eliminate* that risk just yet. In other words, the *entire* project cannot be your *Critical* Component. A good rule of thumb is to choose one substantial component per team member. You might consider assigning two components to an individual if:
  - An individual's primary assignment is less substantial than the components which other team members are analyzing.
  - There is an absolute need to include more components than available team members. If this is the case, discuss the scheduling ramifications with your Instructor.
- Come to a team consensus as to which risk type should be the primary focus of *your* Alternative Selection process. Choose a set of components that all team members agree upon. Be able to articulate that focus.

If you have legitimately attempted to reduce project risk and been reasonable about the workload assigned to each individual and can articulate the team's reasoning for choosing the particular subset of components, you will have made an effective choice.

Use the Block Diagram to visualize the identification process

---

The Block Diagram is a useful tool for guiding your identification process. However, remember that at this stage, the Block Diagram displays the Requirements Document, not the detailed solution. That means the Block Diagram typically displays *subsystems*, rather than *components*. You need to consider what significant components each block/subsystem will eventually require in order to identify Critical Components within a particular subsystem. Adding some detail to the Block Diagram at this point can be convenient. If you do, you will want to maintain two copies: one high-level diagram for use when discussing the Requirements and one detailed-level diagram for use when discussing the Design.

The following examples demonstrate how the Block Diagram will help identify Critical Components. Each of the project teams discussed in these examples considered a different type of risk to choose their Critical Components.

EXAMPLE 12.1: Component Selection to Reduce Technical Risk

The Duty Cycle Analyzer (DCA) was a project to determine the average duty cycle of an LED when an LED was cycled at an extremely high frequency, and its duty cycle could be as short as 1 μs. The most important Specifications for this project was to measure that 1 μs pulse with less than 2% error, meaning that the time resolution of the system had to be at least 20 ns. Also, the system was required to save at least 1 ms of data, resulting in 200 kBytes of data.

While many modern oscilloscopes are capable of doing achieving 20 ns resolution, few commercially available tools are able to simultaneously meet this specification while also recording data for up to 1 ms.

The most significant risk for the DCA project was the technical capabilities of the measurement system. Therefore, the Critical Components identified for this project were all within the LED Measurement block, highlighted by the yellow box in Fig. 12.1.

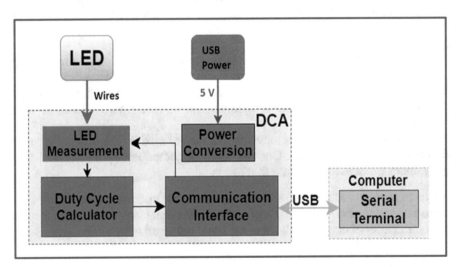

Figure 12.1 DCA component selection

This Conceptual Block Diagram does not provide enough detail for the Alternative Selection process discussion. Therefore, the team created a Detailed Block Diagram for the LED Measurement block. Using another yellow block around this set of blue blocks, Fig. 12.2, added a nice level of consistency when transitioning between the high-level discussion and the detailed discussion.

EXAMPLE 12.1, continued

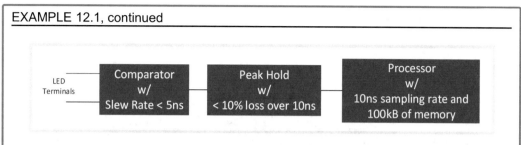

Figure 12.2 DCA component selection details

From this Detailed Block Diagram, we can see that the team focused on three components with challenging specifications to meet, the comparator to determine when the LED pulse went high, a peak hold circuit to maintain that high pulse and eliminate switching noise, and a processor to capture the waveform.

In this example, all the particularly challenging devices *happened* to be contained within a single block. This is somewhat rare. It is more common for the challenging components of a project to be scattered throughout the block diagram. This team did an excellent job realizing the "standard case" did not necessarily apply to them.

EXAMPLE 12.2: Component Selection to Reduce Complete Project Failure

The Raven Mounted Camera Imaging System (RIMCIS) was a Senior Design project tasked with capturing GPS-linked imagery of crops while a farmer drove through a field. This imagery would then be ported into a Graphical User Interface (GUI) and linked to what is known as a "prescription map." The prescription map is a detailed map of planting, irrigation, fertilization, and yield record. The idea behind RIMCIS was to link a visual image of a precise area of the field with the numerical records contained within the prescription map.

The Block Diagram below, Fig. 12.3, displays the two primary Requirements of the project: the RIMCIS within the blue block and the GUI within the red block. Previously collected field data (i.e., the prescription map, in the lower green block) was to be loaded into the GUI along with Image/GPS data collected by the RIMCIS.

EXAMPLE12.2: Component Selection to Reduce Complete Project Failurecontinued

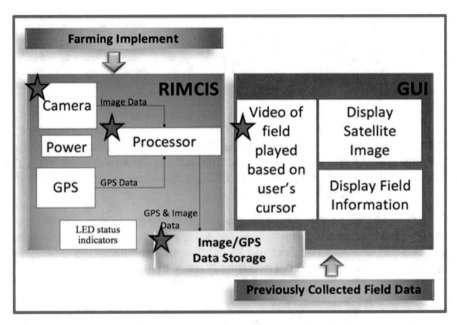

Figure 12.3 RIMCIS component selection

One improvement that should have been made to this Block Diagram is how the Farming Implement relates to the RIMCIS. The direction of the first green arrow makes it *appear* as if the Farming Implement is providing "something" to the RIMCIS.

However, this student team was simply trying to represent the fact that the RIMCIS was attached to the Farming Implement. Perhaps the term "Farming Implement" should have been contained in a larger green box which encompassed the blue RIMCIS block and the arrow eliminated.

This team was primarily concerned with the ability to collect high-quality video footage from a moving vehicle and display it clearly on a GUI. The other features were necessary to produce a marketable product but were not the main focus. Therefore, the team decided to initially focus on a single path through the project that pertained to the primary concerns. This way, if they ran out of time to complete the entire project, at least they could deliver a functional, if not complete, system.

They identified the camera, processor, storage device, and video player portion of the GUI as their critical components. Notice how those topics are denoted with the red stars on the Block Diagram.

After they had completed their Alternatives Selection process, they had a reasonable design which could capture video data, transfer that data to the GUI, and visually display it to a user. During the remainder of the PLANNING process, they completed the design to capture GPS data and link that data, along with the prescription map data to the video data.

EXAMPLE 12.3: Component Selection to Reduce Lack of Knowledge

The Electric Vehicle Photovoltaic Charging Station team, nicknamed Team EVPV, was tasked to offset the electricity required to operate a fleet of university-owned electric vehicles. This design did not directly charge the electric vehicle's batteries since the vehicles were in use during peak sun hours. A battery bank could have been used to store the generated energy during the day and charge the vehicles at night. However, this would have resulted in significant power losses in the storage and conversion process. Instead, the decision was made to back-feed the local power grid during the day and continue charging the electric vehicle batteries directly from the power grid when they were not in use at night (Fig. 12.4).

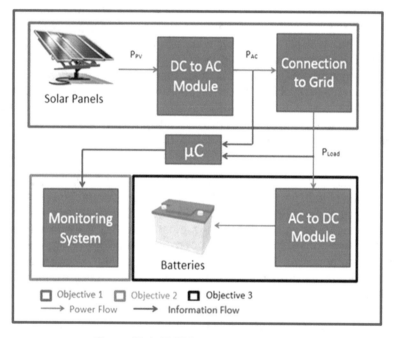

Figure 12.4 EVPV component selection

The challenge for Team EVPV was to *prove* the photovoltaic system was adequately offsetting the energy consumed by the electric vehicle fleet. They did this by monitoring how much electricity was purchased from the local power grid to charge the vehicle's batteries and how much electricity was generated from the photovoltaic solar panels. Hour-by-hour totals were graphically displayed on the university website.

By far the most significant design challenge Team EVPV faced was to design the DC-to-AC module from scratch rather than purchasing a consumer version. The team did not have enough technical background to predict what the output of that module would look like and therefore could not make many valid decisions until they more completely understood that module. Also, it was clear that the Monitoring System would be fairly straightforward to produce and the AC-to-DC module was a prebuilt component of the electric vehicles so it did not require a design effort at all.

EXAMPLE 12.3, continued

So although the project priority was the monitoring system, Team EVPV, with the support of the team's Advisor and Instructor, decided to completely design and produce a working prototype of the DC-to-AC module before making any further decisions.

Dedicating all the team's resources to designing one part of a project in order to reduce the most significant risk can *inadvertently increase* the overall project risk. Care should be taken when selecting this option. Consult with your Advisors before committing to this decision.

Let us take a minute to contrast Example 12.1 and Example 12.3. The focus of Example 12.1 is on a number of technically challenging components. In this example, the DCA team had a number of high-risk components that all *happened* to be located within one conceptual block. However, this is not typical – high-risk components are usually spread across many subsystems. The DCA team was reducing *technology* risk by specifically choosing a number of closely coupled components which all had strict (or high-tech) Specifications. Example 12.3 also presented a case where all the Critical Components were deemed to be contained within a single block but for a very different reason. Team EVPV was *not* concerned whether the technology existed to design a DC-to-AC module; this problem has been solved many times before. Rather, Team EVPV was attempting to reduce their *knowledge* risk by learning everything they could about the DC-to-AC module before making further decisions.

On the surface, it may not appear all that important to distinguish between these two risks. However, the course of action after making the selection *is* different. The DCA team identified technology sufficient to meet the project Specifications and continued on with a "normal" PLANNING process. They incorporated their selections into a system-level design that included a casual component selection (as opposed to an engineering analysis) of nearly every remaining component. Then, they began prototyping. Conversely, Team EVPV identified components within the DC-to-AC module as critical, completed that module's entire design, and then built a fully working prototype. This prototype was then used to guide their decisions about the remainder of the system's components. Only then were they able to complete their system design and complete the PLANNING phase. Team EVPV's methodology was perfectly valid but had to adapt their Project Management structure to that unique decision. Had they not recognized the difference in their risk selection, they may have never realized the need to manage their project differently. As it turned out, they *did* recognize the differences and were extremely successful in modifying the Project Management strategies accordingly. In fact, Team EVPV's team leader, Tyler, was promoted to a Project Manager immediately after graduation when his first employer learned of his Senior Design experience (the promotion came with a $10,000/year raise above the starting salary he was initially offered).

12.2 Engineering Alternatives Analysis

Once the Critical Components have been identified, a detailed Engineering Alternatives Analysis must be performed upon each one. This means completing a thorough search for a variety of alternatives for each component and then quantitatively assessing each possibility.

Not all components must be analyzed in depth; some components may simply be casually selected based on informed decisions and best practices. But the selection *of Critical Components* must be carefully engineered.

> Rigorously analysis of the **Critical** Components within a design is an important risk-reduction technique.

The key difference between an Engineering Alternatives *Analysis* and a casual Component S*election* is that the Alternatives Analysis is driven by comparison *data.* There are a variety of tools that can be used to perform an Alternatives Analysis. In this section, you will learn about three common examples: an Economic Decision Tree, a Radar Chart, and a Pugh Chart.

Economic Decision Trees

An Economic Decision Tree is a particularly useful tool when faced with two relatively complex options and you wish to compare the cost of each. The term "Economic" insinuates that this tool is strictly related to spending cash, and it often is. However, an "Economic" cost can refer to engineering time, testing resources, facility demands, and a variety of other nonmonetary costs.

When analyzing *similar* devices, the remaining design options will typically be approximately equivalent. However, when analyzing significantly *different* design strategies, *every* subsequent design decision will be impacted. The Economic Decision Tree is a tool for analyzing the cascading effect of a particular decision.

> Use an Economic Decision Tree when there are multiple ramifications for your Alternatives Analysis.

For example, if you are simply trying to choose between two microcontrollers such as the MSP430 versus the PIC18, your subsequent design options will be relatively similar regardless of the choice you make. Any necessary Printed Circuit Board (PCB) development will not be impacted other than a slight adjustment to the exact pinout details. The list of potential sensors will be the same for either device. The power draw will be nearly identical. Since both options are similar microcontrollers, subsequent decisions will not be significantly impacted by the initial Alternatives Analysis.

However, if you are trying to choose between a single board computer (e.g., Raspberry Pi) and a microcontroller (e.g., MSP430), your selection process is significantly more complicated. If you choose the Raspberry Pi, you will not have to develop a Printed Circuit Board (PCB) because the Raspberry Pi comes prepackaged. On the other hand, if you select the MSP430, you will have the option of customizing an optimized PCB. If you choose the Raspberry Pi, you will be able to purchase prepackaged sensors specifically designed to work with the Raspberry Pi. If you select the MSP430, you may have the option of implementing uncompensated transducers instead of prepackaged sensors; these will be cheaper to purchase but require additional engineering effort to implement. The ramifications of selecting one product over the other in this scenario will cascade down to many other decisions.

An Economic Decision Tree is useful for analyzing the impact that a particular choice will have on the system-level design. To build the Decision Tree, start by placing the primary decision in a box at the top of the page. From there, draw branches radiating away from the decision box. Each branch of the tree represents a unique Alternative related to the primary decision; label the branch with the Alternative and its cost. Assuming a particular Alternative was selected, determine the next decision that will have to be made. Create another decision box. This box will likely have subsequent branches representing yet more Alternatives. Follow each of these branches to *their* related decisions, labeling the cost of each as you go. Continue until you have considered all the possible Alternatives. Then, sum the cost of each option by following each branch.

Typically, the best engineering decision is to select the lowest cost option. However, that is not always the case. The Engineering Alternatives Analysis is only partly dependent upon cost. If one of the more expensive branches is the best option, it is acceptable to select that Alterative, but you must be able to justify the added expense. Even if you do not select the lowest cost option, the Alternatives Analysis is necessary to make an *educated* selection.

Example 12.4 demonstrates how to build and analyze an Economic Decision Tree.

EXAMPLE 12.4: Economic Decision Trees

This example will demonstrate how to build and analyze an Economic Decision Tree during the Alternative Analysis process. Let us assume you must take temperature measurements in a remote location, store the data until a researcher collects your product from the remote location, and then upload the data to a computer for analysis.

You are trying to decide whether to use a Raspberry Pi or an MSP430 and want to build a Decision Tree to analyze the ramifications of both options.

Start with the Primary Decision and the Two Potential Alternatives

The two alternatives are listed on their respective branches along with any necessary data. For this simplified example, only cost and memory were included. In a thorough Alternatives Analysis, it is probable that additional aspects should be included.

Consider the "Hidden" Costs of Using the MSP4302310

The MSP430FR2310 is a stand-alone chip, meaning a PCB is required.

Let us assume that your Requirements Document states that a minimum of 1Mbytes storage is necessary. The expense of the Raspberry Pi covers this concern, but the MSP430FR2310 is not a storage device; we will need to include a memory chip.

EXAMPLE 12.4, continued

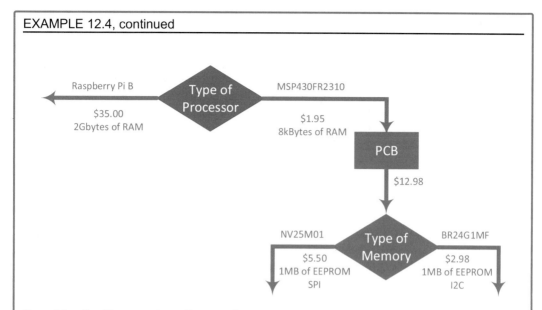

Consider the Temperature Sensor Costs for Each Branch

If we started with the MSP430FR2310 and selected the BR24G1MF I2C memory device, we will not be able to use a I2C sensor (there is only 1 I2C port on the MSP430FR2310); we could also use an uncompensated transducer paired with an Analog-to-Digital port. If we select the NV25M01 SPI memory device, we can still use an SPI sensor (there are 2 SPI ports on the MSP430FR2310), and we have the same I2C and transducer options. If the Raspberry Pi was selected, then there would be a completely different set of options for sensors selections.

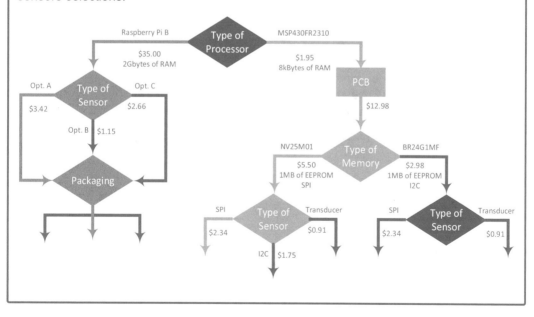

EXAMPLE 12.4, continued

Sometimes Branches Rejoin After a Set of Decisions Are Made

After following the MSP430FR2310 branch through the Memory decision, a decision about Sensors had to be made. There was a slightly different set of Sensor Alternatives depending upon which memory type was selected. Therefore, two branches were required, each with a "Type of Sensor" decision. Conversely, when following the Raspberry Pi Alternative through the Sensor decision, the Packaging had to be analyzed. The type of Sensor selected would probably have not altered any Packaging decisions, in this particular case. Therefore, to save space and effort, the branches could be rejoined into a single Packaging decision block.

Analyze the Economic Decision Tree

After all the Alternatives have been plotted on the Decision Tree, it is possible to analyze the cost of each set of Alternatives. Identify all the paths through the Decision Tree and sum the cost of the path. Two have been highlighted in this example. The first, in green, selects the MSP430FR2310 with a PCB, SPI Memory, and SPI Sensor for a total cost of

$$\$1.95 + \$12.98 + \$5.50 + \$2.35 = \$22.78.$$

The second, in orange, selects the Raspberry Pi, Sensor Opt. A, and the Middle packaging solution. Of course, more information is needed to complete the analysis of this path.

Additional Considerations

This Decision Tree does not account for a number of potential concerns. Take, for example, the additional cost associated with the increased engineering efforts to develop a PCB and implement an uncompensated transducer. A corporation might address this issue by implementing a simple formula, such as follows:

$$\text{Estimated development time} * \text{Engineering fee} = \text{cost}$$

Another concern not represented here may be physical size. This is even more difficult to assign a monetary cost. In this case, a second Decision Tree might be produced with exactly the same branches but assigned sizes, rather than prices, associated with each decision. Both Trees would need to be analyzed before arriving at a final decision. Perhaps a creative metric, such as "cost per area," could be assigned and used as the final determinate.

Radar Charts

Radar Charts are a graphical tool to optimize a "multi-axes," or multidimensional, Alternative Analysis. For example, in embedded system design, a processor must be simultaneously evaluated according to its cost, reliability, general and special computing abilities, and real-time constraints.

Use Radar Charts to perform an Alternatives Analysis when:

1. The focus of the analysis is on a single component.
2. It is more important to compare an Alternative against a set of multidimensional Requirements rather than a list of sequential design decisions.
3. The list of potential Alternatives has already been pared down to a relatively few candidates.

Before moving on, compare the Radar Chart's use case to that of the Economic Decision Tree. The Economic Decision Tree should be used when:

1. The focus of the analysis is on many components along a decision path.
2. It is more important to compare a set of design decisions again a dissimilar set of alternate decisions using only a few Requirements. This suggests that all sets of design decisions can, and will, meet the Requirements.
3. The list of potential Alternatives for each design decision is relatively few.

It is not uncommon that a Radar Chart and Economic Decision Tree are used in tandem. The Economic Decision Tree is used to analyze which design path is the optimal path, whereas the Radar Chart is used to select the different components included along specific paths of the Tree.

Each axis of the Radar Chart represents an important characteristic of a particular component for which you are attempting to select. In order to create a Radar Chart, each Alternative is graphed by plotting the quality of each characteristic along respective axis. The data points on each axis are then connected to create a surface area.

A "Foundation" Radar Chart is first created to represent the Project Requirements related to component being analyzed. This foundation will be used to initially assess whether a potential Alternative is a viable candidate for the project. Then, new charts are produced, one for each Alternative you wish to analyze. Only the Alternatives which produce a chart similar to the foundation chart remain in consideration. After the initial set of Alternatives has been pared down to a few options that all fulfil the Project Requirements, the Alternative with the largest surface area is selected. It is important to down-select the Alternatives based on the Requirements first. "More" is not always "better." The optimal Alternative will produce a chart exactly the same size and shape as the Requirements, whereas an overdesigned component will produce a much larger surface area.

Four different Foundation Radar Charts are displayed in Fig. 12.5, each based on the Requirements related to a microcontroller for different types of projects [29]. Notice that the *power* concern for the Digital Camera project is prolongated compared to the other applications' power axis. This makes sense considering that the Digital Camera is probably required to be mobile, whereas the Microwave, which has the smallest power axis, is connected to the power grid. The real-time constraint axis of the vehicle ESC (Electronic Stability Control) system is significantly longer than the other three projects' axes. Again, this is a reasonable comparison considering you would want your vehicle's stability control system to update every few microseconds, whereas the Digital camera's duty cycle would be measured in fractions of seconds, a 100,000 times slower than the ESC system.

One advantage of using Radar Charts is that they allow for a mixture of quantitative and qualitative analysis.

Typically, detailed Specifications must be accounted for when performing an Alternatives Analysis. However, it is not uncommon for a few qualitative characteristics to be important as well. For example, a Requirements Document might specify that a processor monitors three sensors and draws less than 50 mW of power; these are quantitative Specifications. One of the project Objectives might be to make a product as small and inexpensive "as possible"; these are qualitative characteristics. It would be difficult to incorporate both types of statements in many Alternatives Analysis tools, but the Radar Chart can easily integrate both types in the same graphic.

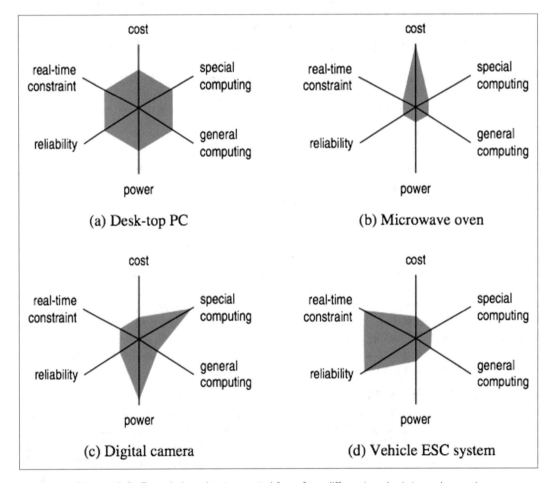

Figure 12.5 Foundation charts created from four different projects' requirements

Example 12.5 demonstrates how to build and analyze a set of Radar Charts for an Alternatives Analysis.

One final note on Radar Charts. The characteristics chosen for the discussion in Example 12.5 (Cost, Special and General Computing, Power, Reliability, and Real-time constraints) are a set of typical concerns for *embedded system* development. These were used as a representative example. You should choose a different subset of characteristics that is applicable to *your* Project Requirements. For instance, you may wish to isolate the number of digital I/O ports and analog-to-digital ports, rather than group all of that capability under "General Computing." You may wish to remove "Special Computing" and replace that axis with different features specific to your project needs. Of course, if the component you are analyzing is *not* a microcontroller, then all six axes might represent completely different characteristics. In fact, there is nothing special about including exactly six axes; Radar Charts typically include anywhere from four to eight axes.

EXAMPLE 12.5: Radar Charts

This example will demonstrate how to build and analyze a set of Radar Charts during the Alternatives Selection process. Let us relate this example to Raspberry Pi B versus MSP420FR430 decision process discussed in Example 12.4.

Why was the **Raspberry Pi B** compared with the **MSP420FR2310** specifically? There are literally 1000s of microcontrollers and dozens of single board computers to choose from. Why these two specifically?

There are two possible answers.

1. If the processor was not initially selected as a Critical Component for this project, simply selecting two reasonable Alternatives would have been acceptable. Maybe these were two options the design team was familiar with.
2. If the processor *was* a Critical Component, then there must have been some detailed analysis performed *before* the Decision Tree was produced.

For purposes of this example, let us assume the processor *is* a Critical Component for this design and discuss how a team might have arrived at the MSP430FR2310.

First, Build a Radar Chart That Represents the Project Requirements

Consider the importance of each characteristic. Make a judgment call on the important of each and place a point on the Chart. Although this judgment call is not "arbitrary," it is still a judgment call that *you* are free to make. Just be sure you can justify and define your decisions.

For purposes of this example, let us assume the following:

1. Measuring data will not require any Real-time Constraints or Special Computing
2. We will need some General Computing and Reliability, but most processors should easily meet those needs.
3. This leaves "low" cost and power as our primary concerns.

EXAMPLE 12.5, continued

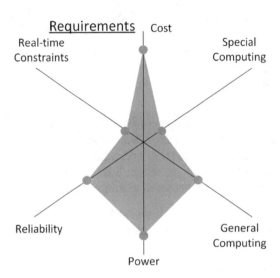

These considerations produce the gray surface area shown in this foundation Radar Chart.

Now, Analyze a Number of Different Alternatives and Create Radar Charts for Each

For this example, the MSP430FR2355 (red), MSP430FR2310 (green), and PIC18F06Q (blue) were analyzed and plotted. The MSP430FR2355 was quickly discarded because its shape does not match the Foundation Radar Chart.

Neither the MSP430FR2310 nor the PIC18F06Q matched the Requirements exactly. The Requirements only called for a single I/O, SPI, and I2C port. Both of these processors provide additional I/O and SPI ports. They are overdesigned in this category, but it might be hard to find a processor with less I/O that will still meet our processing needs. The cost of the MSP430FR2310 is slightly higher than the PIC18F06Q, so it scored a little low on the cost axis. Conversely, the power draw of the PIC18F06Q is slightly higher than the MSP430FR2310. But both of these deviations are minimal, and it appears that either of these devices would make an acceptable choice.

EXAMPLE 12.5, continued

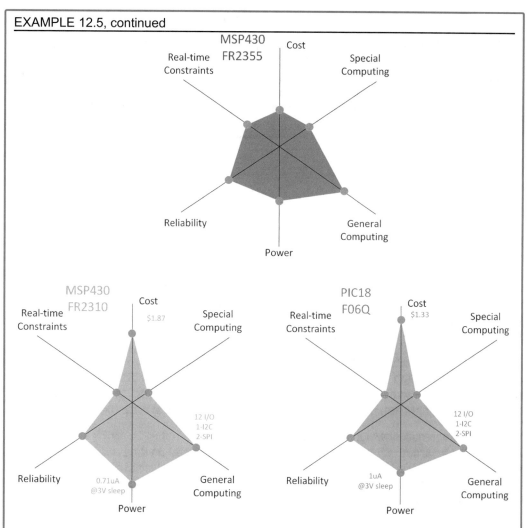

To decide between the MSP430FR2310 and PIC18F06Q, lay the shapes on top of each other. Pick the one that has the larger surface area.

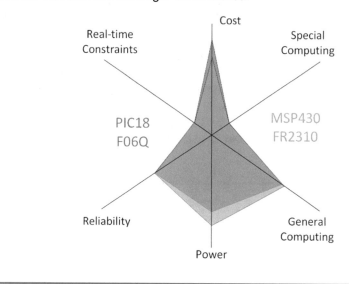

EXAMPLE 12.5, continued

In this case, it appears that the MSP430 FR2310 has a slight advantage and should be selected. The $0.54 savings the PIC18 F06Q provides does not overcome the additional power requirement on this project. For a different project (e.g., one where the product will be mass produced), that consideration may have been more important. If so, the Foundation Chart could be modified to represent that or the weighting you place on that axis could be adjusted.

Pugh Charts

Pugh Charts are another useful Alternative Analysis tool. They should be used when the focus of the analysis is on a single component, similar to when a Radar Chart should be used. However, they have an advantage over Radar Charts. The Radar Chart method becomes cumbersome when there are *many* Alternatives to consider or many characteristics (more than eight axes). Since Pugh Charts can easily be created and analyzed in Excel, or a similar software program, they can easily scale to handle any number of Alternatives and/or characteristics you wish to consider.

To build a Pugh Chart, start by listing all the characteristics you wish to consider in your analysis. Then, assign each a weight based on how important a certain characteristic is compared to the others. Typically, a scale between 1 and 5 is used, but as the number of characteristics increases, it is not uncommon to see a wider scale implemented.

After the characteristics' weights are assigned, rate each Alternative using the same scale. An Alternative is awarded with points for each characteristic based on the product of the characteristic weight multiplied by the Alternative's rating. An Alternative's final score is the sum of all its points. The Alternative with the highest score should be chosen for the project's final design. Example 12.6 demonstrates how to use a Pugh Chart.

EXAMPLE 12.6: Pugh Charts

This example will analyze the same microcontrollers as Example 12.5 but using a Pugh Chart rather than a Radar Chart.

Prepare the chart by listing the characteristics and Alternatives. Then, assign weights to each characteristic. In this case, some of the same characteristics used on Example 12.5 are used again. However, it was determined that Special Computing and Real-time Constraints were not particularly important for this analysis, so they were excluded here. Also, there were too many considerations that fell under General Computing, so that item was broken into three subgroups, number of I/O, analog-to-digital converters (ADC), and number of SPI and/or I2C ports.

EXAMPLE 12.6, continued

For this example, it was determined that

Microcontroller Analysis				
Characteristics	Weight	MSP430 FR 2355	MSP430 FR 2310	PIC18 F60Q
Cost	4			
Number of I/O	3			
Number of ADC	3			
Number of SPI/I2C	3			
Power	5			
Reliability	2			
Final Score:		0	0	0

1. Low power consumption was the most important characteristic and was given a weight of 5.
2. Cost was also very important. It was weighted 4 out of 5.
3. The number of ports all had equal concern, but with a variety of options, these were less concerning than cost or power. These were equally weighted as 3 out 5.
4. Reliability mattered for this project, but most microcontrollers provided enough reliability for this project, so it was not a significant concern. So, it was given a low weight of 2 out of 5.

Now, collect data on each Alternative. Record the data in the top cell of each characteristic's row (blue). This record will help you justify your decisions later. Next rate each Alternative for each characteristic. Typically, the ratings should also be on a scale of 1–5, where a "great" option is rated 5 and a "poor" option is rated 1. If an option is rated 0, the Alternative should be eliminated from consideration. If multiple options receive a 0 in the same category, a decision must be made. Either a new set of Alternatives must be generated, or the category should be eliminated.

Microcontroller Analysis				
Characteristics	Weight	MSP430 FR 2355	MSP430 FR 2310	PIC18 F60Q
Cost		$4.18	$1.87	$1.33
	4	1	4	5
Number of I/O		44	12	12
	3	1	3	3
Number of ADC		12	2	2
	3	1	4	4
Number of SPI/I2C		2+2	1+1	1+0
	3	1	5	3
Power		3uA	0.71uA	1uA
	5	2	5	4
Reliability		??	??	??
	2	0	0	0
Final Score:		23	77	70

EXAMPLE 12.6, continued

The cost of these three microcontrollers was rated first. The cheapest Alternative (PIC18F60Q) was rated a 5 as the best option for the cost category.

Next, the number of I/O category was rated. No Alternative was rated a 5. *More* is not always *better*. Perhaps in this case, the required I/O was 8. Any Alternative with less than 8 would necessarily be rated 0 and removed from consideration. An Alternative with exactly 8 I/O would be rated a 5. In this example, the Alternatives with 10 I/O were rated 3, and the one with 44 I/O was considered a poor choice since it was drastically overengineered and rated a 1.

The ADC, SPI/I2C, and Power categories were rated as well. There was not enough reliability data collected during this analysis to justify a ranking. In this example, it was decided to eliminate that category from the decision. If this had been an important category for project success, a new set of microcontrollers would need to be identified and the analysis process would start over.

The final scores are calculated by the sum of products for each characteristic. In example, the score for the MSP4302310 was calculated by

$$4*4+3*3+4*3+5*3+5*5+0*2=77$$

12.3 Final Alternatives Selection

After analyzing the alternatives for each Critical Component, you must evaluate the top choices of each component for compatibility with all other Critical Components before finalizing your selections. Remember, the purpose here is to select the set of Alternatives which will produce the optimal *system*. If the Alternative Analysis was performed to perfection, the top choices for each component will interface with all other Critical Components seamlessly. However, it is often difficult to account for *every* system consideration when you are focused on an individual component.

Taking the time and effort to carefully verify that the individual Alternatives selected from the Alternatives Analysis are all compatible is an important final step and can eliminate significant problems during the EXECUTING phase. This effort should always be accomplished with support from the entire team. Request that each member present what they believe is the top choice for their assigned Critical Component. Create a diagram or schematic. Pay special attention to any connections between the Critical Components. Provide Specifications such as the following:

- Communication protocol
- Power draw requirements
- Current limitations
- Voltage levels
- Fastener styles
- Package dimensions

Once the Critical Components have been selected and verified to interface with each other well, you can begin selecting the remaining components necessary to complete the design.

Avoiding Common Pitfalls 12.2: Incompatible Alternative Selections

Sometimes what appears to be the top choice of one Critical Component during the Alternatives Analysis will not interface with the primary choice of another component.

For example, let us assume individuals on your team favor the following Alternatives for their respective Critical Components:

1. Processor MSP430FR2310 microcontroller, $1.87
2. GPS transceiver   TESEO-LIV3R GPS module with I2C interface, $12.01
3. Air quality sensor PMSA003I Air Quality Sensor with I2C interface, $39.95

Each device appears to meet Specifications for their respective items on the Requirements Document. All three items are relatively inexpensive. It appears that the team performed an effective Alternatives Analysis.

However, when considered together, there is an issue. The MSP430FR2310 only has one I2C interface. On the surface, this should not be a problem because I2C is designed for one controller to communicate with many devices. Unfortunately, the optimal data transfer rate of the GPS module is significantly slower than that of Air Quality sensor. You must now make some important engineering decisions:

1. You could upgrade the processor to the MSP430FR2355 which has two I2C modules and could individually control the two devices at their optimal rates. This is a 4.3% increase in the total cost of these three items. Is it worth it?
2. Typically, a device will function appropriately if you operate below its optimal data transfer rate, so you could attempt to slow the Air Quality Sensor's data transfer rate to match the GPS module. This would require significant verification testing to prove that the sensor is able to operate below the specified data transfer rate. You may be giving up some functionality if you do this.
3. You could reevaluate the secondary options for the GPS module and/or the Air Quality Sensor. Perhaps another device delivered approximately the same capabilities but had a SPI module. Now you could interface both devices with the MSP430FR2310.

The point here is this: a system-level review of the final Alternatives Selection is a crucial step in this process.

Industry Point of View 12.1: Consider All the Alternatives

Sometimes in your career, you will be asked to solve problems that are outside your scope of expertise. This will require that you think outside the box of your knowledge.

As an electrical engineer, I was asked to evaluate our manufacturing procedures in our plant located in Rochester, MN, because customers had started to complain about the lag time between placing an order and receive the product.

Our containment units held up to 48 batteries, but a customer could order any number (up to 48) of batteries to be installed in the containment unit. Once the order was placed, the units were manually configured with the desired number of batteries and then put through a series of performance tests.

Industry Point of View 12.1, continued

The testing process required three weeks regardless of how many batteries were installed.

Having never been involved in the manufacturing end of the business, I had to research many alternative processes before I could propose a solution. I spent time with our manufactures to learn their process. I also consulted with colleagues in different companies to learn about their processes.

We could not build-to-suit faster than three weeks, and we initially thought we could not prebuild products since there were far too many variations to keep in stock given the size of our storage facility.

We simulated a number of standard alternative processes to find a solution that would reduce the manufacturing timeline. We considered purchasing a large storage facility so that we could prebuild every possible configuration and maintain a large stock of products. We considered limiting the number of configurations we would sell and maintain very limited stock of pre-built products. We even consider reducing the testing we conducted to speed up our existing process. All the alternatives met the primary goal of speeding up the process but had drawbacks as well.

We eventually started thinking outside the box. We developed a new process wherein fully configured units were pre-built and tested. Then, when a customer's order came in, any unnecessary battery was removed and immediately installed in another unit. Once correctly figured, the unit could be immediately shipped since all the necessary testing had been performed.

By thoroughly considering many options and analyzing the results as well as the greater impacts, we identified a solution that optimally solved the issue. Had we immediately implemented one of the first ideas, we never would have identified the optimal solution. Engineering design takes time and careful consideration before jumping into implementation.

12.4 Chapter Summary

In this chapter, you learned about the Alternatives Selection process. During this process, alternatives for a number of the system's Critical Components are analyzed, and the optimal alternative is eventually selected. Here are the most important takeaways from the chapter:

- The first step in the Alternatives Selection process is to identify the system's Critical Components. Although a team has the freedom to identify any aspect of a design of a Critical Component, the choice should be focused on reducing project risk.
- Appropriate choices for the Critical Subsystem include the following:
  - The most expensive components
  - The components with the tightest (most difficult) Specifications
  - The components which will impact many future design decisions
- Engineering Alternatives Analysis are based on a qualitative analysis of data. Pugh charts are probably the most common tool used to perform this analysis on Senior Design

projects, but the Decision Tree and Radar Chart are two other useful tools that you can employ.

- Rigorous Engineering Alternatives Analysis should be performed on each Critical Component, as well as any other component which will significantly impact the system design. The remaining components (such as resistors, wire gauge, or packaging material) may be reasonably selected without a direct comparison to a set of alternatives.

- The final set of Alternatives must be *integrated*. This means that the top choice for any given Critical Component *may* not be the optimal choice for the system. The optimal choice for each Critical Component is the Alternative which will best integrate into the system design.

12.5 Case Studies

Let us look at our case study team's Alternative Selection processes. We will start by looking at the Critical Component Selections, and then one of the related Alternative Analyses will be discussed for each team. Finally, the team's final Alternative Selection for the related Critical Component will be reviewed.

Augmented Reality Sandbox

Figure 12.6 displays the AR Sandbox team's Block Diagram, this time with their choice of Critical Components identified by gold stars. The team identified the Projector, Sand, Base Structure, PC, and Corn Models as the five Critical Components for the five students.

- **What went well:** The team reasonably determined that a high-quality projector and the specific type of sand were Critical Components to consider in order to display a high-quality image. Since the project would eventually be displayed in public, the aesthetics of the base structure was also deemed critical. So, three of their five Critical Components were well chosen.
- **What could have been improved #1:** The team originally identified the PC as a Critical Component (the project was constrained to use a PC for compatibility with other measure exhibits). Often, choosing the optimal computer *is* critical. However, in this case, the computer itself did not matter so much. What *was* critical was the computer's graphics card. The computer should have been considered a subsystem which contained a number of components. Then, by selecting one of those components (e.g., the graphics card), rather than the entire subsystem, the Alternatives Analysis would have been much more focused and therefore easier to complete.
- **What could have been improved #2:** The fifth Critical Component selected was the software models of the corn. While this initially seemed to be a daunting aspect of the project, the predictive crop yield algorithm proved to be a much more significant challenge. Predicting which component would require the most effort during the early stages of the project would have been difficult, so the team is not necessarily "to blame" for this selection. However, consider the fact that the algorithm produced the model. If the algorithm does not work, then the entire Objective of teaching children about agriculture will not be possible. Conversely, perhaps if the corn models were not produced well, an alternative method of conveying the algorithm's results could have been developed. When selecting your own Critical Components, consider which components will require the most effort as the AR Sandbox team did, but also consider which components are the most necessary and which can possibly be omitted if necessary.

Once the Critical Components were identified, each student performed an Alternatives Analysis on one component. Logan was tasked to analyze options for the Projector.

Table 12.1 presents his results in a Pugh Chart. On the far-left column, we can see the characteristics important to this component along with requirements from the Requirements Document. Initially, Logan analyzed over a dozen alternatives and narrowed the options down to the three choices presented here.

This was a difficult Alternatives Analysis because only one projector completely met Requirements, but that item cost nearly twice the amount of the next-most-expensive alternative. Logan appropriately accounted for this disparity in his initial analysis, but the expensive item still scored the highest point value and was initially selected as the preferred alternative. Let us consider how Logan assigned the point values presented here.

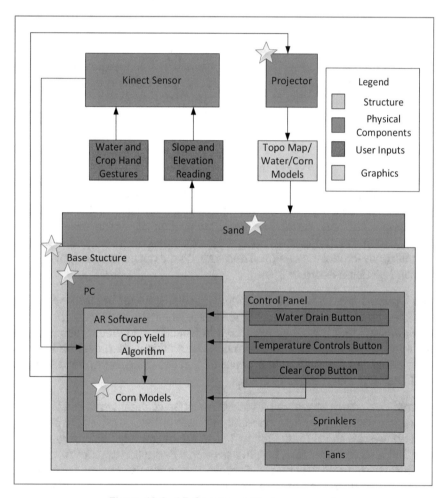

Figure 12.6 AR Sandbox critical components

Table 12.1 Alternatives Analysis – Projector

Critical Component: Projector			
	Option 1	Option 2	Option 3
Lumens (bright)	4000	3000	6000
	3.33	2.5	5
Aspect Ratio (4:3)	4:3	16:10	4:3
	5	3	5
Throw Ratio (0.9)	1.3	0.9	0.9
	1	5	5
Cost (effective)	$699.99	$553.99	$1675.00
	4.5	5	1
Total:	13.83	15.5	16

The Projector was required to be as "bright" as possible so that the projected image would be visible on the field of sand, but there was no Specification defining exactly what that meant. Logan gave the brightest alternative 5pts. Then, use the ratio 5/6000 to linearly score the other two alternatives. Equation 12.1 provides an example of how the Option 2 alternative score was calculated.

$$3000 \text{ lumens} \times 5/6000 = 2.5 \tag{12.1}$$

The Requirements Document defined that the Projector's aspect ratio was 4:3. Therefore, the two projectors which had a 4:3 ratio scored 5pts. An aspect ratio of 4:3 was defined by the desired dimensions of the sandbox. Although Option 2 would not meet the initial Specifications, Logan thought that a change request could be made because a 16:10 ratio would not significantly impact the design. Therefore, he did not immediately discard the Option 2 alternative and gave it a score of 3pts.

Similarly, the Option 1 alternative did not meet Specifications for the throw ratio. The Throw Ratio was defined by the characteristics of the sensor which the team was constrained to use. While Logan guessed that a work-around for this Specification would also be possible, the effort to modify this Specification would be much more difficult than altering the aspect ratio. Therefore, he penalized Option 1 with a score of 1 pt (as opposed to penalizing the aspect ratio score of Option 2 with a score of 3pts).

Scaling the cost characteristic was a bit trickier because of the inverse relationship between the characteristic and the score (i.e., cheaper is better or less-is-more). To address this, the cheapest option was awarded 5pts. Option 1 was approximately 25% more expensive, and Option 3 was 200% more expensive. Logan assigned a deduction of 0.5pts for every 25% cost increase. This resulted in assessing a 0.5pts deduction for the Option 1 (giving Option 1 a score of 4.5pts) and a deduction of 4pts for Option 3 (giving Option 1 a score of 1 pt).

These decisions resulted in awarding Option 3 the highest overall score, and so Logan initially chose Option 3 for his Alternative Selection. However, the team modified the choice during the final integration step of the Alternative Selection process. The cost of Option 3 was 33.5% of their total budget, and the Alternatives Selections made by the other team members were going to require approximately 60% of the budget (meaning that the Critical Components alone would demand 93.5% of the total). The team presented a change request to the

Sponsor asking that the wording of the Specification state "the aspect ratio of the projected image must be 4:3," *as* opposed to stating, "the aspect ratio of the projector must be 4:3." Then, a 4:3 shutter was developed to limit the projected image. This was an intuitive solution that produced the desired result without exceeding the project budget.

Smart Flow-rate Valve

Figure 12.7 displays the Smart Flow Rate Valve team's Block Diagram again, this time with the Critical Components highlighted in orange text. The team first showed potential to be a Performance team during the Critical Component identification process. Their selections indicated a thorough consideration of many possibilities. For example, while ordinarily a motor and the primary sensors of any control system should be considered Critical Components, the team recognized that the Sponsor had specified a strong preference for these component selections already. The team was not necessarily constrained to those preferences but would require a strong justification for a deviation from the Sponsor's preferences. Therefore, the team felt that; although a review of alternatives for these components was necessary, the formal Alternatives Selection process should start with the previously unconsidered aspects of the project. This decision was supported by the Sponsor and resulted in the Housing, MATLAB GUI, Power Converter, and Valve/Encoder being selected as the Critical Components. Another reason this team's Alternatives Selection process was chosen for this case study was how well the team integrated the mechanical and electrical aspects of the project.

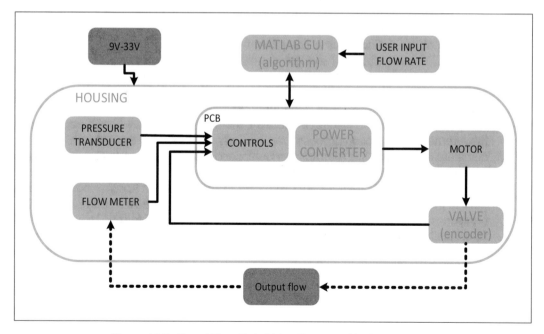

Figure 12.7 Smart Flow Rate Valve Team's critical components

- What went well #1: Christian, a mechanical engineer, was assigned to select the Valve. Keith, an electrical engineer, was assigned to select the Encoder. Knowing that this would be one of the project aspects which would require the mechanicals and electricals to most closely interact, the team decided to have each student perform an individual analysis on their respective components but also perform a joint analysis on alternatives which manufactured an Encoder with the Valve already. Then, the two students performed an integration of their three analyses before contributing to the official team integration.

- What went well #2: Gannon and Mason were both assigned the Housing as their Critical Component. Under normal circumstances, the academic constraints of Senior Design would not allow two students to select the same Critical Component. However, this Critical Component was integral to the team's design, and it was substantially more complex than most Senior Design Alternative Analyses. The two students brought this issue to their Advisor and ask for approval to perform the analysis together. Then, during the Alternatives Analysis process, they each focused on different aspects of the component and produced three-dimensional sketches of their individual Alternative Selections which demonstrated that they had individually contributed to the process. Then, they merged the best ideas together into an intermediate Alternative Selection. Finally, they worked together after the team's system integration discussions to produce the final Alternative Selection.

- What could have been improved: The team planned to perform an additional Alternatives Analysis on the sensors and motor which the Sponsor had suggested after they completed the Alternatives Analysis on the Critical Components. However, Gannon and Mason both used the preferred sensors and motor in their respective Alternatives Analysis for the Housing. During the Alternatives Analysis for the motor, a DC motor was identified which may have been a substitute for the initially preferred motor. However, by that point in the process, a substitution would have meant significant changes to the Housing and the Valve/Encoder selections. A Critical Component, by definition, is a component upon which many other decisions must be based. The selection of the motor impacted the Housing and vice versa; therefore, the motor should have been considered "critical." However, it would not be fair to find fault with the students for their decision process on this issue. Practically *every* component of this project was dependent on every other component, meaning that every component was "critical." In industry, more time might have been dedicated to the Alternatives Selection process than what can be allotted in the academic cycle to completely integrate the Alternatives Selections. Alternatively, additional engineers may have been tasked to this process. Recall that both time and resources are legs of the Triple Point Constraint theory discussed in Sect. 1.2. Yet another consideration here is that spending too much time and/or resources in the Alternatives Selection process is also possible. In this case, the initially preferred motor met all Specifications and did not negatively impact any other decision. An engineer must balance the desire to produce "the perfect solution" with the absolute need to provide "a solution which meets specifications"; sometimes "good enough" is sufficient.

Thein was assigned to analyze the power converter. This Critical Component was necessary to convert the external supply voltage to the desired voltage or voltages necessary for the Smart Flow Rate Valve system. The Alternatives Analysis quickly became complex considering how many design options were involved. Here is a summary:

1. The preferred components previously selected by the Sponsor were all 12 V components, but microcontrollers typically run off 5 V or 3.3 V.

   • Should the Power Converter provide one 12 V output? This would mean that the control system must be 12 V analog hardware rather than performed in software.
   • Should the Power Converter provide two outputs, one at 12 V and one at microelectronic levels? If so, should the microelectronic level be 5 V, 3.3 V, or both?

2. An even more difficult challenge was presented by the range of potential inputs, which was defined to be anywhere between 9 V and 33 V. This is a wider input range than what most commercially available DC-to-DC converters can accept.

   • Should a high-end, customized Power Converter that can handle this wide range be selected? Doing so will likely be expensive, require significant physical volume, and potentially have to be custom-made resulting in a time delay.
   • Should there be two Power Converters, one for the lower half of the range and one for the upper half? This solution would require a switching circuit to power the appropriate converter under specific conditions. A second power converter and the necessary switching circuit would require additional cost and volume.

Thein's Alternatives Analysis was too complicated to start with a Pugh Chart. A Decision Tree, Fig. 12.8, was used to first select the best power topology. Then, the necessary components could be identified for each potential topology and an Alternative Analyses could be performed on each. The Alternative along the green path was ultimately selected.

• What went well #1: Without a detailed design, the exact cost of the "Additional Switching Circuitry" or the "Analog Controls" was not known. Thein made reasonable estimates and documented his decisions and then consistently applied those estimates for all Alternatives which required those values.
• What went well #2: Thein did not select the cheapest Alternative. In fact, when considering the $5000 Project Budget, even the most expensive Alternative was reasonably priced and would have been an acceptable selection. The Alternative that required the smallest physical space was ultimately selected; however, even this was not the primary justification for the final selection. The final selection was made based on the risk analysis of the Alternatives. There was only one Alternative which did not present a significant Project Risk (noted in orange text).

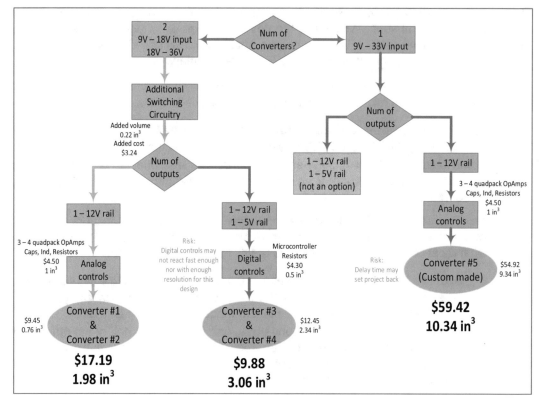

Figure 12.8 Alternatives Analysis – Power Converter

Robotic ESD Testing Apparatus

Figure 12.9 displays the RESDTA team's Block Diagram. The Critical Components are highlighted by the red boxes. Because the team's Block Diagram was focused on high-level Objectives, three of the blocks had to be modified from the original Block Diagram (Fig. 8.11) by including additional details. For example, the "STRUCTURE" block now specifies both "DUT mounting" and "Moving ESD tip" and the "MOTORS" block now specifies "X/Y direction motors" and "Z direction motor."

- What went well #1: The general guideline for a Senior Design team is to select one Critical Component for each student to work on. However, this team insightfully realized that they had more Critical Components (7) than students (5). Moreover, they realized that the solution to "move the ESD tip" would affect nearly every other subsequent decision. Therefore, they coined a term specific to their project by calling the ESD movement an "Ultra-Critical" Component. The entire time contributed to the

Alternatives Analysis for this so-called Ultra-Critical Component then divided up the remaining Critical Components and performed a more "typical" analysis on those. The initial team analysis actually generated additional details that needed to be incorporated into the Requirements Document which were then used to guide the Alternatives Analysis process for the remaining Critical Components. This process indicated that the team had transitioned from a real team into a performance team. They creatively thought outside the box when applying the Project Management "rules" in order to best manage their unique project. Then, supported each other's work and eventually produced a phenomenal result.

- What went well #2: After determining the concepts for how to move the ESD tip around, the team divided the remaining Critical Components. Leah was ultimately in charge of the control system, but Kaitlin wanted to be involved with this subsystem as well. Leah took on the responsibility of the X/Y Direction Motors Alternatives Analysis, and Kaitlin was responsible for the Z Direction Motor. The two teammates reviewed each other's Alternatives Analysis before either was finalized. Then, they worked together on the Motor Drive Alternatives Analysis. Next, they integrated a variety of options for all three components. When Leah and Kaitlin eventually contributed to the team's system integration, they presented the three alternatives which were the top results of their integrated analysis.

- What could have been improved: After the team selected a structure for moving the ESD tip, Easton was assigned with a responsibility for identifying a solution for mounting the DUT (which had to be manually placed with a $\pm$ 0.005" tolerance). Kelsi needed to analyze safety enclosure systems. Both performed their analysis well. However, the team had already identified (and labeled) the Motors as a joint electrical and mechanical issue. Yet, no mechanical input was considered when Leah and Kaitlin performed their Alternatives Analysis on the motors. Fortunately, Kelsi identified and shared some concerns about the motors during the system integration step that the electrical engineers had not previously considered. Additional modifications had to be made to the initial motors' Alternatives Analysis before a final selection was agreed up. There is nothing negative about having a team member challenge the Alternatives Analysis presented during the integration step – that is not the point being made here. What was strange about the RESDTA team's division of labor is that they identified a block as being a joint electrical/mechanical issue but then assigned two electricals to work on it. Perhaps Kelsi should have been assigned some of the responsibility for this analysis in the first place. Alternatively, Kelsi could have been included when initially deciding upon the analysis's characteristics to help guide Leah and Kaitlin but then have been tasked to analyze the enclosure after the electrical engineers got started on the actual analysis. Doing so would have incorporated Kelsi's expertise into the motor's analysis without significantly increasing her workload.

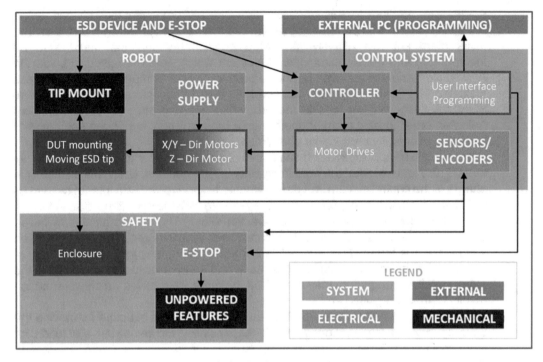

Figure 12.9 RESDTA Team's critical components

Chapter 13: **Preliminary Design Process**

At this stage of the PLANNING phase, the Critical Components have been identified, analyzed, and selected through the Alternatives Selection process. It is now time to produce the Preliminary System Design. The Preliminary System Design process will be carried out much like the Engineering Alternative Analysis process. In that process, each Critical Component was selected, but before the selection was finalized, the entire list of alternatives had to be considered from a system-level point of view to ensure compatibility. In the Design process, individual subsystems will be identified and designed. Then they must be integrated into a system-level design.

> During the Preliminary Design process the Critical Subsystems are designed and a framework for the complete system is developed.

The outcome of this process is not expected to be a *finalized* system design yet. There will likely still be many factors which will not be completely understood by the end of this process. That is alright for now, these remaining details will be investigated and resolved before the culmination of the PLANNING phase in the following process.

Therefore, another important outcome of the Preliminary System Design process is to *identify* the unknown factors that are preventing the completion of the design. These will be resolved in the Preliminary Prototyping process (Sect. 14.4).

In this chapter you will learn to:

- Identify subsystems and select the Critical Subsystems
- Design the Critical Subsystems
- Develop the framework for the complete system design
- Identify technical issues preventing the completion of the system design

C.J. Mettler, *Engineering Design*, https://doi.org/10.1007/978-3-031-23309-8_13

13.1 Identifying Subsystems

The first step in the Preliminary System Design process is to identify *all* the subsystems which must eventually be designed. This step is an important difference between "design" and "development." A development process might start by producing the project's input subsystem without considering the project's processing subsystem. The project's processing subsystem might be developed next, but it would be severely constrained by the now-existing input design. The project's output subsystem might then be developed, but it would necessarily be based on *all* the decisions made previously. Each decision severely limits the options for the remaining subsystems. The development process does work sometimes, but in a *rigorous* design project, it is no more effective than the adventurer traveling to the Himalaya without procuring the proper permits to achieve their goal as discussed in the introduction to PART IV: THE PLANNING PHASE. The *entire* system (i.e., plan) must be considered before any part of the system is finalized.

A system is typically divided into discrete subsystems where each subsystem produces a particular "feature." A "feature" might be as complicated as producing an entire Objective or as simple as ensuring a particular Specification is met. The size of a subsystem should be inversely proportional to the complexity of the subsystem. Large subsystems are acceptable if the complexity of the entire subsystem is relatively simple. For more complex issues, subdividing the system into small subsystems is more effective.

The primary reason for dividing up the system is to divide the *workload* of the remaining Design Life-Cycle into manageable chunks. Even a Performance-Team would struggle to manage their project if they were instructed to *"just complete the project"* in a single step. At the other end of the spectrum, it is just as difficult to manage a project at the single-component level; there are too many details to manage effectively. Breaking the project into subsystems allows the team to effectively manage the project design.

For any given project, the number of subsystems, and the exact functionality of those subsystems, is somewhat arbitrary. This section will assist you in identifying appropriate subsystems for your project and then assigning those subsystems to the team members.

Identifying Subsystems

Any complex system can be divided into multiple subsystems. The exact divisions are somewhat arbitrary. There is not a single "correct" way to do this. There are, however, some guidelines which will aid you in this process.

Each subsystem should explicitly carry out a "single purpose."

"Single purpose" does not mean one-and-only-one function. A function can be relatively simple or extremely complicated. The more complicated a function is, the more narrowly-defined the related subsystem should be. Typically, each function of the system is defined by a Requirement. Occasionally, a subsystem can represent an entire Objective, assuming that Objective is a relatively simple one. It is rare, but not impossible, that a subsystem could represent a single Specification. Example 13.1 presents three possible ways the same features of a particular project might be grouped into different subsystems.

EXAMPLE 13.1: Identifying Subsystems

For this example, consider the common Objective: Monitor an input signal. Let's assume that the particular signal for this example is an acceleration signal.

A subset of potential Requirements related to this Objective might be:

1. Capture the raw signal
2. Process the raw signal
3. Analyze the processed signal

One option would be to simply define a subsystem "Measure the Acceleration Signal." The related portion of the Block Diagram might look like this (of course, the exact "Output" would depend upon the specific project requirements):

"Monitor Input Signal" could be considered a single function. However, since many steps will be required to accomplish this function, it would be a fairly large subsystem. This is often the case when defining a subsystem at the Objective-level. However, a large subsystem does not necessarily correlate to a significant engineering effort. For example, the final design of this subsystem may simply entail connecting a pre-package sensor to a signal-analyzer with built-in signal processing capabilities. There would be almost no engineering work required to complete the design.

However, in a more complicated design, this large subsystem may be difficult to manage. It may potentially be better to define multiple subsystems for this Objective; perhaps "Monitor Input Signal" should be broken down to two subsystems: "Signal Processing" and "Data Processing." The Block Diagram would be modified to look like this:

These subsystems still define a signal function, but the definition is more precise than "Monitor Input Signal." These subsystems might require the following tasks:

Signal Processing	Data Processing
1. Collect Data	1. Compare to preset thresholds
2. Filter Noise	2. Determine correct output
3. Apply Gain	

It is possible to divide the project into too many subsystems. For example, the "Signal Processing" block contains three tasks just like the "Monitor Input Signal" block. The "Signal Processing" subsystem *could* be broken down into three additional subsystems, "Collect Data," "Filter Noise," "Apply Gain," and the related Block Diagram would look like this:

EXAMPLE 13.1, continued

This level of detail may be necessary for *certain* projects. For example, if a team is attempting to design an Accelerometer from scratch and the filter is a 10th order filter, then each of these blocks may be necessary.

However, on *most* Senior Design projects, the Accelerometer will be a sensor that was purchased and the Filter/Amp combination will only require a few components. In that case, dividing the "Signal Processing" subsystem into more specific subsystems is unnecessary and will hinder the project. In this case, the "Accelerometer" block would be a single component, which is, by definition, *not* a subsystem. In fact, the arrow "Accel Sig" and the block "Accelerometer" would physically be the exact same thing.

Occasionally it is beneficial to document the specific tasks required by a particular subsystem on the Block Diagram without necessarily identifying those specific tasks as subsystems. That diagram might look like this:

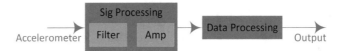

Here we see that the subsystem "Signal Processing" is composed of two pieces, the "Filter" and the "Amp." The Project Management structure should focus on the subsystem, but the extra detail may prove beneficial for some technical discussions.

One final note about the "Accelerometer" block: Recall that a subsystem cannot consist of a single component or process. However, a subsystem may consist of a single component AND single process. For this example, perhaps the team is concerned with the *calibration* of the accelerometer. Then, a reasonable subsystem may consist of the sensor component and the calibration process.

Here are some general guidelines for defining subsystems:

- A subsystem must contain a minimum of two components and/or processes. A single component or process cannot comprise the entire subsystem.
- By definition, a subsystem performs a single purpose. Keep in mind, however, that a single purpose may require multiple steps (or functions).
- The number of subsystems should roughly relate to the size of the design team. A good rule of thumb is to define 2–3 subsystems per team member. However, this rule is extremely dependent on the project.
- A starting point for defining subsystems is to build a subsystem around each Critical Component. This means there is only one Critical Component within a particular subsystem. There are almost always more subsystems than Critical Components. So, this guideline is only a *starting* point for defining the subsystems.
- Consider the following project concerns which may not always be explicitly defined in the Requirements Document but might need to be represented by a subsystem:
  - How will each component of the project be powered?
  - How will the final product be contained, packaged, or protected?
  - How will the product attach or interface to its operating environment?

The responsibility of each Subsystem must now be assigned to specific team members.

> On most Senior Design teams, it is expected that each student will be solely responsible for at least one subsystem throughout the remainder of the project.

This will allow each student to demonstrate they are individually capable of designing, building, and testing an engineering product. The remainder of the Subsystems should be divided among the team in the most effective way possible. How this is accomplished will, of course, differ between teams.

The most common implementation of distributing the remaining subsystems among the team members is to simply ensure that each member has an equal number of subsystems. However, it is also common that at least one subsystem will require significantly more time and effort than the others. So, care should be taken when distributing the assignments that one member does not receive an unequal share of the workload.

Example 13.2 demonstrates how a high-functioning team used this strategy to set up their initial assignments and again when the assignments required modifications.

EXAMPLE 13.2: Assigning Subsystems to Individual Members

The Block Diagram discussed in Example 8.7 was originally developed by a four-member team, Heather, Jacob, Menisha, and Zach. The diagram is shown here.

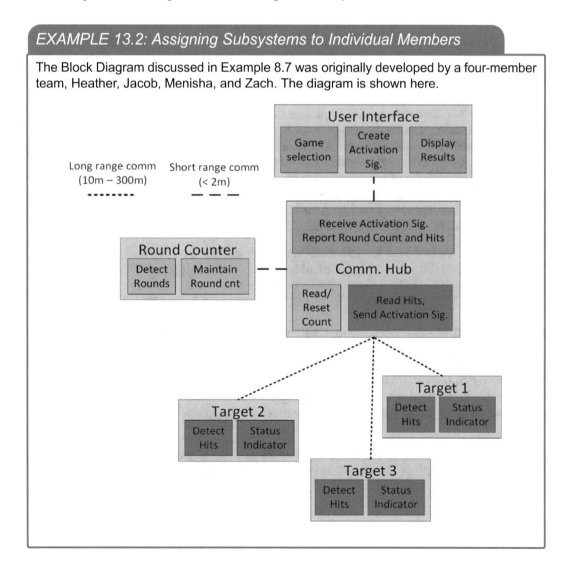

EXAMPLE 13.2, continued

The project's Requirements Document consisted of four Objectives. The team decided to break the project into four large subsystems shown in this block diagram. Each student was assigned one subsystem. Heather was responsible for the User Interface, Jacob for the Round Counter, Zach for the Targets, and for the Communication (Comm.) Hub.

This division of labor would likely have been successful. However, Menisha had to withdraw from the project due to personal health concerns very early in the Design Life-Cycle. Therefore, the team redistributed Menisha's responsibilities evenly among the remaining three members by dividing the large Comm. Hub subsystem in to three smaller subsystems. Heather developed the ability on the Comm. Hub to Receive an Activation Signal and Report Round Counts/Hits to the User Interface. Jacob programmed the ability to Read/Reset Counts. Zach programmed the ability to Read Hits and Send Activation Signals to the Targets. It is worthwhile to point out that in this modified approach, each team member took responsibility for the portion of the Comm. Hub most directly connected to their existing project responsibilities, likely increasing the overall effectiveness of the team.

Initially, this modified division of labor seemed to make sense. However, the User Interface was already the most labor-intensive part of the project. Additionally, the ability to communicate between the User Interface and Comm. Hub was the most difficult portion of the large Comm. Hub subsystem. Before long, it became clear that Heather had been tasked with a disproportionate workload.

Fortunately, this team managed their project extremely well and essentially operated as a Performance-team. Ultimately, they chose *not* to modify the team responsibilities even after recognizing the unbalanced assignments; they recognized that their team (and therefore their project) would operate more efficiently if Heather maintained the leadership of the User Interface and communication with the Comm. Hub. So instead, Jacob and Zach modified their schedules so that they were able to finish their subsystems well ahead of schedule. Then, with Heather taking the lead, the three team members worked together to complete her two subsystems on time.

This strategy worked well for this particular team largely due to Heather's advanced leadership skills. Her leadership contribution salvaged this project which eventually won a national level professional design pitch competition.

Another method of distributing subsystems is to assign one Critical Subsystem to each member, and the remaining subsystems to "small-groups." A "small-group" is a subset of team members who work closely together on a particular task. This method is particularly useful when one subsystem is used to join two or more Critical Subsystems. The joint effort of the members is often more effective than a single member's efforts. Example 13.3 discusses an alternative strategy the team discussed in Example 13.2 could have implemented.

When assigning the design responsibilities among your team, make sure you keep in mind the primary purpose of dividing the system into subsystems: to make the management (and successful completion!) of the project as efficient as possible.

EXAMPLE 13.3: Assigning Subsystems to Small Groups

This example presents an alternative to that chosen by the team discussed in Example 13.2. After recognizing that the team had an unbalanced workload, they may have also performed effectively had they immediately reorganized.

The subsystems were probably defined appropriately as they were depicted by the colored blocks in Example 13.2. However, the ability to read Hits and Counts (blue and green blocks within the Comm. Hub subsystem) were essentially the same process. Once Zach had figured out how to Read Hits, he probably would have been able to implement the same design (albeit with a few modifications) to read Counts. Assigning similar tasks to one team member would have been more efficient than two members attempting to independently design essentially the same function.

Assigning Zach to both these subsystems would have freed Jacob up to directly assist Heather with the Comm. Hub's subsystem ability to communicate with the User Interface. A joint responsibility on that task would have been more effective than an individual assignment. Heather had the knowledge of what needed to be accomplished but did not have the time due to the complexity of her primary responsibility. Jacob did not have the intricate knowledge of the User Interface but would have had the extra time to dedicate. Working together may have been the most effective way of designing this particular subsystem.

13.2 Critical Subsystems Design

The next process in the PLANNING phase is to design the Critical Subsystems.

Critical Subsystems are centered around a Critical Component.

Each team member should be responsible for designing a Critical Subsystem. Typically, the team member who performed the Alternatives Analysis on a particular Critical Component should be assigned to the related subsystem. In rare cases, more than one Critical Component is located within a particular Critical Subsystem. In that case, one member should design the Critical Subsystem and the other member should design the next-most-important subsystem.

Since every project is so drastically different, it is not possible for this textbook to explain how to complete the design of a given subsystem. For that, you should rely on previous course work. This section is meant to provide guidance on the final deliverable of Critical Subsystem Design process.

The output of this process should be either:

1. A Finalized Subsystem Design with complete documentation or
2. A nearly-Finalized Subsystem Design with nearly-complete documentation *and* a detailed plan on how the missing details will be finalized.

Finalized Subsystem Designs

The purpose of the Critical Subsystem Design process is to produce a Finalized Subsystem Design for each of the most important subsystems. Many students confuse a "design" of a subsystem with the basic "ideation" of the subsystem. A design often starts with a basic ideation of a solution but does not end there. The ideation simply describes the primary concepts at a high level. A design, on the other hand, specifies every detail of the final product. This means that a technician should be able to build the design without having to make any further engineering decisions. Moreover, the design should be verified somehow. Ideally, the design should be simulated in some sort of engineering software package such as MATLAB, PSpice, Powersim, ANSYS, ProE, or SolidWorks. If direct simulation is not possible or appropriate, then the design should be verified using theoretical equations or some other engineering analysis.

In essence, this is similar to the INITIATING phase of the project. The client has a general ideation of a project. That ideation is formalized in the official Problem Statement. Paring the ideation and formalization of concepts continue throughout the project. The Block Diagram is a useful visual tool which can be used to discuss the ideation of the Requirements; the Requirements Document formalizes all those details. The pattern repeats yet again during the Preliminary Subsystem Design process. Ideation of the solution at both the system and subsystem levels is almost certainly helpful in understanding and designing the solution. However, the ideation of the subsystem alone is not sufficient to complete the design; it is the starting point.

Anyone can develop the ideation of a project. In order to complete an engineering design, however, an engineer must consider each design decision in great detail. Often, as the engineer considers one aspect of the design, additional questions arise that must be answered before the design is considered "finalized." One way to determine whether the design if finalized or not is to consider whether two independent fabrication facilities could be contracted to build the design and produced *exactly* the same product. The more aspects of the design which have not been specified by the engineer increases the likelihood of the two facilities producing different products.

Example 13.4 discusses the difference between the *ideation* of a subsystem and the *design* of a subsystem; it also demonstrates how each high-level decision often generates additional secondary questions.

EXAMPLE 13.4: Design vs. Ideation of a Subsystem

Consider a simple containment subsystem. Perhaps a particular project requires a box with a lid in which the product will be packaged. The specifications may state that the box be 20 cm (width) × 25 cm (length) × 10 cm (height). The constraints limit the design to be metal and have a lid.

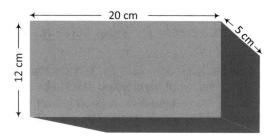

20 cm

5 cm

12 cm

Even though the Requirements Document has defined certain aspects of the design that must be met, this description is still a very basic ideation. Any contractor with a fabrication shop would be able to fabricate a box from the ideation provided above. However, if this was the extent of the instructions provided to three different contractors, each of the three contractor's solution would be different. Consider all the outstanding decisions that are required to synthesize the box which have been left undefined:

- What type of metal is this?
  Weather-proofing requirements may dictate galvanized steel, but that choice will affect whether the box can be welded.
- How will the joints be connected?
  If welding is not possible, perhaps riveting the corners is necessary. However, riveted corners are not tightly sealed from all weather conditions.
- Will the joints need to be sealed?
  If so, what type of compound will be used?
- How will the lid attach?
  One long hinge for extra stability or two small hinges to save weight?
- How will the lid close?
  A tight seal? If so, what fabrication tolerances would be required so that the seal is "tight enough?" Or, will the lid close with a lock and how "unbreakable" must that lock be?

Each undefined property of the solution requires that the contractor make a design decision. Each decision that a contractor makes increases the likelihood of producing a solution that does not exactly meet the Project Requirements.

The value an engineer adds to a project is the ability to consider all these decisions at once and determine the *combination* that will produce the best design. A complete subsystem design would therefore include details addressing all of the above questions, and not leave any important issues for the contractor to "figure out."

Avoiding Common Pit Falls 13.1: Produce a "Finalized" Design

It is common for students to assume they have completed the Critical Subsystem Design process only to be told at the PDR they have more issues to address. Rarely will a student *choose* to submit an incomplete design, but they simply did not consider all the details due to a lack of experience.

So, how do you *know* when you are done if you have never produced a similar design before? Admittedly, this is difficult to know for certain, but there are at least three steps you can take to reduce this risk..

1. After you believe the design is complete, explain to a team member how you would build the design. Start with what parts would need to be ordered and then describe how the design would be constructed. A good teammate will attempt to identify any missing information that would need to be determined before completing the process.
2. Assuming both you and a teammate agree the design is complete, "build" the design in a simulation package (or sketch it on paper). In doing this, pay close attention to whether you are using any knowledge you might have because you have been working on the project, but which is not included in your design and documentation. If so, consider whether this information should be included in your design?
3. Finally, ask an Advisor to review the final documentation, including the simulation results. This review should not be during a general team meeting. Instead, schedule a technical meeting specifically regarding this subsystem so that the entire focus of the meeting is on your design.

Nearly-Finalized Subsystem Designs

While the purpose of the Critical Subsystem Process is to produce a *Finalized* Subsystem Design, it is likely that some of the details will be difficult to specify at this early stage. The next significant process in the PLANNING phase is called the Preliminary Prototyping process. During the Preliminary Prototyping process, small test cases are built, specific hardware is analyzed, and initial software features are investigated. The purpose of these Preliminary Prototypes is to investigate the unknown details of the Critical Subsystem designs. Therefore, it is acceptable to have a few minor gaps in the design at the end of the Preliminary System Design process – but only if these gaps are clearly acknowledged by the design team and a plan is in place to address them.

It is important that any remaining gaps in the design at the of the Preliminary System Design process should be *minor* concerns. There is a significant difference between simply producing the basic ideation of the design versus producing a nearly-Finalized Subsystem Design with a minor gap. Additionally, in order to be acceptable as the output of the Preliminary System Design process, a nearly-Finalized Subsystem Design must include a plan on how the Preliminary Prototype process will provide necessary information to complete the design. Example 13.5 presents an example of an acceptable nearly-Finalized Subsystem Designs compared to the simple ideation of the design.

EXAMPLE 13.5: Nearly-Finalized Subsystem Designs

This example discusses the difference between *basic ideation* of a solution and a *nearly-Finalized* Subsystem Design. Perhaps, one of the Subsystems that must be designed for a particular project is an amplification circuit. Let's assume the Critical Component for this subsystem was an OpAmp, and the LM741 part number was selected. Consider the following two stages of the design process:

Ideation of Design: An OpAmp will be used to amplify the signal.

Nearly-Finalized Design: A non-inverting OpAmp, Fig. 13.1, will be used to amplify the signal. The OpAmp will be powered with a +5 V rail-to-ground connection. The amplifier's output will be connected to a Microcontroller's input which should draw approximately 10 µA which is well below the LM741's limit of 40 mA.

The required slew rate and gain of the design cannot be known until the signal is analyzed during the Preliminary Prototyping process. The LM741 has a slew rate limitation of 0.5 µV/s. This is assumed to be sufficient for this application but will need to be tested. The required gain is unknown at this time but can easily be determined during the Preliminary Prototyping process as well. Once the required gain has been determined, values for R_F and R_S will be calculated by (13.1).

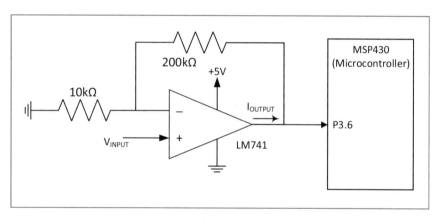

Figure 13.1 Non-inverting OpAmp

$$A = 1 + \frac{R_F}{R_S} \tag{13.1}$$

The *design* description expands upon the basic *ideation* of the solution by considering all the important parameters of any OpAmp circuit. The power rails were defined and the two most common non-idealities (Slew Rate and Output Current Limits) were addressed. The output current requirements were reasonably estimated using a logical argument that can be backed up by the personal experience of anyone who has taken a microcontrollers course. An assumption was also made about the slew rate performance. This is acceptable at this stage as long as the assumption is recognized and documented. The resistor values are also unknown, but an equation is provided to prove they can easily be calculated once the required gain has been determined.

Design Documentation

The Finalized (or nearly-Finalized) design should be documented using a short description and a detailed schematic or figure. The figures should be annotated with component values. A table of exact part numbers and purchasing information should be included as well. Any equations or theory used to determine parameters should be included. At minimum, this should be documented in the Engineering Notebook. It will also be presented in multiple presentations and reports in upcoming Design Life-Cycle processes. So, it is worth producing a high-quality figure and formatted equations in the Engineering Notebook now.

The type of figure used to document a particular subsystem should be chosen to effectively communicate the details of the design to all stakeholders, not just the ones who are intimately familiar with the technology involved in the design; remember to consider your entire audience. The most common types of schematics used for the output of this particular process are:

- Circuit diagram
- Three-dimensional structural diagrams
- Free-body diagrams
- Flowcharts

Avoiding Common Pit Falls 13.2: Circuit Diagrams

It is common to interchange the term *Pin Diagram* for *Circuit Diagram*, particularly if a design contains multiple Integrated Circuits (IC). Both types of figures are important but have different purposes. The correct method to document a circuit at this stage is a **Circuit Diagram**. Pin Diagrams are used for circuit fabrication.

The following Fig. 13.2 is a Pin Diagram. This is a necessary step in PCB manufacturing. This step depicts exactly how specific pins of multiple chips are connected. It is common to occasionally draw an arrow leaving a particular pin and label the pin number that arrow should connect to rather than draw the complete connection. This technique is used on the Pin Diagram below on pin 6. The arrow pointing toward U7.36 indicates that pin 6 of IC U1 should connect to pin 36 of IC U7. Typically, passive components, such as resistors and capacitors, will be labeled with part numbers rather than component values. The ICs are often numbered without a part number. Without prior knowledge of the part numbers, component values, and pin layouts, this figure does not convey much information to a general engineering audience.

Avoiding Common Pit Falls 13.2, continued

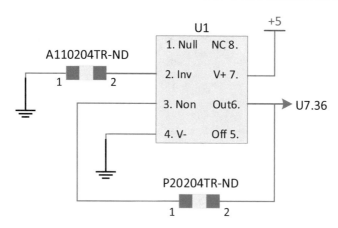

Figure 13.2 Pin diagram example

This next Fig. 13.3 is a Circuit Diagram; it uses standard symbols for basic circuit components whenever possible. Most engineers, particularly in electrical engineering, would quickly recognize this circuit as a non-inverting amplifier and could mentally apply the appropriate Eq. (13.1) to determine that the gain would be

$$A = 1 + 200/10 = 21V/V$$

The LM741 is a rectangular IC with eight pins, not a triangle with 5 connections. However, to best communicate the design of this circuit, the standard OpAmp symbol (the triangle) and common formatting (the non-inverting topology with a feedback resistor) should be used in a presentation discussion.

Conversely, the MSP430FR2355 is a 48 pin IC with millions of transistors. There is no possible way to conveniently represent the complete circuit details on a single presentation slide, and doing so would not improve an audience's understanding of YOUR design. In this case, the rectangular block can be used to represent the complicated IC. However, this is still not meant to be a physical representation of the circuit. The MSP430FR2355 is a *square* package and pin P3.6 is actually located on the *right* side. But, just like the Block Diagram, it is preferred to present the inputs on the left and outputs on the right. So, the physical location of pin P3.6 is ignored, and instead it is placed on the left side of the MSP430's block.

Moving the location of P3.6 on a Pin Diagram would result in a PCB that was incorrectly fabricated. Moving the location of P3.6 on a Circuit Diagram improves the ability of the audience to understand the design. Generally, you will need to produce *both* types of figures during the design process and use the appropriate figure for its intended purpose.

Avoiding Common Pit Falls 13.2, continued

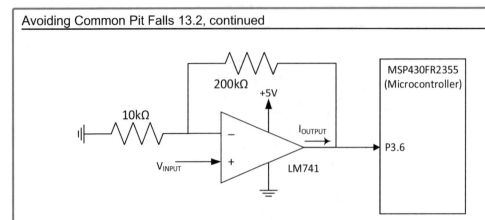

Figure 13.3 Circuit Diagram Example

Students occasionally question whether there are significant communication differences between the Pin Diagram and Circuit Diagram. Consider for a moment, how long it took *you* to determine that the Circuit Diagram was an OpAmp. If necessary, would you have been able to determine the gain to be 21 V/V? Now consider, how long did it take you to realize that the Pin Diagram shown above is the exact same circuit as depicted in this Circuit Diagram? Even if you annotated the values of the ICs used, it is doubtful that a general engineering audience would recognize the circuit and equation as quickly.

Along with the specific design details for each Critical Subsystem, the output of the Preliminary Design Process includes documentation of how any remaining issues will be resolved during the upcoming Preliminary Prototyping process.

During the Preliminary Prototyping process, a small portion of each Critical Subsystem's design will be built and analyzed. The primary purpose of doing this is to investigate any remaining issues that will assist you in finalizing the design. At this point, all that is necessary is to identify what parts of the design will be prototyped and to document why that prototype will help complete the final system design.

To be clear, it is not necessary to build and test the *entire* subsystem at this point. It is doubtful that there is enough time left in the PLANNING phase to build *everything*. Instead, focus on what small piece needs to be prototyped before the final paper design is completed. Section 14.4: will assist you in this process.

Industry Point of View 13.1: Systems Engineering

I am employed as a Systems Engineer at a large aerospace design firm located in Minneapolis, MN. Systems Engineering is an exciting career field that I was completely unaware of in college. So, I enjoy using my membership in Society of Women Engineers (SWE) and Girls in Engineering, Math, and Science (GEMS) to promote this career field as often as I can.

Industry Point of View 13.1, continued

System Engineering is the art of adding value to a product by optimally combining the necessary components and subsystems. My design team consists of Mechanical, Aeronautical, Software, and Electrical engineers. Together, we define a project's Requirements. Once the system has been defined, each of my engineers become the Project Manager of a team responsible for a particular subsystem. Part of the reason I love this career is that while we work with cutting-edge technologies, we also build strong interpersonal relationships. We ensure project success by maintaining overlapping responsibilities – each member of my team is primarily responsible for a particular subsystem but can also step into a management role for at least one other team – the team responsible for designing the subsystem most closely related to their primary responsibility.

We assign Critical Subsystems to our most experienced team members. To define a Critical Subsystem, we perform a detailed statistical analysis. Any system with a potential for failure higher than 12% is defined as Critical, and any system with a 6% failure probability is defined as near-Critical.

During our Preliminary Design process, each Critical Subsystem is designed as thoroughly as possible. Then, a Senior Engineer on a different design team is assigned to validate all design calculations and decisions. A team must have carefully documented all of this or the Senior Engineer will simply reject the design because I cannot afford to pay my Senior Engineers to redevelop or decipher poorly documented designs. That is my Junior Engineers' responsibility.

Once all the designs have been completed, we hold individual Preliminary Design Reviews for each subsystem. My entire team, along with the technical leads of each subsystem, must be in attendance for each review.

After all the PDRs have been completed, this large group meets to review all the Preliminary Prototype Proposals and determine how to best spend our limited PLANNING phase budget.

13.3 Preliminary System Designs

Once all the Critical Subsystems have been Finalized, or nearly-Finalized, the Preliminary *System* Design must be developed. This includes, at minimum, the ideation of any remaining subsystems and/or any required interconnections between subsystems. Of course, any effort you put into developing these ideations into designs now will pay large dividends later in the project.

Before moving onto the Preliminary Prototyping process, a Preliminary Design Review (Sect. 13.4) will be held. During this event, the Preliminary Design will be presented to the stakeholders. This will include the details of the nearly-Finalized Critical Subsystem Designs, the ideation of the complete System Design, and a plan for how the Preliminary Prototyping process will assist in the finalization of that completed System Design. It is important that the ideation of the complete System Design is as detailed as possible by the time it is presented at this review.

13.4 Chapter Summary

In this chapter: you learned about the Preliminary Design Process. This included identifying subsystems, choosing Critical Subsystems, and producing the Preliminary Design. Here are the most important take-aways from the chapter:

- A subsystem is a set of components that produces a single feature. How precisely a team defines "single feature" depends on the complexity of that portion of the design.
- A Critical Subsystem is a subsystem which contains a Critical Component. These subsystems should be designed first.
- While the ultimate goal of the Preliminary Design process is to produce Finalized Designs for each Critical Subsystem and the Ideation of the Complete System, it is acceptable at this stage to have produced nearly-Finalized Critical Subsystem Designs.
- Nearly-Finalized Designs include as much design detail about the subsystem as possible, an explanation of what remains to be designed, and a plan as to how that issue(s) will be resolved. This is dramatically different than a subsystem ideation.

13.5 Case Studies

Let's look at how our case study teams divided their project into subsystems.

Augmented Reality Sandbox

Figure 13.4 shows the AR Sandbox team's Block Diagram for purposes of discussing the project's Critical Subsystems. The Critical Subsystems were defined as:

1. Structure: This Critical Subsystem included two Critical Components – the base structure and the sand. The Structure Subsystem also included the mounting solution for the Kinect sensor and the projector.
2. AR software: This Critical Subsystem was responsible for detecting sand elevations and then converting that information into a projectable image. The subsystem was composed of the Projector (a Critical Component), the AR Code, and the Kinect sensor. It also had to interface with the PC which was defined as another Critical Component.
3. Crop Yield Software: This Critical Subsystem used the predictive Crop Yield Algorithm to produce the appropriate Corn Model. This subsystem also was responsible for overlaying the corn model on the topographical image. The subsystem included the Corn Model Critical Component and had to interface with the PC.
4. Control Panel: This subsystem did not include a Critical Component since all the Critical Components were contained in the previous three subsystems. The Control Panel included user interface buttons and the communications with the PC.

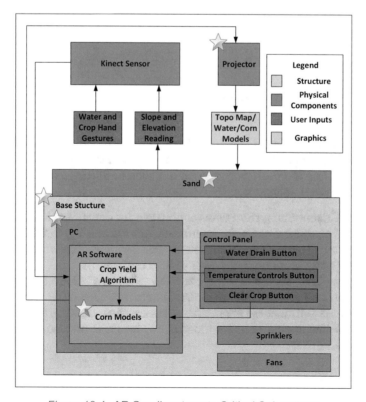

Figure 13.4 AR Sandbox team's Critical Subsystems

- What went well: The students astutely recognized that while selecting the best type of sand and projector to display the topographical image were critical decisions to make during the Alternatives Selection process, there was no further design work associated with those components during the Preliminary Design process. Wrapping those components into Critical Subsystems along with other Critical Components was a good idea. However, doing so meant that the team needed to designate an additional subsystem as "critical" so that each student was assigned one Critical Subsystem. In this case, while the Control Panel did not include a Critical Component, the team decided that this subsystem was the next-most-important subsystem to design and therefore designated it as a Critical Subsystem.
- What could have been improved #1: The team only identified four Critical Subsystems for the five team members. Two students were assigned responsibilities within the AR Software subsystem. This created confusion and problems with both the software development and the academic course structure. The two students were able to navigate these issues and ultimately produced a high-quality result, but they would have had fewer difficulties had the AR Software subsystem be further defined into two smaller subsystems. The AR Software subsystem was responsible for collecting information *and* displaying the image. A subsystem should be responsible for *one* feature only. Perhaps in this case, the team could have defined a Data Collection Software subsystem and an Image Projection subsystem that interfaced together to complete the AR Software's responsibilities.
- What could have been improved #2: We have already mentioned that the decision to designate the PC as a Critical Subsystem was not the best choice (Sect. 12.5). Here we can see another reason why. Normally, a Critical Component should be encompassed within a Critical Subsystem. Occasionally, a Critical Component may interface between two subsystems. Unfortunately, in this case the PC actually encompassed both the AR Software and Crop Yield Software subsystems. By definition, a *component* cannot encompass a *system*. Designating the PC as a Critical Component continued to plague the AR Sandbox team throughout the Preliminary Design process.

Smart Flow-rate Valve

Figure 13.5 shows the Smart Flow-rate Valve team's Block Diagram for purposes of discussing the project's Critical Subsystems. The author has taken the liberty to rearrange the diagram and coordinate the color of each subsystem as discussed in Sect. 12.5. The Critical Subsystems were defined as:

1. Housing: This Critical Subsystem was responsible for containing the entire project. The subsystem was also responsible for interfacing into the existing farmer's sprayer.
2. Power Converter: This Critical Subsystem was responsible for converting the input voltage to 12 V and delivering that power to the components.
3. Control System: This subsystem included the *"Controls"* block on the PCB and the process-control code (i.e., algorithm) located in the *MATLAB GUI*.
4. Valve: This subsystem included the motor, valve, and encoder. The subsystem received a command from the analog controls located on the PCB and adjusted the flow valve accordingly.

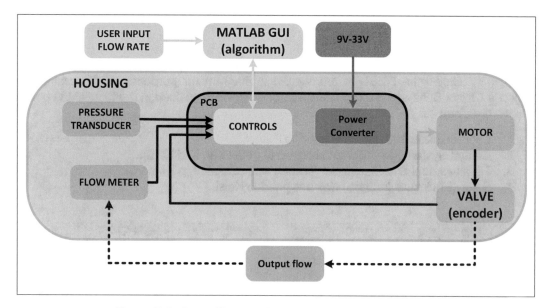

Figure 13.5 Smart Flow-rate Valve Team's Critical Subsystems

- What went well #1: Gannon and Mason continued to work well together on the Housing subsystem. Each took responsibility for a different aspect of the subsystem and produced independent work. This is a rare case of how two students worked on the same subsystem and independently contributed to the success of the project. This probably is not a model for most teams to follow.

- What went well #2: Christian and Keith initially were assigned the Valve and Encoder, respectively, as their Critical Components. They each analyzed Alternatives for their individual component and worked together to consider single-package Alternatives which included both features. Ultimately, they chose independent products. During the Preliminary Design process, the team chose to define one of the Critical Subsystems as the "Valve Subsystem" and include both Critical Components (i.e., the Valve and Encoder). The Valve Subsystem was assigned to Christian, the mechanical engineer, since it was expected that the most complicated aspect of this subsystem would be interfacing with the Housing subsystem. This freed Keith to assist Tyler with the Controls Subsystem – a task which could not have been successfully accomplished by one student within the allotted Senior Design schedule.

- What could have been improved: Gannon and Mason truly worked on different aspects of the Housing subsystem. Further subdividing this subsystem would have been an arbitrary decision forced upon them by the Senior Design curriculum; doing so would have *added* to the complexity of the design and caused confusion. But, this is a rare case. Keith and Tyler also worked together their large subsystem, the Controls. However, the Controls subsystem could have easily been divided into two distinct subsystems in a number of ways; one possible division could have been hardware vs. software. In fact, the workload was essentially distributed this way anyway; Keith developed much of the software, and Tyler was responsible for most of the hardware. The criticism being discussed here is not about the actual workload distribution – in fact, the students handled that perfectly well. Rather, the issue here was the fact that the Controls subsystem was not officially subdivided. There was no reason this subsystem could not have been subdivided further and doing so would have reduced the complexity of managing and documenting workloads.

Robotic ESD Testing Apparatus

Figure 13.6 shows the RESDTA team's Block Diagram for purposes of discussing the project's Critical Subsystems. The Critical Subsystems are outlined in the highlighted boxes and were defined as:

1. User Interface (red) was assigned to Patrick.
2. Control System (green) was assigned to both Leah and Kaitlin.
3. Tip Movement and Placement (purple) was assigned to Easton.
4. Safety Enclosure (orange) was assigned to Kelsi.

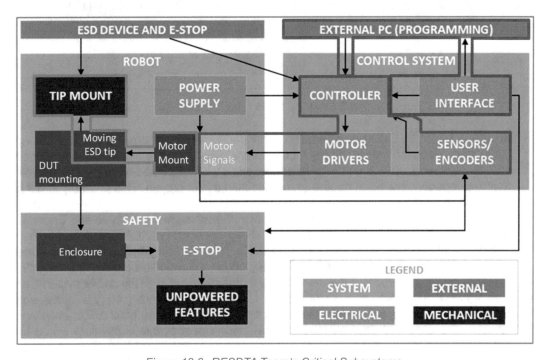

Figure 13.6 RESDTA Team's Critical Subsystems

- **What went well #1:** The team corrected the oversight encountered during the Alternatives Analysis by dividing the MOTORS block into a Mechanical and Electrical component. The Motor Signals were defined as part of the Control subsystem and the physical motors and required mounting systems were considered part of the Tip Movement subsystem.
- **What went well #2:** These subsystems expressly define which *signals* were included with which subsystem. For example, the arrows point to and from the "EXTERNAL PC" block are defined within the User Interface subsystem. Since the USER INTER-FACE block and EXTERNAL PC block are both clearly defined within this subsystem, it is intuitive that the signal between the blocks is also contained within the subsystem. The more interesting definition here is that the signal between the EXTERNAL PC and CONTROLLER blocks is defined to be within the User Interface subsystem.

When a signal is located between two subsystems like this, it is easy to forget to clearly define who is responsible for that signal. In this case, the signal in question is clearly defined as part of the User Interface's responsibility.

- What could have been improved: One of the defined subsystems was the Tip Movement subsystem. This subsystem was composed of the Tip Mount and gears and motors required to move the Tip Mount into place about the DUT. The term "tip mount" referred to mounting the tip of the ESD device (i.e., ESD gun) onto the test apparatus. Originally, the team was considering dismantling the ESD device and only using the "tip" of the gun. However, the team realized that dismantling the ESD device created a safety concern and ended up mounting the entire gun. Remember, the Block Diagram was created during the INITIALIZING phase when the team was focused on the problem, not the solution. The team astutely labeled the Block Diagram focused on the problem of getting the tip of the ESD device (not the ESD device itself) into the proper location. However, when defining the Tip Movement subsystem, they did not encompass the ESD device into the subsystem. Whether the design solution included just the tip or the entire gun, this component should have been considered with the Tip Movement subsystem. When the team finally included the entire ESD gun into the RESDTA design, they realized that the gun was too heavy for the mount they had designed. Fortunately, this was a simple fix, but reprinting the 3D mount caused a time delay that almost caused the team to miss their final milestone. By incorporating the ESD device into the Tip Movement subsystem, they may have avoided this issue.

Chapter 14: **Preliminary Design Review**

The culmination of the Preliminary Design process is an event called the Preliminary Design Review (PDR). At this event, the design team presents their solution to the Problem Statement for the other Stakeholders' approval. At this point in the Design Life-Cycle, the Stakeholders have not risked any significant resources (money). The next process, however, will require the first substantial risk.

> The purpose of the PDR is to verify that all Stakeholder's agree that the preliminary design process has produced a potential solution worth pursuing further.

The purpose of the PDR is similar to the purpose of the Project Kickoff Meeting (Chap. 9). At the PKM, the design team's objective was to convince the Stakeholders to accept a risk. That risk was relatively small with the potential of substantial rewards. The extent of the monetary risk at that point was a few weeks' worth of engineering resources (salaries). This is a relatively insignificant cost. The more concerning risk at the PKM is the "opportunity cost" of pursing the project. The Stakeholders must consider whether the project is a priority above other possible opportunities. So, while it is important to produce quality work at the PKM, the burden of proof is relatively low.

Risks encountered after the PDR are more significant. This process will typically require the purchase of hardware and equipment. This means that the design team will have a higher burden of proof at the PDR than at the PKM. Your priority will be to convince the Decision Makers that you have a valid solution and that the most important aspects of the solution have already been developed, thereby limiting the financial risk of continuing the project.

A well-presented solution will allow Stakeholders to leverage their knowledge and experience to improve the design. In fact, a common outcome of a PDR is a list of improvements to the design that should be considered during the final stages of the PLANNING phase. You should actively seek this assistance! Effectively presenting your design, will allow Stakeholders to provide you with detailed and thorough feedback.

There are three potential outcomes of a PDR. The ideal outcome is that the Stakeholders are convinced that your design has a high likelihood of success and agree to continue project support. Alternatively, if the Stakeholders are not satisfied with your design, they may ask you to repeat an aspect of the Preliminary Design phase. If the Stakeholders are truly unhappy with the design, they may cancel the project outright. This is an extremely rare occurrence, and one that the Design Life-Cycle was designed to avoid.

In this chapter you will learn:

- How to prepare and present a Preliminary Design Review.
- How to accept and address feedback from the review panel.

© The Author(s), under exclusive license to Springer Nature Switzerland AG 2023
C.J. Mettler, *Engineering Design*, https://doi.org/10.1007/978-3-031-23309-8_14

14.1 Formatting a PDR

Just as the Project Kickoff Meeting format can be different depending on the preferences of the primary Decision Makers, so can the exact format and content of a PDR differ slightly from one organization to the next. The exact format is not what is important here. What is important are the fundamental concerns of all PDRs. These include:

* Do you have a realistic solution to the Problem Statement?
* Does this solution directly address the Requirements Document?
* Have the critical aspects of the design been addressed in enough detail so as to reduce the risk of expending more effort and funds?

Figure 14.1 depicts the general flow of a typical PDR. The event starts off with the Problem Statement; this includes the Motivation, Goal, and Objective Statements. This discussion ensures that all Stakeholders understand the problem that is to be addressed by the project. The High-level Design Concept is explained next; typically, the Block Diagram is used here for visualization purposes. Each of the nearly-Finalized Critical Subsystem design are presented in detail. The System-Level ideation is then presented; this includes how the Critical Subsystems interact with each other as well as any remaining subsystems. Finally, an estimated budget is presented.

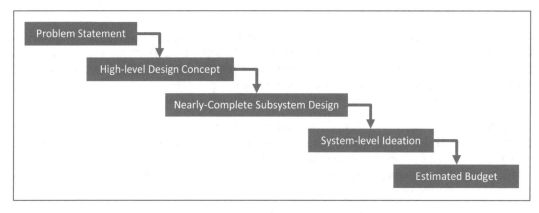

Figure 14.1 Typical topic-flow of a PDR

The most important formatting issue to consider is:

The flow between the Design Concept, Subsystem Designs, and System Ideation should flow in such a way that someone with very little experience with the project will easily understand each topic as it is presented.

14.2 A PDR Template

This section describes one method of presenting a Preliminary Design Review (PDR) effectively. Keep in mind that the purpose of performing a PDR is to convince the reviewers that you have developed a solution that will successfully address the Problem Statement and fulfill the Requirements Document, and that you have made significant progress in reducing the project risk. The topics this method will cover are:

- Problem Statement
- Necessary Background
- Project Design Concept (including the Block Diagram)
- Critical Subsystem Design Solutions
- Preliminary Prototype Proposals
- System-Level Ideation
- Estimated Budget

Assuming that the primary Decision Makers are somewhat familiar with the project at this point is reasonable. However, you should not expect them to remember any of the *details* regarding the project; you can expect that they know who *you* are and can be quickly brought back up-to-speed. Therefore, the PDR template presented here does not start with an "Introduction of the *team*" as the Kickoff Meeting did. Instead, the PDR begins with an "Introduction of the *project*." This is accomplished by using the Problem Statement.

You may wish to state your team's name and welcome the audience to "your Preliminary Design Review" on the cover page of the slide deck, but this should require no more than a few seconds. The real introduction comes on the next slide.

Problem Statement

Problem Statement

Problem Motivation:
[text]

Project Description:
[this is the goal and objectives...you can use a bulleted list for the obj to reduce text]

2

Start your PDR with the Problem Statement. This should be essentially a polished version of the same introduction that you gave at the PKM and that you wrote in the Charter.

Select a team member to present this who is very comfortable presenting. This needs to be delivered in an enthusiastic, confident manner, without using notes or reading from the slide. When you convey confidence and excitement about the project, the Stakeholders are more likely to have similar feelings and be more apt to support the project. Have confidence when you present this slide – you know this material more thoroughly than anyone else in the audience.

Necessary Background

Present any necessary background that the audience requires to fully understand the material. This is not the *entire* background developed for the Charter. It is a subset of the most pertinent information about the project. Focus on ensuring the audience fully understands the Problem Statement.

You may wish to place this slide before the Problem Statement depending on how complicated your Problem Statement is.

Necessary Background

- Focus ONLY on what is necessary to understand THIS presentation.
- MUST include figures!

3

Project Design Concept

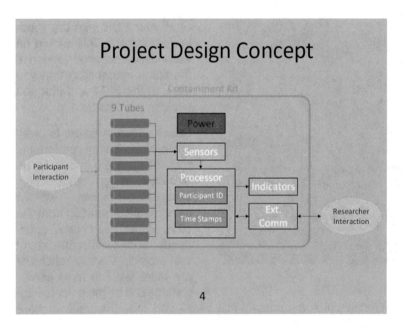

Use the Block Diagram to describe the solution. Although this should be a high-level discussion, it should provide the audience with enough information to understand what the solution is and how it will work. Starting with an "input" and walking the audience through the design until they understand the "output" is often the most effective delivery.

Make sure to highlight the Critical Subsystems on this slide and explain which aspects of the Requirements Document each subsystem addresses. Do not attempt to cover the entire Requirements Document, just refer to the highlights.

Design Solution

Present each of the Critical Sub-system Designs, one at a time. Start the discussion of each sub-system by relating the subsystem back to the Project Design Con-cept (Block Diagram). Reshowing the Block Diagram at the start of each new discussion is a useful technique to ensure the audience correctly remembers where each subsystem fits into the overall solution concept.

This discussion should be highly detailed. Include clear figures and equations so that reviewers are able to understand and analyze

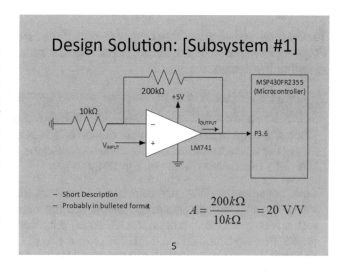

the design. When appropriate, simulate the design and present the simulation results. Dem-onstrate that the results (or at least the *expected* outputs) fulfill the Specifications defined in the Requirements Document.

Although the Design Solution discussion should be detailed, it is not expected that you have produced a Finalized Design yet. It is acceptable that there are details which still need to be addressed. If this is the case, ensure that you are clear about what is complete and what still needs to be finalized.

Preliminary Prototype Proposals

The next process of the PLANNING phase is called the Preliminary Prototyping process. Preliminary Prototyping was mentioned in Sect. 13.2: and will be covered in more detail in Sect. 14.4: That material should be reviewed before presenting the PDR.

After presenting the nearly-Finalized design solution for a particular Critical Sub-system, you will propose a Preliminary Prototype. In order for the reviewer to approve your proposal, they require a clear understanding of the prototype's benefit.

The Preliminary Prototype Proposal should include an explanation of what remains to be designed and specifically how the prototype will directly assist in completing the design effort.

Subsystem #1 Preliminary Prototype

- Remaining aspects of the design:

- What will be built to address these aspects:

- What data will be collected:

- How will that data be used:

7

It is often tempting to build a portion of the solution, improve it by trial-and-error, and continue to tweak the solution until it fulfills requirements. However, that is the antithesis of *design.*

> Instead, the Preliminary Prototype is used to produce and collect data which can be used to complete the Critical Subsystem design.

To convince Stakeholders that proper design procedures will be used, the Preliminary Design Proposal includes an explanation of what data will be collected and how that data will be used to further the design.

Avoiding Common Pit Falls 14.1: Circuit vs. Pin Diagrams (again)

Choosing a figure to represent your design efforts is a crucial choice. You should carefully select the figure that most clearly communicates the design to the reviewers. To illustrate this, let's reconsider Circuit vs. Pin Diagrams.

Let's consider the difference between a Circuit vs. Pin diagram from a reviewer's point of view. The two circuits presented below are the same circuit. Most electrical engineers would recognize the circuit on the left as an amplifier circuit, while few could immediately realize that the circuit on the right is the same circuit. Of the few who might realize what the right-hand circuit is, essentially no one would know the specifics of a A110204TR-ND (that is the 10kΩ resistor, by the way). So, by presenting the Pin Diagram, you have already introduced some ambiguity into your presentation.

The reason this is important is that the reviewer's job is to critically *check your work*, not just listen to you speak. If they spend the entire discussion interpreting the figure, they do not have time left to verify whether your design is correct.

In fact, while you were reading this, have you noticed the design flaw presented on both of these slides? Can you find it? Which slide made it easier to detect?

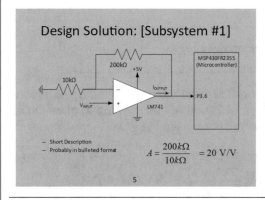

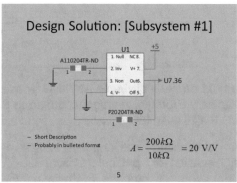

Analysis:

The amplifier circuit is structured in the non-inverting configuration (see the input entering the positive terminal). The equation presented is not correct for a non-inverting configuration. The equation should be:

Avoiding Common Pit Falls 14.1, continued

$$A = \frac{200\ k\Omega}{10\ k\Omega} + 1 = 21\ \text{V/V}$$

Reviewers can, and commonly do, find flaws like this in a PDR. But, only if the design is presented in such a way that the reviewer can perform this level of technical review. It is very unlikely that a review would find that flaw in the Pin Diagram.

System Ideation

Ideation of Complete System Design

- System-level explanation of any and all remaining design concerns.

- This slide should contain a figure (either a more detailed block diagram or a circuit schematic).

- The purpose of this discussion is to convince Stakeholders that all aspects of the required solution have been considered and integrated.

8

After each nearly-Finalized Critical Subsystem has been presented and their respective Preliminary Prototype Proposals have been discussed, the remaining System-Level issues are addressed.

It is *likely* that there are additional subsystems which must be designed which were not initially considered "critical." It is *certain* that at least the interfaces between subsystems must still be designed.

The Stakeholders need to know that you have considered all aspects of the System-Level design, even if you do not have all the details planned out, to demonstrate that the individual subsystems are compatible at a System-Level. Also, any subsystem which was not initially deemed "critical" should at least be ideated at this time.

Estimated Budget

The final aspect of the PDR is to convince Stakeholders that your preliminary design can be accomplished within the provided budget.

Of course, with only a portion of the solution designed, an exact budget is not yet expected. However, since the critical aspects of the system *have* been addressed, you should be able to produce a reasonable estimate.

Estimated Budget

- Show a table

9

If the Critical Subsystems were chosen well, most of the expensive components required for the solution should have been analyzed in the Alternatives Selection process. Additionally, you should have a reasonable idea of many other important aspects of the project. The remaining concerns should be relatively minor and therefore should be relatively inexpensive.

On this slide, you should present a budget (in table format) that clearly indicates what costs are finalized and what costs are estimates. The more items that must be estimated, the larger the risk to the project success. Therefore, it is wise to leave some margin of error in the budget. A good rule-of-thumb is that the "planned" budget at PDR should not exceed 75% of the total budget. Of course, the exact amount of necessary margin-of-error depends greatly on the specific project. The more complete your System-Level design is, the less error you need to account for.

14.3 PDR Feedback

Your presentation should not exceed half of the allotted time of the PDR (check with your Instructor for the exact format). The remainder of the allotted time should be dedicated to a discussion with the review panel. It is this discussion that produces the real value of a PDR. During this time, you have the full attention of a panel of experts (or at least Stakeholders with more experience than the design team typically has). These Stakeholders are in attendance specifically to provide you with suggestions that might help you improve your outcomes and advice that might help you avoid problems. You might as well make good use of that time!

This section will help you make the most of that feedback. There are at least four broad categories that your feedback will fall into, each will be discussed below.

Before discussing the specific types of feedback, it should be mentioned that a "Review" is just another name for a formal "Meeting."

> This means there should be a notebook entry related to the PDR.

Carefully record the feedback you receive during the Review/Meeting in a Meeting Minutes entry so that you will remember to address ALL of it in the coming processes.

Presentation-Quality Critiques

Criticism about your presentation is the one type of PDR criticism that should be avoided. Any time spent discussing the *presentation* is time NOT spent discussing the *project*. The best strategy to avoid these issues is to prepare the PDR in advance, practice, and seek informal feedback before the real event.

Your Project Advisor is a valuable resource. Not only will most faculty members have experience presenting technical topics, but this is also a chance to get their support on the design details before the graded event. Part of the value of a PDR is that you present the entire project in one cohesive presentation. Even the most involved Advisors will be unable to see *every* possible issue with a project during the weekly advisor meetings. By practicing the PDR, you will give your Advisor the "full picture" and perhaps preemptively catch some of the issues which may have otherwise been uncovered during PDR. Therefore, it is strongly suggested that you use the advisor meeting immediately preceding the scheduled PDR to run through the PDR in its entirety.

Your classmates are another valuable resource. Often, when you become intimately familiar with a topic, you fall into the trap of thinking because "it is obvious to you; it is obvious to everyone." Even your Advisor may miss those types of shortcomings since they discuss the project with you on a regular basis. On the other hand, your classmates understand the expectations of the class but are not as familiar with your project as are you and your Advisor. Their ability to understand your project is a good indication of how well your Instructor and Sponsor will understand it. Of course, to be of much value, your classmate must be willing to *objectively* critique your PDR, but critiquing friends can be challenging. It is tempting to "go soft" on the critique. But, a *harsh* review is more valuable to the presenters at this time. Learn to push each other toward excellence during these help sessions.

Eliminating all comments related to the presentation is extremely difficult. Each reviewer has their own personal preferences, and it takes a bit of experience to *perfectly* format a presentation for a particular reviewer. So, a few comments related to the presentation should not be considered "negative." Just take a few notes and incorporate that reviewer's requests in the next presentation.

Clarifying Questions

Often reviewers will start the PDR discussion by requesting that you clarify a topic that you just presented. This request, in a way, is providing you feedback. The request suggests that there is some room for improvement in the way you presented that topic. Often, the most effective way of improving that part of your presentation is to use a better figure in future presentations or documentations.

Answering a clarifying question occasionally requires a thoughtful answer. Sometimes the response the reviewer requires in not a direct answer to the question they have asked. This occurs when there is a fundamental misunderstanding between the speaker and the reviewer. Consider for a moment: Have you ever asked a professor a question and then received an answer more in-depth than you were expecting. The professor recognized that you may have had a deeper issue that was preventing you from understanding the specific topic that you inquired about. This happens all the time to reviewers. So, before answering a clarifying question, pause and consider *why* the reviewer is confused about this topic. Then, articulate an answer that addressed the expressed and hidden misunderstanding.

This issue can be very difficult to detect in the moment. One key indicator that a deeper misunderstanding is present is when you attempt to answer the surface-level question, and the reviewer still seems confused. Perhaps the reviewer asks a follow-up question that indicates they did not understand your answer. Occasionally, students may feel that the reviewer is "just being difficult," but that is almost never the case. What is really happening is that the reviewer and the presenter are "speaking around each other."

Sometimes, team members who are not directly involved with the conversation may have an easier time detecting this issue than the speaker and/or reviewer. A supportive team member will actively listen to the conversation and perhaps interrupt the discussion if they see indications of this type of miscommunication. Example 14.1 provides an example of this type of miscommunication.

Technical Improvements

Ideally, most of the PDR discussion will revolve around technical improvements that could be made to the project. It is useful to think of technical issues in terms of "levels."

Example 14.1: Clarifying Questions

A 2021 Senior Design team sponsored by Schweitzer Engineering Laboratories faced this issue during their PDR. In their Problem Statement, they explained:

The Montana Smart Energy Solutions Laboratory (MSESL) monitors the power grid that feeds the engineering facilites at Montana State University. Unfortunately, when a power outage occurs, the MSESL equipment also looses power, and therefore is unable to monitor the power outage. To improve the ability to monitor potential power outages a PV+battery microgrid would be designed to carry the MSESL load.

The team then explained that their objectives were:

1. *Collect energy using PV.*
2. *Storage energy in a battery bank.*
3. *Design a switching system to switch between grid-tied power and battery backup power when required.*
4. *Install a power monitoring system (called a PMU) in a SECONDARY facility.*

The reviewer asked a clarification question: What is a PMU?

The students gave a perfect technical answer: *A PMU is a device that measures a quantity called a "phasors" Phasors provide information on the magnitude and phase of an AC signal.*

The reviewer asked a follow-up question: Why would you want this information?

The students gave another accurate technical answer; they even had a diagram prepared that explained all the benefits of phasor measurements.

The reviewer was still confused: Why would *this team* need to know *that* information at a *secondary* facility? How does that relate to carrying the MSESL load?

The students were also confused, and perhaps a bit frustrated. They had just provided the reviewer with perfectly accurate technical information.

Example 14.1, continued

Analysis:

The reviewer was trying to understand how the functionality of the PMU was going to support THIS project so as to understand why installing the PMU was an Objective. The students were answering "what is a PMU" not "why is this an Objective."

The cause for confusion stemmed from the fact the PMU was never meant to directly support the PV+battery microgrid! The installation objective was a separate task the sponsors required in exchange for financially supporting the microgrid project.

Eventually, one of the students recognized that when the reviewer was asking "what does the PMU do" they were really asking "what does the PMU do to support the project Goal," and provided the answer "nothing, it is the cost of doing business."

To avoid this confusion for your own projects, keep in mind two concepts:

1. *When a reviewer asks multiple follow-up questions, you are probably not addressing the real issue.*
2. *All questions should be considered in the context "for this project goal."*

Fundamental course work introduces students to "first-level" concepts. Technical elective courses begin developing a deeper understanding of technical concepts, sometime these are called "second-level" concepts. Project-based courses, like Senior Design, introduce students to even deeper level issues; these tend to arise in system-level designs when multiple concepts must work together to achieve the ultimate goal. Senior Design, in particular, may even push you further that than when project concerns include non-technical issues like marketing, reliability, reusability, or aesthetics.

As a senior engineering student, you are expected to understand most of the first-level concepts and have the ability to research and learn second-level concepts without much direct support from your Advisors. Additionally, you are expected to *recognize the need* to consider even deeper level concerns, but not necessarily know exactly what that means for your specific projects. That is where the Advisors support the project.

Example 14.2 provides examples of different levels of concerns for an amplifier design.

Example 14.2: "Deeper Level" Concerns

Let us put the term "deeper level" concerns into context by considering a simple amplifier design.

Examples of *first-level* concerns:

- In most Introduction to Electronics courses, students learn that an amplifier can be designed using prepackaged OpAmps.
- The designer must select a non-inverting vs. inverting configuration and, in either case, apply the correct equations to select the feedback resistor.

Example 14.2, continued

Examples of second-level concerns:

- An OpAmp's power configuration (rail-to-rail vs. rail-to-ground) is an important consideration that is only briefly mentioned in your first Electronics course.
- When implementing a OpAmp, the designer must consider the OpAmp's non-idealities. These might include voltage clipping and output current limitations. These issues also will only be briefly mentioned in an introduction course.

Examples of deeper level concerns:

- An OpAmp's slew rate may affect the potential sampling rate (data resolution) at the system level.
- The tolerance of the feedback resistor may affect the accuracy of the data measurements.
- The OpAmp's packaging profile may affect a PCB layout.

Electrical Engineering students were most likely required to complete an Electronics course prior to enrolling in Senior Design. Therefore, you would be expected to know how to choose the correct configuration and calculate gain for this amplifier design.

You would also be expected to remember to consider the power supplied and basic non-idealities. You might not have performed this task perfectly, but you are still doing just fine assuming you at least attempted to address those issues.

You would not be expected to recognize the resistor tolerance issues at PDR. If that was an issue, your reviewer's job is to remind you about those types of issues. You would be expected to investigate that topic in the next design phase and have an answer prior to CDR.

When you receive feedback relating to technical improvements during the PDR, first consider what level of concern the feedback is addressing.

- If the feedback primarily revolves around first- and second-level issues, that is an indication that you may need to dedicate more effort to your project. However, it is rare that the feedback revolves around these levels for teams that have dedicated a reasonable amount of effort to the project thus far.
- If the feedback primarily revolves around deeper level issues, you are doing a great job – no matter how many critiques and criticism you receive! The *purpose* of the PDR is to get as much useful input as possible; the more the better! Reviewers are only able to focus on deeper level issues if:
  - You have presented well.
  - You have met the expectations of addressing first-level issues.

  Occasionally, when a team receives a lot of criticism they feel "defeated" or even "attacked." Please do not allow yourself to feel this way! The fact that you received a lot of feedback on deep-level technical issues means *you are doing a great job* and that the reviewers are *highly invested and personally interested in helping you!*

When you received feedback on Technical Improvements and have correctly characterized which level of improvement you are dealing with, you can begin to take the appropriate actions to improve your project results. Here are a few suggestions:

- First- and second-level issues:
    - It is likely that you need to spend more time on the project design effort.
    - Review your design details with your teammates. Typically, verbalizing your ideas or plans is the best way to identify shortcomings in your work.
    - Ask for help! Your Instructor and Advisor are there to help.
- Deeper level issues:
    - Train yourself to think in stages. First consider, "is my design going to work on its own?" Then ask, "what happens when it is connected to the 'next' subsystem?" and "how does this affect the system's results?"
    - Train yourself to think about the User's or Sponsor's point of view. *How will your design be used by someone else?*
    - Present the work to an Advisor and specifically ask for a technical review. This may need to be accomplished outside normal advisor meetings.

Big Picture Considerations

Another type of common feedback at PDR relates to "big picture" considerations. Recall that the PDR is *primarily* focused on the technical details of your Critical Subsystems. However, the System-Level Ideation discussion is also important. You are not expected to have the entire system designed at this stage, but you cannot present a set of mutually exclusive Critical Subsystem that are unable to be integrated into a system.

Regular team discussions regarding "integrating" subsystems are necessary to avoid producing subsystem designs that cannot be integrated during the upcoming processes. You have already experienced this during the Alternatives Selection process. You may recall that Alternatives Selection process was conducted as follows:

- Teams defined the Critical Components of their designs.
- Individuals researched and analyzed Alternatives for a specific Critical Component.
- Teams integrated the individual Alternatives into a final set of selections.

Many of the integration concerns are *deeper level* concerns as discussed above. You are not necessarily expected to have identified *every* integration issue, but you are expected to have considered how the Critical Subsystems will integrate to some extent. By the time the PLANNING phase concludes, you will need to have addressed this integration in detail. The more you can address now, the easier completing the PLANNING phase will be.

If you are asked a "big picture" question that you have previously considered,

Answer the question by referring to a system-level figure, perhaps the Block Diagram.

If you are asked a "big picture" question that you have not previously considered,

Do not "make up" an answer.

Instead, admit you "have not considered that yet." Then, if appropriate, you could carry on a discussion of potential solutions. In fact, you may wish to ask the reviewer for their opinion or suggestion on the topic rather than provide a "guess" yourself.

Industry Point of View 14.1: Use Evidence to Improve Communication

My first job out of engineering school was with a large company with offices in Milwaukee, WI. I was assigned to work on a project that had previously completed the STARTING and INITIATING phases. The Requirements Document had been approved and I was tasked with designing a minor subsystem during the Preliminary Design process.

Within the first month, I suspected that the technology we were using would not be able to meet the project's Specifications. I addressed my suspicions with my team leader during our weekly team meeting. Unfortunately, I had come to the meeting without any documentation – I assumed the meeting was "just a discussion." So, considering that the Requirements had been developed by a team of Senior Engineers, my team leader assumed the mistake was in my calculations.

Over the next few weeks, I continued to work the problem with no success. The technology simply could not meet the timing specifications defined in the Requirements Document. I again addressed the issue with the team, but this time I brought my timing diagrams and calculations with me. I even spent the night before rehearsing what I would say. When I presented my findings in a professional presentation, the team leader finally understood my concerns.

The team leader called an emergency review with the decision makers, and I again presented my findings. I had never presented to an executive before, so a few of the more experienced engineers coached me to improve my presentation. By the end of the meeting everyone understood the flaw in the Requirements Document and agreed that we would have to revisit the INITIATING phase.

I was included in the efforts to modify the Requirements and got to see firsthand how many cascading effects the single flaw had caused. We performed a comprehensive review to verify that there were no other flaws in the Specifications; there were not. Then, we rewrote the flawed Specification. That change had to then be analyzed to ensure that it had not affected the feasibility of any other item on the Requirements Document; it had. Turns out, many items had to be adjusted to accommodate the correction to the one issue.

That event was a significant step in my engineering career. I finally understood how precise I needed to be when I presented my work. I couldn't just talk about it, I had to prove it. My team leader was very supportive, but the executives were not particularly inclined to believe me at first – doing so meant that they had to disbelieve their senior engineering team's work. It was a hard sell. But, the data doesn't lie. When presented with detailed diagrams and organized calculations, there really was no doubt. I have tried to remember that lesson whenever I have presented since then.

Fortunately, the mistake was caught during the Preliminary Design. We had yet to dedicate many resources to the project. The Requirements Document was corrected and a new design was developed. The product was very successful.

14.4 Chapter Summary

In this chapter, you learned about the Preliminary Design Review. The PDR is a milestone encountered approximately halfway through the PLANNING phase. At PDR, the design developed thus far is presented for a technical review. The purpose of the PDR is to identify flaws or omission in the technical design so that they may be addressed before the end of the PLANNING phase. Here are the most important take-aways from the chapter:

* The PDR should include a top-down discussion of the design. This starts with the Problem Statement include a high-level description of the design concept and provide the details of the Critical Subsystem Designs.
* The Critical Subsystems are not expected to be Finalized just yet; Stakeholders understand that you will now need to spend part of the Project Budget to create prototypes that will assist you in finalizing the designs. Prototype proposals are pitched as part of the nearly-Finalized Design discussions.
* Reading Chap. 15 before proposing your Preliminary prototype is highly advisable.
* The PDR includes a discussion of the System-Level Ideation to convince reviewers that the design will not be limited during future process by mutually exclusive Critical Subsystems design.
* A PDR is NOT analogous to a regular-course "test." The presenters and reviewers *are on the same side*; both have a goal to make the project more successful. A reviewer's job is to "poke holes" in your design to find the flaws and/or omissions early in the Design Life-Cycle so that the design team may address them before significant resources have been allocated.

Chapter 15: **Preliminary Prototyping Process**

We are now entering the Preliminary Prototyping process of the PLANNING phase. During this process, the Critical Subsystem Designs will be finalized and the System-Level Ideation will be advanced into a nearly Finalized Design. An event called the Preliminary Prototype Demonstration (PPD) will eventually be conducted as the culmination of the Preliminary Prototyping process. During the process proceeding the PPD, called the Critical Design process, all remaining details of the PLANNING phase will be addressed to produce a Finalized System Design. An event called the Critical Design Review (CDR) will be conducted to complete the PLANNING phase and allow entrance into the EXECUTING phase.

This chapter is focused on the Preliminary Prototyping process. A Preliminary Prototype is a device or experiment that will be used to gain information or data that can be applied to the Critical Subsystem Designs. The purpose of this process is to gain enough information so that the Critical Subsystem Designs can be finalized.

In this chapter, you will learn:

- The purpose of the Preliminary Prototyping process.
- How to identify a useful Preliminary Prototype.
- How to format a Prototype Proposal.
- How to present a Prototype Demo.

C.J. Mettler, *Engineering Design*, https://doi.org/10.1007/978-3-031-23309-8_15

15.1 Purpose of the Preliminary Prototype

Before discussing the Preliminary Prototyping process, let us review the entire flow of the PLANNING phase. Each box in Fig. 15.1 represents an important design process; each diamond represented a review which must be conducted in order for the project to advance to the next process. The lightly shaded boxes are processes which have already been conducted, and the filled boxes (starting with the Preliminary Prototyping Process box) are upcoming processes.

As you recall, during the Alternatives Selection process, the system's Critical Components were identified. Alternatives for each component was analyzed and finally selected. During the Preliminary Design process, Critical Subsystems were designed around each of the Critical Components. An ideation of how these subsystems would interface and interconnect at the system level was developed. Then, a Preliminary Design Review was conducted. Assuming you have successfully navigated the PDR, it is time to move on.

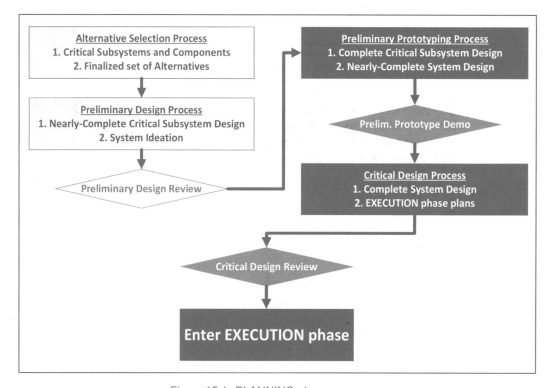

Figure 15.1 PLANNING phase processes

During the Preliminary Design process (Chap. 13), nearly Finalized Critical Subsystems were designed. Recall that a nearly Finalized design is not the same thing as a solution ideation; refer to Example 13.4 for details. It is anticipated that, in many cases, a design cannot be finalized during the Preliminary Design process due to a lack of information about specific components or characterizations about particular signals.

The purpose of the Preliminary Prototype process is to develop missing information required in order to complete the Finalized Critical Subsystem designs.

During the Preliminary Prototyping process, a prototype will be constructed and analyzed for each Critical Subsystem. A prototype of the *entire* subsystem is not necessary at this point of the Design Life-Cycle. Under normal circumstances, the prototype should be constrained to include only the portion of the subsystem that will produce the necessary data required to complete the subsystem design.

Selecting the optimal prototype will require planning before you get started. *This should be conducted before the PDR!* Occasionally, it is reasonable to build the entire subsystem, but often this is too large of a task for the time remaining in the PLANNING phase. Typically, the best prototype is a single, and focused, feature of the subsystem. Infrequently, the best prototype is not even a part of the actual design, but rather it is some sort of testbench designed to produce a desired dataset.

> In general, the specific prototype you choose to build must produce enough information so that you are able to finalize the design of your subsystem.

This is a crucial decision that will affect numerous aspects of the design process. If you are unsure about this selection, seek out advice from your Instructor.

Avoiding Common Pit Falls 15.1: Selecting a Preliminary Prototype

Starting to build your subsystem during the Preliminary Prototyping process can be tempting. The thought of *finally* starting to synthesize your ideas into reality is an exciting prospect. Unfortunately, starting this process too early, before the design is complete, will lead to setbacks and frustration during the EXECUTING phase. Do not rush, we are almost at the end of the PLANNING phase and the appropriate time to start building is just around the corner.

Instead, the focus here should be finalizing the Critical Subsystem *design*. Conduct this mental exercise before selecting your prototype.

1. Identify aspects of your subsystem which are still unknown. If you asked two different manufacturers to build your design, which aspects may not be produced in identical fashion?
2. Identify why those aspects have not already been designed. If possible, address those aspects now and finish the design.
3. Design an experiment or test that will provide the missing information. This should include the physical apparatus required to conduct the test – this is the heart of your prototype.
4. Plan out a schedule to conduct this experiment using your Gantt Chart. Be sure to allot time for procuring components, building the prototype, conducting the experiment, collecting data, and analyzing the results.
5. Finally, consider whether your schedule is reasonable for the allotted time. If you have additional time, you may choose to increase the scope of your prototype. If the timeline seems too aggressive, you will have to find a way to reduce the scope.

15.2 Identifying Useful Prototypes

As previously mentioned, identifying the best prototype requires thought and planning. You want to use your time well. The primary focus must be collecting data to complete the design work, but a secondary effort includes getting a head start on the EXECUTING phase by constructing a portion of the system design. However, do not be tempted to get a head start on a future phase before the current phase's concerns are completely addressed.

Example 15.1 discusses a variety of proposals presented at PDR in 2021. Two of the proposals were clearly intended to verify rigorous design efforts. One of the proposals indicates some basic design decisions that still needs to be made; this is not the purpose of the Preliminary Prototyping process.

Industry Point of View 15.1: Subsystem Prototypes Reduce Risk

Even in mega-corporations, expenditures are carefully tracked. Large expenditures must be justified with proof that there will be a return on investment. When a new concept is proposed, there must be data that shows there is a very high probability that the company will recoup the money spent.

In the early 1990s, as the disk drive industry made significant advances in speeds and storage capacity, the need to protect the read/write system from outside noise became imperative. This had never been done before, but the new storage technologies required all new solutions for nearly every aspect of a drive.

Our lead mechanical engineer, Robert, designed a "shield" that would fit over the read/write head. He presented simulation results that looked promising at a design review. Nobody was positive that the shield would work, but the simulations were convincing, so the team leader decided to move forward by building a prototype.

Now, as the procurement engineer, it was up to me to work with external vendors to source the part. There were many unknowns. Will the material actually protect the read/write head from interference? Is the shape of the shield correct? Will the weight of the shield effect the performance of the arm assembly?

With all the unknowns, we did not commit large sums of money to the design. Instead, I found a company that was willing to partner with us to produce a prototype of the design. When they looked at the spec drawing, they were concerned about the dimensions of the shield wall. They had never made a part using the necessary material with such small dimensions. Thinking that attempting to build such a shield might advance their own technical expertise, they agreed to share some of the risk in manufacturing the part.

After 3 weeks, they delivered a few shields for us to test. The shields worked perfectly for attenuating noise. However, the shields weighed too much. The arm assembly bowed from the weight which caused the reader head to fly too close to the spinning disk. Test results showed we needed to cut the weight of the shield in half.

I went back to the vendor with new specs for the shield. Since the first prototype was partly successfully, they again agreed to share some of the risk in developing the next prototype. When we tested the new shield, it worked perfectly.

I think this is an example of what makes my company so successful. Our mission is to push the limits of technology. But, if we take careless risks, a single mistake could be catastrophic. We have a specific design life cycle, similar to what you are studying now, that is meant to prevent any one mistake from resulting in a huge loss. By simulating before prototyping, and by initially prototyping in small batch, before committing to large manufacturing contracts, we ensure that mistakes are also relatively small and contained.

In this case the risks paid off. The vendor was awarded a lucrative contract and our product was a great success for us. But, if the shield had not proven to be a successful solution, we had salvaged enough of our budget to have attempted at least one more alterative design before putting our project at risk.

Example 15.1: Useful Preliminary Prototypes

This example provides the three prototypes presented by a 2021 Senior Design team at their PDR. This team made two excellent proposals and one that needed work.

Prototype Proposal #1: (Subsystem: CAN communication with Microcontroller)

Predetermined Controller Area Networks (CAN) signals will be sent to the microcontroller and interpreted. The CAN signals will be stepped down from the required 5 to 3.3 V (the microcontroller's bus voltage).
Justification: It is unknown whether the voltage conversion between the CAN signals and the microcontroller will create errors in the data transmissions.

Analysis: Prototype Proposal #1 was an excellent idea. The proposed prototype directly allowed for analysis of a potential problem, and if successfully completed a portion of the system would be produced in advance of the EXECUTING phase.

Prototype Proposal #2: (Subsystem: Data logger)

Find a real-time clock and interface it with a microcontroller. Record data every second for a minute and verify 60 consecutive seconds were recorded
Justification: It is unknown which real-time clock should be used.

Analysis: There are a number of issues with this proposal. The phase: "find" a real-time clock, is an indication of an incomplete design. This is an aspect that should have been addressed during the design phase using datasheets. Also, verifying that 60 "seconds" were recorded is not a rigorous data analysis. Perhaps what the student actually meant was "timing data would be collected to verify that the real-time clock updated every second +/– 10 ms."

Prototype Proposal #3: (Subsystem: Protection Relay)

A signal from a processor will trigger a relay. The relay should disconnect the "output." The time between sending the signal and the output deenergizing will be monitored with a scope. Data will be compared to Spec 4.3.2

Example 15.1, continued

Justification: All design calculations suggest that the timing data will meet spec. However, this is such a critical spec, it needs to be verified before integrating the output subsystem.

Analysis: This is another excellent prototype. It is clear what data will be collected and why. It is clear that the purpose of this experiment is to verify the design work previously performed, as opposed to a trial-and-error experiment.

15.3 Exceptions to the Rule

Occasionally, one or more Critical Subsystems can be completely finalized after the Preliminary Design phase. In those cases, the rules presented for the Preliminary Prototype do not make much sense and therefore can be modified.

First, before assuming that you are able to apply the following rule-expectation, ensure that you have, in fact, completed a Finalized Critical Subsystem design. Consult with your team and verify the entire team is in agreement. Also consult with both your Advisor and Instructor; verify that they too agree you are ready to move on without a Preliminary Prototype.

In industry you would likely, skip the Preliminary Prototyping process for this subsystem and a portion of your team would move on to the next process. Either additional subsystems would be designed or System-Level integration issues would be addressed while you wait for the remainder of your team to "catch-up." However, in academics (where all students should be treated equally), having part of the team move on may not be the best course of action. Many different compromises can be made here. Exactly what this concession looks like is negotiable, but ultimately your course of action should be chosen to provide the greatest benefit to your project. Here are some options to consider:

- Fabricate your Critical Subsystem
  One option would be to fabricate your critical subsystem. This must be completed eventually anyway. It is typically a task assigned during the EXECUTING phase, but, as you learned in Sect. 11.3, one effective Schedule Management technique is to assign future tasks to a team member if and when a team member has a lull in their workload. This allows the team to get ahead in the long run.
  In this case, a prototype will still be built, but the purpose is different. The purpose of the prototype in this exception is to collect data and start the Project Verification process (Chap. 18). You will use that data to verify your system design meets Specification.
- Develop a non-Critical Subsystem
  Rigorous design is not always required for every aspect of a project – particularly in Senior Design. Often, non-Critical Subsystem can be simply "developed." This means that there is no Alternatives Selection process, nor is there a formal Preliminary Design process conducted. Instead, you can simply build a working subsystem for a simple (i.e., non-critical) feature.
  Data can still be collected and used to start the Project Verification process.
  Care should be taken when implementing this option. Although a non-rigorous development process may be acceptable, it is common that a combination of design and development techniques would produce a better result for your project. Discuss these options with an Advisor or Instructor before moving on.

- Design a new portion of the system

  This option could include a paper-design of a subsystem or an integration of two previously designed subsystems. This is probably the most realistic approach for an industry project, but since it is difficult to produce data for this option, it can cause issues in the academic Design Life-Cycle. Perhaps simulation data could be used in lieu of collecting data from a physical prototype. Discuss this option with an Instructor to negotiate the grading details before proceeding.

15.4 Formatting a Prototype Proposal

The Preliminary Prototype Proposal presented at the PDR is...a proposal. This means you must provide the reviewers with enough information to accept, reject, or modify the plan of action. The reviewers will be interested in:

- What information is lacking that currently prevents the completion of the design?
- What do you plan to build that directly addresses that lack of information?
- What data will be collected using the apparatus built?
- How will the data collected be used to improve the design?

Knowing this is what the reviewers are looking for will give you an indication of how best to format your proposal. Recall that you will be presenting this proposal immediately after presenting your nearly Finalized Critical Subsystem design during the PDR. Section 14.2 presents a template for your PDR presentation slides. This section expands on that information specifically for the two slides related to the Prototype Proposal.

The format and the amount of information about the proposals discussed in Example 15.1 require improvement before presenting the proposals at a PDR. Figure 15.2 shows how you might effectively format one of those examples. In these two slides, you can see that each of the reviewer's primary concerns are explicitly addressed.

Notice the effective use of figures to explain the proposal. Rather than simply describe "what will be built," a circuit schematic was provided. Figures are an important part of the proposal to clarify what the prototype will look like. Additionally, when discussing "what data will be collected," a graphical representation of the pertinent data was provided. Of course, this is not *real* data, since the apparatus has not yet been constructed, but rather it is a conceptual representation of what the presenter expects from the data.

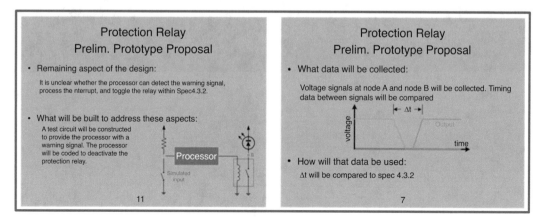

Figure 15.2 Formatting a Preliminary Prototype Proposal

15.5 Template for Prelim: Prototype Demos

At the end of the Preliminary Prototyping process, the prototype and results will be presented to the Stakeholders. This is not necessarily a formal review like the PDR or CDR, but it is still an important milestone. At the Preliminary Prototype Demo (PPD), you must demonstrate that you produced something useful and explain how the design will now be improved. Producing "something useful" hopefully means an "operational feature of the system design." However, a working apparatus that does not produce the expected results also provides useful information. What you must demonstrate at the PPD is that you have built something that provided useful information that will advance your design.

This section describes how to present a Preliminary Prototype Demo (PPD) effectively.

> The purpose of performing a PPD is to demonstrate to Stakeholders how you have continued to make progress on the project solution since the PDR.

The topics this method will cover are as follows:

- Introduction (including the Problem Statement, Background, and Design Concept)
- Prototype Introduction
- Prototype Demonstration
- Complete Subsystem Design Solution
- Nearly Complete System Design

Introduction

The introduction of the PPD includes the Problem Statement, Necessary Background, and Project Design Concept slides. These slides should be essentially the same as what was presented at the PDR (albeit improved if necessary).

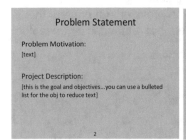
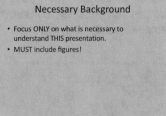
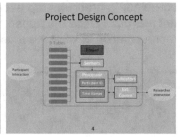

Prototype Introduction

Carefully introduce each proto-type. Remember that the reviewers will not necessarily remember your proposal. Before presenting the prototype itself, explain the specific problem your prototype is meant to address. This should directly relate to the "unknown aspect" of your design.

Using figures that relate back to the Block Diagram and/or your subsystem's nearly Finalized design is a good idea.

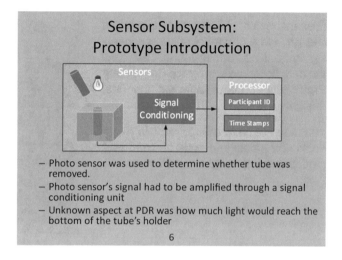

Sensor Subsystem: Prototype Introduction

– Photo sensor was used to determine whether tube was removed.
– Photo sensor's signal had to be amplified through a signal conditioning unit
– Unknown aspect at PDR was how much light would reach the bottom of the tube's holder

6

Prototype Demonstration

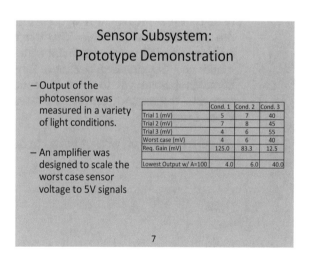

Sensor Subsystem: Prototype Demonstration

– Output of the photosensor was measured in a variety of light conditions.

– An amplifier was designed to scale the worst case sensor voltage to 5V signals

	Cond. 1	Cond. 2	Cond. 3
Trial 1 (mV)	5	7	40
Trial 2 (mV)	7	8	45
Trial 3 (mV)	4	6	55
Worst case (mV)	4	6	40
Req. Gain (mV)	125.0	83.3	12.5
Lowest Output w/ A=100	4.0	6.0	40.0

7

Next, you should discuss the demon-stration itself. Sometimes performing a live demonstration is convenient, but that is not always necessary or feasible. Regardless, the *focus* of this discussion should be on the data and related analysis.

Be clear where the data came from and how it was collected (note: explaining where the data come from was not provided in the example shown here). Be sure to *analyze* and *interpret* the data for the audience.

Finalized Subsystem Design

Present your Finalized Subsystem Design with a communicative engineering figure. Ensure that you discuss how the data collected during the Preliminary Prototyping process affected the final design.

You may not have time to explain every detail of the design. Focus on what was modified or improved since PDR.

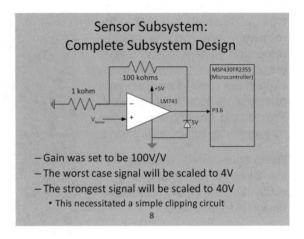

Nearly Finalized System Design

Present your nearly Finalized System design. Present this with figures that are as detailed as possible. This might include a nearly complete circuit schematic *and* 3D drawing package. The items that have yet to be addressed can be referenced on an updated Block Diagram.

Be sure to focus on what *remains* to be designed.

15.6 Chapter Summary

In this chapter and Chap. 12, you learned about the Preliminary Prototyping process. During this process, prototypes are developed for each Critical Subsystem. Here are the most important takeaways from this chapter:

- Care should be taken in choosing the prototype which will most benefit your project.
- Under normal circumstances, the prototype should be used to collect data that can be directly used to advance and complete the Critical Subsystem design.
- If you have already produced a Finalized Critical Subsystem design, consider what replacement efforts will be most beneficial to your project. Discuss these with your Instructor, and possibly your Advisor, before committing to any course of action.
- The Preliminary Prototype (or replacement efforts) are proposed for approval at the PDR. This ensures that all Stakeholders agree that your efforts will be beneficial to the project. Assuming your proposal is accepted, a small portion of the Project Budget may be released to support the prototyping efforts.
- The results of the Preliminary Prototyping process are presented at an event called the Preliminary Prototype Demonstration. If time allows, actually showing a physical demonstration at this event is a nice touch. However, the focus should be on the data collected and the impact to the design efforts. If time or circumstance does not allow for the demonstration, it is acceptable to present and analyze a summary of the data collected.

15.7 Case Studies

Let us look at our case studies Preliminary Prototypes.

Augmented Reality Sandbox

Tucker and Logan worked together to develop a joint Preliminary Prototype. Tucker built a prototype of the AR Sandbox structure, including the Kinect sensor and projector mounts, Fig. 15.3. Logan implemented open-source AR Software to detect the surface of the sand and produce the topographical image, Fig. 15.4.

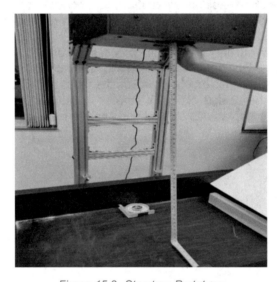

Figure 15.3 Structure Prototype

Figure 15.4 Mapping Prototype

- What went well: This was a rare case when it was acceptable for two students to work so closely together on a Preliminary Prototype. Logan needed the structure to be built in order to test the interaction between the Kinect sensor and projector. Tucker needed an image projected onto the sand in order to analyze the dimensions of the sandbox and the quality of the sand. What is important to remember is that although they worked together to build a complex prototype, each student had well-defined responsibilities and each student independently collected data to investigate different aspects of the prototype.

- What could have been improved: Not much could have been improved on the prototype itself. The students did an excellent job fabricating their Critical Subsystem designs. However, Tucker could have improved the data collection and analysis of his prototyping process. His primary concerns were whether the sandbox would contain the fine-grain sand well enough and whether the sand would reflect the image clearly. In both cases, there was little data collection necessary – the system was simply shown "to work." Rather than discuss this situation with an Advisor, Tucker decided to unofficially support his teammates. But without official communication about the situation, it appeared to his Advisors that he had "checked out" for a couple of weeks. Fortunately, Moe identified this concern early and took steps to better manage the team's workload. Tucker became *formally* involved with the data collection of a different aspect of the project. On the one hand, this is an example of one member of a Performance-team watching out (i.e., caring for) another member. On the other hand, this is an area for individual improvement on Tucker's part. Throughout the project, Tucker consistently helped his teammates "get the job done," but often at the cost of not completing his own course assignments. Tucker should have realized that he was *expected* to collect data from his prototype, and when he realized that no data was *necessary* for this prototype, the correct course of action would have been to communicate this unique situation with his team, including his Advisor. Tucker's desire to help his teammates was admirable and is part of why this team can be characterized as a rare Performance-team. The step that Tucker overlooked was *communicating* his reasonable modification to the process.

Smart Flow Rate Valve

The Smart Flow Rate Valve team's Valve subsystem is a perfect example of a nearly Finalized Design. The design of this subsystem was relatively easy after the components had been selected. All the physical and electrical connections were easily established. Then, the parameters necessary to articulate the DC motor such that the connected valve would turn to the desired position were determined without trouble. The design was essentially complete, but what Christian did not know was "when placed into the Sponsor's system, what flow rate would be achieved given a particular valve position." Since the control output-parameter was flow rate, not valve position, this information was vital to complete the system design.

Interestingly enough, the "missing information" that Christian needed to procure during the Preliminary Prototyping process would not necessarily be used to complete his (i.e., the Valve) subsystem. Rather, this information was needed to characterize the Valve subsystem's output which the Controls subsystem designers required to complete *their* Preliminary Design.

Yet again, this group demonstrated Performance-team characteristics during this process. Although Christian had already produced an essentially Finalized Design, he prototyped his subsystem to gain information that would support his team members. That allowed the Control subsystem designers to prototype a different aspect of the project.

Christian's prototype implemented the selected valve, in this case a ball valve with a flow-optimizing disk – Fig. 15.5 – into a liquid delivery system. Since this was a Preliminary Prototype, the motor was not yet purchased. Instead, Christian manually turned the valve and collected flow rate data at each position, Fig. 15.6. Using this data, a transfer function could be determined that related angular position of the ball valve to flow rate in terms of gallons-per-minute, GPM. Since the transfer function across the DC motor already related input-voltage (to the motor) to angular position of the valve, a complete transfer function that related input-voltage to flow rate across the motor-and-valve was possible.

Figure 15.5 Ball valve with optimizing disk

- What went well: Although the data collected was not needed to complete the Valve subsystem design, the data was necessary to complete the Control subsystem design. In fact, this data was the most important data of the Preliminary Prototyping process for this team. Christian and Keith, although assigned to different subsystems, had continued to work closely throughout their respective Preliminary Design processes to identify this need and determine the best method by which the need could be met.
- What could have been improved: Fig. 15.6 does not indicate how much data was actually collected at each position, how many positions were sampled, nor what the margin of error in the data was. These are important components to any data collection. Section 18.7 discusses data collection and presentation; you may wish to review this section now since the data collection topic applies to the Preliminary Prototyping process as well.

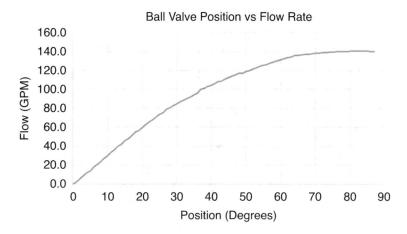

Figure 15.6 *Preliminary Prototype data – ball valve*

Robotic ESD Testing Apparatus

The RESDTA team had a challenging situation to navigate during the Preliminary Prototyping process. All students were required to produce a Preliminary Prototype. However, the estimated cost of building the Structure subsystem's prototype was $13,000, and the customized machining was scheduled to take 6 weeks.

There are times in Senior Design where a student simply cannot produce the needed prototype in the allotted time and within the allotted budget. However, this situation does not negate the need to produce a risk-reduction prototype of some sort. In fact, this situation makes the risk-reduction prototype all the more necessary. The challenge is to identify a useful prototype that can be produced given the time and budget constraints dictated to you.

Easton reasonably argued that an advanced 3D model of the Structure subsystem was the most useful contribution he could make to the project. Most of the time, a 3D model is produced during the Preliminary Design phase. However, due to time constraints, those models are often static, meaning that components which are intended to move in the physical product are not animated to move in the software model. With additional time and effort, the 3D model can be upgraded to an articulating model (i.e., one that has animation) which can then be used for further analysis. Easton proposed to produce an articulating model of the Structure and then to perform a stress analysis on the model to determine parameters such as the frictional force created along the slide rails and the deflection inflected upon the structure while in operation. The amount of friction created on the rails would then be used to determine whether the motors were sized correctly or not. The amount of deflection the system experienced would be used to predict the tolerance of the structure's movement.

Figure 15.7 displays an image of Easton's Preliminary Prototype. Originally, only the back two blue upright structures were part of the design and the mounting arm (which suspended the ESD gun above the DUT, i.e., the dark blue box) was supposed to be mounted as a cantilever without the front blue upright support.

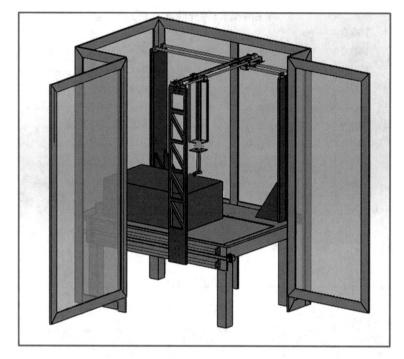

Figure 15.7 Structure subsystem model

Table 15.1 displays comparison data between the displacement on the arm when mounted as a cantilever versus supported using the third blue upright as shown in Fig. 15.7.

Table 15.1 Displacement data comparison

Cantilever beam with force at end		Simply supported, center force	
Force (lb)	Displacement (in)	Force (lb)	Displacement (in)
1	−1.38	1	−0.09
2	−2.75	2	−0.17
3	−4.13	3	−0.26
4	−5.51	4	−0.34
5	−6.88	5	−0.43
6	−8.26	6	−0.52
7	−9.64	7	−0.60
8	−11.01	8	−0.69
9	−12.39	9	−0.77
10	−13.77	10	−0.86

- **What went well:** Easton developed his Preliminary Prototype at no cost – which is often the case when the prototype is a software model. The model generated important data about the structural integrity of the design. That data drove an important design change. Easton updated the design to include the third supporting upright shown in Fig. 15.7. Easton appropriately integrated this change into the system-level design. Most notably, he recognized that the estimated force required to slide the arm along the back rail was now significantly reduced with the arm supported by the newly added upright. This meant that the motors used to articulate the arm along the back rail could be much small (and cheaper) than originally expected.

- What could have been improved: Table 15.1 claims to display "comparison data." However, the reader was left to make the comparison themselves. Displaying the table was the first step when presenting data. However, this is the raw data necessary to make the comparison, not the actual comparison. Results should have been included. Perhaps in this case a Percent Difference column should have been included. Alternatively, this data could have been graphed to present a visual comparison. Next, the results should be analyzed. The analysis of Table 15.1 is straightforward – "the arm is predicted to experience only 6% of the displacement when mounted with a support as compared to mounted as a cantilever." Only now can conclusions, such as changing the design, be drawn. Section 18.7 discusses data presentation in much more detail.

Chapter 16: **Critical Design Process**

The information gained during the Preliminary Prototyping process is used to complete the system design during the last major process of the PLANNING phase. This process is called the Critical Design process and it culminates with the Critical Design Review (0). Recall that during the Preliminary Design process (Chap. 13), a number of Critical Subsystems were designed, although it would have been difficult to finalize the design of many subsystems at that stage. Engineers need time to analyze functionality, interpret data, and investigate interactions to develop a complete understanding of their design challenges before finalizing a design. This enhanced understanding was developed during the Preliminary Prototyping process (Chap. 15). Now, during the Critical Design process, that understanding is applied to the project with a goal of completing the System-Level design.

> During the Critical Design process a complete System-Level design will be produced.

A challenging design will likely require engineers to revisit the Preliminary Design, Prototyping, and Critical Design processes many times before a complete system design is produced. The scope of a Senior Design project is typically limited such that only one, or perhaps two, cycles through these processes are necessary.

In fact, regardless of how many cycles through the Preliminary Design, Prototyping, and Critical Design processes are required, the design will not actually be considered truly "finalized" until the entire Design Life-Cycle has been completed; this includes the EXECUTING phase. Issues identified during the EXECUTING phase often require a design team to revisit the PLANNING phase yet again. However, this would be nearly impossible to implement during a typical Senior Design experience where the PLANNING is conducted during semester one and the EXECUTING is conducted during semester two. The purpose of the PLANNING phase is to reduce the probability of this cycle.

The design produced at the culmination of the PLANNING phase is commonly referred to as the "Finalized System-Level" (or simply the "System-Level") design. This is a bit of a misnomer since we cannot *guarantee* that no additional issues will be uncovered. However, the term is used to indicate that we have done everything we could do "on paper" to avoid encountering an expensive and time-consuming issue after we start fabrication.

> The status of the System-Level design at the end of the Critical Design process should be such that every aspect of the first prototype is ready to be built and tested.

C.J. Mettler, *Engineering Design*, https://doi.org/10.1007/978-3-031-23309-8_16

The Critical Design process actually encompasses four smaller processes to produce the Finalized System-Level Design. These processes include the technical description of the solution, instructions on how to produce that solution, a plan to verify (i.e., validate) the final results, and a risk analysis of what might hinder the successful completion of the design.

In this chapter, you will learn to:

- Identify a complete System-Level design.
- Develop an Execution Plan.
- Develop a Verification Plan.
- Perform a rudimentary Risk Analysis.

16.1 Overview of the Design Life-Cycle

Before discussing the Critical Design process, it will be beneficial to review a few details of the Design Life-Cycle. Figure 16.1 portrays the processes of the PLANNING phase flowing into the processes of the EXECUTING phase. You probably recall that early in the PLANNING phase you selected a number of Critical Components (ovals) and performed a detailed Alternatives Analysis upon them. Preliminary Subsystems (rectangles) were developed around the Critical Components to produce desired functionality. Of course, each subsystem required a variety of additional components (circles) besides the Critical Components.

Recall that the Preliminary Prototyping process (not shown) supported the completion of the Preliminary Design process and enabled the development of the Finalized System design. It is probable that the Finalized System design also required additional components and possibly even additional subsystems (squares).

The Finalized System design will be revealed at the Critical Design Review (CDR). The purpose of this review is to convince Stakeholders that the design is complete and as much of the potential Project Risk has been eliminated or at least reduced as possible. Besides the technical design, there remain two important details to consider during the PLANNING phase that are also discussed at CDR: Can the solution be fabricated, and how will the solution be verified.

Assuming successful completion of the CDR, the majority of the Project Budget will finally be released to the team so that the System-Level prototype can be fabricated. This is a significant risk to the Stakeholders, so the CDR must convince them that the project design was well thought-out and a solid plan is in place before the EXECUTING phase starts!

The System Prototype is produced during the execution of the Fabrication and Verification processes. These processes typically coincide with each other to a large extent. The output of the Fabrication process is the System-Level prototype itself. The output of the Verification process is a dataset which convincingly proves that this prototype has successfully addressed the entire Requirements Document. The prototype and related data are presented at the Final Design Review (FDR). Only upon successful completion of the FDR can the design team truly claim a "Final" Design has been produced.

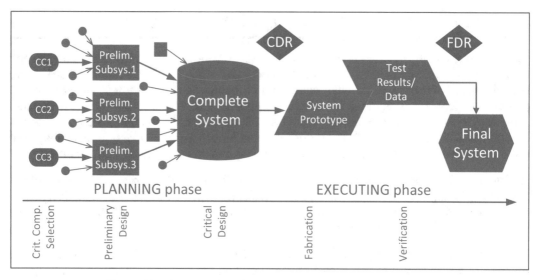

Figure 16.1 Details of the PLANNING and EXECUTING phases

Currently, we are entering the Critical Design process of the PLANNING phase in order to prepare for CDR. During this process, we must produce the Finalized System-Level Design, the Fabrication Plan, and the Verification Plan. The following sections will help you complete these tasks.

16.2 Finalized System-Level Designs

It would be nearly impossible for this textbook to instruct you on how to complete your unique System-Level Designs. The wide range of engineering projects encountered during Senior Design requires an engineering-student to *understand* the Design Life-Cycle concepts (particularly at this stage) and to personally apply those concepts to their individual System-Level Design. This section is not meant to instruct you on *completing* your design, but rather it is meant to assist you in *verifying* that the design is, in fact, complete.

The concepts related to a "Finalized" versus "nearly-Finalized" Subsystem Design discussed in Sect. 13.2 directly apply to the System-Level Design as well. Recall that it is rare for the engineering team to personally fabricate the System-Level Prototype. Typically, the construction is outsourced to machine shops, PCB fabrication houses with pick-and-place capabilities, or tooling specialists. To ensure that these specialists manufacture exactly what is required, 100% of all engineering design work must be completed and documented before requesting someone builds your design.

Unless you have experience designing products similar to the one currently under design, it can be challenging to determine whether all necessary engineering decisions have been made. Here is a checklist that can help you make that determination:

- Has every item on the Requirements Document been addressed?
- Will every Specification-solution directly address their related Requirement?
- Will every Requirement-solution directly solve their related Objective?
- Will every Objective-solution directly solve the Project Goal?

- Are the exact part numbers and suppliers known for *every* component?
- Have the connections between every component been addressed?
- Has the physical containment solution been addressed in enough detail?
- Are there any loose connections (or wires) that need to be bundled?
- Are there any fragile components that need to be protected?
- Are there any dangerous components that need to be shielded?
- Has the integration between all software features been considered?
- Are there ANY remaining design decisions to be made?

One potentially helpful thought-exercise at this stage is to consider: *given the documentation of the Finalized System-Level Design, could a different team of engineers – technically qualified but with no other knowledge of the project – successfully construct the intended solution?* If not, what information or instructions are missing to enable this? Note that in this scenario, the second team of engineers does not know the Project Requirements and so cannot be expected to "figure out" any important details that are missing; they will simply implement what is given in the documentation handed to them. If this hypothetical scenario would realistically have a chance for success, then your design *might* be complete.

16.3 Fabrication and Verification Processes

The majority of the EXECUTING phase simultaneously includes both the Fabrication and Verification processes. These two processes are, to a large extent, conducted simultaneously as shown in Fig. 16.2. The general idea is to build small, test small. This means that we verify the functionality of individual components first. Assuming those components meet Specification, they are assembled into subsystems, and the subsystems are verified to meet Requirements. Once those subsystems are verified, they are integrated together into the complete system where the Objectives and Goal may be verified.

The detailed coordination of the Fabrication and Verification processes requires a thought *plan* before they can be successfully enacted. This is unique in the Design Life-Cycle. So far, we have been able to enter each new process in a linear fashion, one right after the next. However, the Fabrication and Verification processes plans must be developed before entering them to ensure that the team functions efficiently. These plans are prepared during the final stages of the PLANNING phase and presented at CDR.

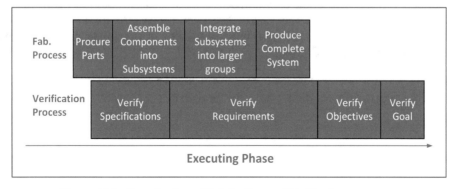

Figure 16.2 Coordination of Fabrication and Verification processes

Besides the increased level of coordination required to navigate the Fabrication and Verification processes, there is another important reason these processes are considered during the PLANNING phase. Just because a concept (i.e., the Design) can be produced on paper does not necessarily mean that idea can be physically realized.

The Fabrication Planning process is conducted during the PLANNING phase to ensure that the design can be physically synthesized. One example where this can become an issue relates to a system's packaging. 3D printers have made prototyping intricate details commonplace; 3D printing a single iteration of a concept is relatively quick and easy; printing 10,000 iterations of the same concept is likely resource-prohibitive. Injection molding is an extremely expensive method of prototyping a single iteration, but once the mold has been produced, printing numerous versions requires essentially no additional cost. So, a design team may 3D print the initial prototype but plan to mass produce the part using injection molding. Unfortunately, it is quite easy to 3D print a concept that has no realistic chance of being replicated by an injection molding process. Some design teams must consider whether the System-Level Design prototype has any chance of being scaled to a production scenario. This is just one example, but the point is this:

> Ensuring that the design can be fabricated, and done so within the allotted Project Schedule, is an essential consideration of the PLANNING phase.

Similarly, the Verification process requires some forethought. Each item on the Requirements Document will need to be verified with test data. This is a time-consuming and intricate process. A detailed plan is required to ensure that all the necessary test equipment will be available at the desired point in the EXECUTING phase timeline.

Another reason for the plan is to coordinate efforts between team members. Perhaps you desire to put a particular subsystem under a verification test at a particular point in time. Simultaneously a different team member desires that the same subsystem should be physically integrated with another subsystem. This might mean that one of the team members must compromise their schedule, it might mean that two identical subsystems should be developed, and it might mean that the two team members need to coordinate their efforts. There is not necessarily a "right" course of action here, but the situation should be accounted for in the "plan."

Additionally, the Verification plan is presented at CDR in order to procure support from all Stakeholders. It is not uncommon that a team intends to conduct a particular test, but another Stakeholder is not convinced that the data collected will actually provide conclusive evidence of the intended results. The rigor, environment, and/or testing methods might need to be modified in order to gain support from all Stakeholders. By procuring that support *before* the tests are conducted, you have yet again reduced the risk to resources and time that you'll expend during the EXECUTING phase.

> Ensuring that the design can be conclusively verified within the allotted Project Schedule is another essential consideration of the PLANNING phase

16.4 Fabrication Planning Process

The purpose of the Fabrication Planning process is to develop a plan for synthesizing the design. Occasionally, the term "fabrication" leads students to only focus on physical hardware construction. However, the term is applied much more broadly here. Fabrication can apply to electronic hardware, physical packaging, and software – or even data structure – development. The point here is "how will you **physically realize, or synthesize,** your design."

The Oxford Dictionary defines synthesize as:

> **Synthesize** – to combine a number of things into a coherent whole.

In other words, the Fabrication Plan essentially describes how you will build or produce each piece of the design and then combine all the pieces together in order to translate the concept into reality. Additionally, the plan describes the optimal timeline of when each stage of the Fabrication process will be conducted and by whom. It is also a good practice to include when each test of the Verification process will be conducted on this timeline. Notice, the Fabrication Plan does not include details of *how* the Verification process will be conducted, only *when* the tests will be conducted. The details of *how* the tests will be conducted will be specified in the Verification Plan (Sect. 16.5).

The best way to document the Execution Plan is to use the Gantt Chart. Example 16.1 presents a Fabrication Plan; this plan is based on the project described in Example 11.1: Alt. Selection and Subsystem Design Gantt Chart Example 11.1.

EXAMPLE 16.1: Execution Plan Using a Gantt Chart

The team introduced in Example 11.1 completed their PLANNING phase of the Hoop-Strain monitor. At their CDR, they presented a System-Level Design along with an integrated Fabrication Plan. This example analyzes a small portion of that plan, Fig. 16.3; notice that the taskbars are all highlighted *purple* to indicate these are EXECUTING phase action items.

In Example 11.1 Scott was assigned to design the strain sensor subsystem while Chris was assigned to the processor subsystem. However, during the Preliminary Prototyping process, Chris supported the strain sensor subsystem design by interfacing with a specialized tooling company to develop the sensor mount while Scott simulated the necessary signal conditioning circuitry.

In their Fabrication Plan we can see that Scott is to begin fabricating the strain sensor subsystem while Chris plans to return to his primary responsibilities related to the processor subsystem. The team allocated responsibilities effectively so that gains were made on both subsystems while the mounting fabrication was outsourced.

A variety of Specification-level tests were incorporated throughout the initial fabrication tasks. These are indicated by the Milestones and specified with the Requirements Document line item, rather than the team member's name. Notice that some tests address more than one Specification. Sometimes this is acceptable, other times it is not, depending upon

---

EXAMPLE 16.1, continued

the specific situation. At this point, it is not clear whether the proposed tests will sufficiently address these specifications. That is why the Verification Plan (Sect. 16.5) is necessary.

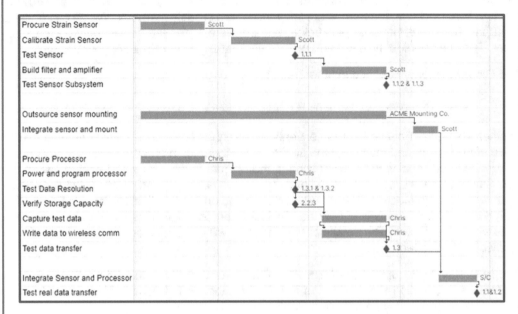

Figure 16.3 Fabrication Plan example

The important thing to note in this example is that Chris and Scott worked together to ensure that the sensor mounting system would arrive in time for Scott to integrate the three systems before the necessary test. Additionally, they ensured that work was continuously produced even while waiting for the tooling company to fabricate the mount. This level of coordination is necessary to complete the EXECUTING phase within the allotted timeframe.

Another look at this team's Fabrication Plan shows that as individual components were assembled into subsystems, a variety of Specification-level tests were conducted. As more items were integrated, it became possible to verify certain Requirements. What is not shown here is that once the entire system was fully integrated into a working prototype, the Objectives were carefully verified as well.

Although a general guideline states that Specifications are tested at the component level and Requirements are tested once multiple Subsystem are integrated, this is not neces-sarily a "rule." Notice that while Chris was still initially developing the processor subsystem, he planned on verifying Requirement 1.3. It is common that a few Requirements can be verified at the initial Subsystem level. It is also not uncommon that a few Specifications are tested after larger subsystems have been produced.

Avoiding Common Pit Falls 16.1: Organized Fabrication Plans

It is important to verify all Requirements Document items *in a logical order.*

It is important that all Specifications under a particular Requirement are verified before attempting to verify the Requirement, and all Requirements under an Objective are verified before validating the Objective.

Notice that in Example 16.1, Requirement 1.3 included two specifications, 1.3.1 and 1.3.2, both of which were scheduled to be completed during the second week of the EXECUTION phase, whereas Req1.3 was schedule in week 3. That is the proper order. However, the team planned to verify Req1.1 and 1.2 during week 4, but there is no documented plan to verify any of Requirement 1.2's Specifications. The team actually proposed that the test for Requirement 1.2 would "cover" all the Specifications under that Requirement. This is rarely an acceptable plan and in fact, caused this team considerable difficulties at the end of their Design Life-Cycle.

We should also point out that the Specifications under Req1.1 were not scheduled to be completely tested before Req1.3 was to be tested – this is acceptable, but all Specifications under Req1.1 must be tested before testing Req1.1. The point is that the rule does not state "ALL Specification be tested before *any* Requirements," just that "the Specifications for a particular Requirement must be previously tested before *that* Requirement."

Formatting Gantt Charts to Document the Execution Plan

As you develop the Fabrication Plan within your Gantt Chart, you will probably realize that this is much more intricate than the scheduling you were required to do thus far. Developing a useful Gantt Chart requires some patience and creativity. Remember, the Gantt Chart is a presentation tool as much as it is a design tool. If your chart is so cluttered that it is difficult to present and/or understand, it is not a very useful document. Take the time to arrange the tasks in such a way that the chart clearly describes the schedule you intend to embark upon.

16.5 Verification Planning Process

The Verification Plan is an important supplement to the PLANNING phase documentation. The Fabrication Plan describes *when* tests will occur, but not *how* they will be performed. The Verification Plan describes each test in detail so that all Stakeholders agree as to whether a successful completion of a particular test will provide convincing data that prove the design was successful. This may seem like a trivial process, but it is far from trivial. Any educator will attest to the challenges of developing *effective* assessment tools, and many students have felt the frustrations of knowing course material more proficiently than test results show. Developing effective assessment tools (i.e., designing a test) requires some forethought and effort. Unlike most of your other course work, during the Verification Planning process, *you* will have significant input on how you (i.e., your project) will be "tested."

Strangely enough, the purpose of the Verification process is NOT to prove the design was successfully completed. Rather, the purpose of the Verification process is to find the limitations and shortcomings of the design. This means that the tests must be rigorous and aggressive.

> The key to developing a proper Verification Plan is to assume there is something wrong with your design and to aggressively look for it.

Unfortunately, the realities of the academic world often restrict this process to some extent. There are tight budget constraints, limited resources, and nonnegotiable timelines that must be accounted for. According to the Triple Point Constraint concept discussed in Sect. 1.2, this may severely limit the "quality" of any endeavor. The trick for Senior Design is to define rigorous enough tests that will convince Stakeholders the project was successfully completed but which can be accomplished within the scope of the project. This section will help you do that.

What needs to be tested?

The short answer to "what needs to be tested" is: "the Requirements Document." This means that every Specification, Requirement, and Objective is carefully and rigorously tested, in order. "In order" does not necessarily mean that every single Specification is tested before any Requirement is tested. Figure 16.4 depicts this concept. The Specifications (ovals) under Requirement 1.1 (rectangle) are laid out in the most common manner: Specification 1.1.2 is performed after 1.1.1 and before 1.1.3, once all the Specifications have been tested, and Requirement 1.1's test is subsequently performed. However, testing does not *have* to be performed in this order. Specifications under Requirement 1.2 are scheduled in the order of 1-3-2, and there was quite a delay between testing 1.2.3 and 1.2.2. In fact, the test for Req1.3 was conducted *before* Spec1.2.2. The Specifications under Requirement 1.3 show yet another possibility. Specification 1.3.2 required that 1.3.1 was performed first, but then 1.3.2 and 1.3.3 could be verified simultaneously (rarely can two Specifications be tested with a *single* test, but often multiple tests can run *concurrently*). The exact order of the Requirements' testing does not matter, as long as all Specification tests have been conducted for a given Requirement before that particular Requirement is tested. All Requirements must be tested before the related Objective (diamond) can be verified.

> It is critical that all Objectives-level tests are conducted on the *complete* system.

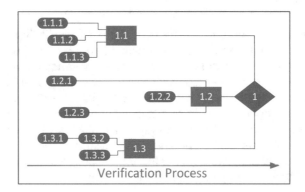

Figure 16.4 Conceptual timeline of the Verification process

Simply validating each Specification does not necessarily translate into a guarantee that the related Requirement was met. Similarly, meeting all the Requirements does not mean that an Objective was fulfilled. Therefore, unique tests must be defined for each item on the Requirements Document. The Objective shown in Fig. 16.4 would require 13 unique tests (9 Specification tests, 3 Requirements tests, and 1 Objective test).

Types of Tests Used in the Verification Process

Test Engineering is a unique and challenging career field that requires drastic amounts of creativity and flexibility. There is no one "right" way to test a Specification or verify a Requirement. Table 16.1 provides a few ideas of how to test each type of Requirements Document item. Although most of the tests should be Rigorous Laboratory Tests or Field Experiments, using Datasheets, Focus Groups, and Sponsor Approval can be effective and inexpensive methods of testing certain features.

Table 16.1 Appropriate tests for Requirements Document items

Req Doc Item	Applicable Types of Tests
Specifications	Datasheet Analysis
	Rigorous Laboratory Testing
Requirements	Rigorous Laboratory Testing
	Field Experiments
Objectives	Field Experiments
	Live Test Cases
	Focus Groups
Goal	Live Test Cases
	Focus Groups
	Sponsor approval

Documenting the Verification Plan

Two methods should simultaneously be used to document the Verification Plan. First, a summary table, Fig. 16.5, is a useful tool that provides a high-level overview. Unfortunately, many test descriptions will be too detailed to include within the table. Therefore, a written narrative that describes the tests in detail is also required.

Figure 16.5 is an example of a Performance Verification Plan summary table. The table is organized similarly to the Requirements Document, except rather than working top-down as you did during the INITIATING and PLANNING phases, the Verification Plan will work bottom-up. When developing the Problem Statement and System Design, the sequence of your focus should have been:

Motivation → Goal → Objectives → Requirements → Specifications.

When verifying the results, the order flips and you work from the small details toward the big picture in this order:

Specifications → Requirements → Objectives → Goal.

The V-model of the Design Life-Cycle shown in Fig. 1.7 effectively depicts this concept. A quick review of that figure would be useful at this time.

Performance Verification Summary {Subsystem X}				
Num	Name	Test description	Pass Criteria	Planned Date
1.1.1	Spec 1	How will this be tested	What is required to pass this test	When
1.1.2	Spec 2	How will this be tested	What is required to pass this test	When
1.1	Requirement 1	How will this be tested	What is required to pass this test	When
1.2.1	Spec 1	How will this be tested	What is required to pass this test	When
1.2.2	Spec 2	How will this be tested	What is required to pass this test	When
1.2	Requirement 2	How will this be tested	What is required to pass this test	When

Figure 16.5 Performance Verification Summary template

Verification Summary Table

The Verification Summary Table, Fig. 16.5, provides an overview of the Performance Verification Plan for one particular subsystem. The **Num** column organizes the table by Requirements Document item number. The **Name** column presents the Requirements Document text; so, the entry "Spec 1" should be replaced with the actual Specification or at least a summary of that item. The **Test Description** column should provide the test description, but often this is too long and detailed to fit into a table-cell; therefore, this entry can be limited to a shortened description or even just a name. The full description will be provided in the supporting narrative – discussed below. The **Pass Criteria** column documents what a successful test requires; again, this may be a summary of what is provided in the narrative. The **Planned Date** column lists the date of the Project Schedule (i.e., Gantt Chart) in order to maintain coordination between the Fabrication and Verification processes.

If every item in the Requirements Document was included in a single Verification Summary Table, the table would become quite large and cumbersome to present. Therefore, the table is often subdivided by subsystem. Since all of our previous and future presentations have started with a design summary (i.e., Block Diagram) and then provided details for individual subsystems, dividing the table by subsystem will format the Verification Plan into a structure that will match the presentation formats.

Recall that the presentation formats typically end with a System-Level discussion. Therefore, when the Verification Summary Table is subdivided by subsystem, the Objectives-Level tests are not included. Rather, a final table is included with all the tests performed at the System-Level, Fig. 16.6. This table should include all the Objectives, as well as any of the Requirements that were not previously tested. In rare cases, this table may also include a limited number of Specifications. However, if the Verification Plan was appropriately developed and coordinated with the Fabrication Plan, there should not typically be any Specifications left to test at this point.

Performance Verification Summary {Project Name - Complete System}				
Num	Name	Test description	Result	Date
1	Objective 1	How was this tested	What is required to pass this test	When
2	Objective 2	How was this tested	What is required to pass this test	When
3.4	Requirement X	How was this tested	What is required to pass this test	When
3	Objective 3	How was this tested	What is required to pass this test	When

Figure 16.6 System-Level Verification Plan summary

Test Descriptions

The amount of detail required for each Test Description will depend upon the specific tests. For example, if a Specification is being verified using a datasheet, the Test Description might simply be "Datasheet Verification." However, *most* of the Test Descriptions should read similar to a lab manual from one of your technical courses and be far too lengthy to document in the summary table. These detailed Test Descriptions should be documented in a written narrative. Each test description should be given a unique name which then can be linked to the summary table's **Test Description** cell.

The written narrative should address the following issues for each rigorous test:

- Name/Req. Doc. Item Number: Label each test description with the Requirements Document item number and a unique name. Both of these should directly match the respective cells in the summary table.
- Setup/Environment: How will the test be set up? Will the circuit be breadboarded? What nodes will be measured? How will it be configured? Will the packaging be 3D printed or simply modeled in software? Is there any test-hardware that must be interfaced with a software-feature under test? If so, is that hardware a simple test case like an LED or the system's actual hardware. Where will the test be conducted? Test step-up descriptions should include a figure.
- Equipment: What equipment will be used? Are you measuring voltage with a cheap multimeter or monitoring it with an Oscilloscope? Will you require an environmental chamber or will naturally occurring conditions suffice? Is the necessary equipment standard lab equipment at your university or will you be using your personal equipment, or perhaps some sort of specialized equipment is necessary? Will scheduling time with the required equipment be challenging for any reason? The test description should address any issues related to the equipment necessary to conduct the planned test.
- Procedure: Briefly describe the step-by-step process that will be applied to conduct the test. Keep in mind that this documentation is not for YOU. Remember, the synthesis of the Critical Design (including the Verification Plan) will often be *outsourced*. In terms of the Verification Plan, the tests are often outsourced to technicians who are experienced with performing similar tests. While they are often more competent than the average engineer when it comes to laboratory procedure, they will need guidance on what to test and how –

they may not be privy to the "big picture" for your specific project – that must be provided by the engineer. The test description should include enough information that an experienced test engineer could complete the test for the engineering team.

- Data Collection: What data will be collected and how will it be analyzed? Will the data be compared across many trials, or against specific standards or Specifications? Will extrapolating or interpolating data be necessary? The test description should explain what data will be collected and how it will be analyzed.
- Repeatability: How many successful trials are required to "prove" the design was successful? Perhaps only one or two trials are required to prove that a particular Specification was met, but many trials may be necessary to demonstrate an overall Objective was met. For example, a single lab test might demonstrate that data is transmitted using the correct formatting, but 1000 tests are required to prove that data will be reliably transmitted in a "real" operating environment with a variety of noise conditions. The test description should document the number of trials of each test that will be required and justify why that number is sufficient.

Pass Criteria

As mentioned already, the **Pass Criteria** is one item documented in the summary table. As you probably have ascertained, the Pass Criteria defines what it means to "pass" the test. What students occasionally overlook is that passing a test should directly relate to the Requirements Document. Pass Criteria are usually easy to define for all the Specifications; however, there are sometimes minor considerations that must be reviewed at this time. Example 16.2 illustrates this issue.

EXAMPLE 16.2: Examples of Pass Criteria for Specifications

In this example, Pass Criteria are defined for the Specifications provided for the tiny home discussed in Example 7.4.

- Req: The tiny home must be towable by a pickup truck
  - Spec: Weight of the tiny home must not exceed 7500 lb
    - Pass Criteria option 1: Weigh less than 7500 lb
    - Pass Criteria option 2: Weigh 7500 lb $\pm$ 50 lb
  - Spec: Width of the tiny home must not exceed 8.5 ft
    - Pass Criteria option 1: Width less than 8.5 ft
    - Pass Criteria option 2: Width 6 inches less than 8.5 ft or less

---

Analysis:

What exactly does it mean to meet a Specification? A well-written Requirements Document should already include a tolerance or required safety factor for each specification. However, occasionally these limits may not be documented or open to some interpretation in the context of a given test. Defining the Pass Criteria for each test, even when it may seem obvious, provides one last chance to verify that all stakeholders are on the same page. This is a relatively easy, but necessary task.

Defining Pass Criteria for Requirements and Objectives usually require more coordination with the Sponsor than for Specifications. Consider for a moment, the reason Specifications were necessary was that they further define the Requirement; the reason Requirements are necessary was that they further define the Objective. This inherently means there is more ambiguity with a Requirement than a Specification, or with an Objective versus a Requirement. This also means that there is more creativity required when verifying higher levels of the Requirements Document. Moreover, depending on how a particular test is defined, the acceptable Pass Criteria may be different. Example 16.3 illustrates this.

16.6 Rudimentary Risk Analysis

The last piece of the Critical Design process is called Risk Analysis. At CDR, Stakeholders want to know what might go wrong. Decision Makers *hate* to be surprised. If they are taking a risk, they want it to be an *informed* risk. So, they are essentially asking you to "look into your crystal ball" and predict the future – a difficult task to be sure. Fortunately, Project Management includes a process called Risk Management to help you do this.

An effective Risk Management strategy is developed using statistically analyzed volumes of data to determine the probability of a risk occurring. Then, a variety of mitigation techniques are considered. This may even include developing an entire alternative design and analyzing which design has the highest probability of success. Once the mitigation strategy is defined, benchmarks are set to indicate if and when the project must enact the mitigation. Clearly, such a process is beyond the scope of a Senior Design project.

However, many Senior Design projects benefit from performing what we will call a Rudimentary Risk Analysis (or Risk Analysis for short). This technique includes identifying potential risks late in the PLANNING phase and considering mitigation alternatives that might be taken during the EXECUTING phase. Often, simply by being cognitively aware of a potential risk, the project team naturally avoid the risk. If a risk does occur, then having a *predetermined* plan of action will save the team time if they have to adapt – at least the mitigation strategy has already been discussed and determined, it just has to be enacted. This is not exactly formal Risk Management as defined by PMI [30], but it is a good introduction to the topic and fits within the scope of a Senior Design project.

EXAMPLE 16.3: *Examples of Pass Criteria for Requirements*

In this example, Pass Criteria for a Requirement related to the Tiny Home presented in Example 7.4 are discussed.

* Req: Must be towable by a pickup truck
  Related test description: *Tow prototype of tiny home with a pickup for 100 miles.*
  – Pass Criteria option 1: Truck survives 100 miles of towing.
  – Pass Criteria option 2: Average fuel consumption is reduced by less than 5%.

Analysis:

An experienced reviewer should consider a test's Requirement, Test Description, and Pass Criteria comprehensively. None of these statements are obviously wrong when considered individually, but there are some holes that need to be addressed when reviewed systematically. Consider:

EXAMPLE 16.3, continued

- *Is it necessary to specify which pickup truck? Would testing this Requirement with a 2018 Dodge Ram with a 6.4 L Hemi and heavy towing package convince an owner of a classic 1950 Chevy Truck with an LS3 V8 engine that you have met this Requirement?*
- *Is a 100-mile road test sufficient to prove that this Requirement has been met? Does it matter which 100 miles? Would driving over the flatlands of South Dakota convince a Sponsor who lives in the mountains of Montana?*
- *If the Pass Criteria is simply that the truck "survives the road test," who is going to supply the test vehicle? If there is a guarantee that the truck does survive, is the testing rigorous enough? If this is a rigorous test and the truck might not survive, who will be putting their vehicle at risk?*
- *Perhaps simply testing whether the truck survives is not a great option. Perhaps there is some metric such as fuel consumption that provides the desired information. But, is using fuel consumption a valid metric to determine whether the truck is able to tow something? Is a change of 5% a reasonable indicator of success?*

There is no "right" answer to any of the questions posed here. The point that is being made by this example is that (1) the Verification Plan often poses many questions depending on the point of view the review takes and (2) all Stakeholders should agree upon the answers before the test is conducted.

This section will cover how to properly define a risk and potential mitigation options for dealing with the risk should it occur.

Definition of a Risk

A **Risk** is defined as an uncertain event (positive *or* negative) which would have an impact on at least one project Objective. A **Threat** is defined as a risk which produces a negative impact, and an **Opportunity** is a risk which produces a positive impact. For the purpose of Senior Design, let us limit the discussion of risks to those that threaten the outcome of the project.

> A Risk is any uncertain event that impacts a project; in Senior Design , the focus will be on negative risks (i.e., Threats).

There are four categories of risks, Example 16.4 describes these categories and provides some examples.

EXAMPLE 16.4: Four Categories of Risks

There are four categories of risks. This example will present the four categories and provide examples of each.

- **KNOWN RISK** – Events which have been identified.
  - Weather concerns: There is a 92% probability Bozeman, MT will receive between 63″ and 82″ of snow fall each year.
  - Component tolerances: 5% resistors are accurate to their stated values to within +/− 5% with 3 sigma probability (meaning 99.7% of the resistors will stay within the +/− 5% bounds).

  Known risks are quantified and predictable. Therefore, they are also the easiest risk to mitigate.
- **UNKNOWN RISK** – Events that cannot be predicted.
  - COVID-19: No one could have predicted the flu epidemic that hit society in March 2020. Engineering businesses were not prepared to mitigate this risk.
  - Inaccuracies in component datasheets: It is reasonable to trust a manufacturer's datasheet, but sometime there are errors in the documentation which can cause drastic setbacks.

  Predicting unknown risks is not possible. Therefore, mitigation of these risk is also not directly possible. We should be aware that unknown risks may exist and be watchful for indications of their effect on our projects.
- **Business Risks** – Risks that directly affect a business's profit or ability to conduct business.
  - Competition Successes: A competitive product hits the market before your product release date.
  - Fuel prices: The transportation industry's profit is directly related to the price of gas.

  Most companies face similar business risks; therefore the amount of predictive-data is enormous. An engineering degree is not required to understand or manage these types of risks; engineers may rely on best practices for the type and size of their business to do so. This is one aspect that employees with an MBA would support.
- **Insurable Risk** – Risks that can be insured
  - Flood or Fire: Most insurance companies will assume this risk for you.

  Occasionally a company will "take out" an insurance policy for certain types of risks. This is a realistic concern, but not one that directly applies to Senior Design.

The risks associated with Senior Design projects typically include Known and Unknown risks. An Unknown Risk is not a risk which a student simply did not think about, or was uneducated about. An Unknown Risk is a risk which *nobody* could have predicted. So, effectively, the risks you should be concerned about are limited to Known Risks. Known Risks can be technical or non-technical. You should focus on the technical risks.

Finally, notice that none of the risks in the previous example were related to the engineer's ability. If there is a chance that an engineer is unable to perform a task, a manager's mitigation strategy might be to simply reassign that engineer; another might be to send the engineer for additional training. Your Risk Analysis should be focused on the *technical issues* related to your project.

Articulating Risks in a Meaningful Way

Risks must be specific and concise to be mitigated. To articulate a risk, three issues must be identified and documented:

- Event
- Cause
- Impact

In order to mitigate a risk, the risk's Event, Cause, and Impact must be articulated. Each risk should be stated in the following convention:

If "X" (event) occurs then that could cause "Y" (cause) which could possibly result in "Z" (impact to project)

Example 16.5 provides instruction on properly formatting a well-articulated rick.

EXAMPLE 16.5: Articulating a Risk

An articulated risk describes the risk's Event, Cause, and Impact. Here are some examples:

Risk 1:
> If I miss class, then I may not learn the material which could possibly result in my failing this class.

Risk 2:
> If my PCB has a connection mistake, then my project schedule will be delayed by at least 2 weeks and my budget will increase $50; this could result in not completing my project on time.

Risk 3:
> If the ADC is slower than expected, then the sampling frequency will be reduced, which could result in signal aliases.

Analysis:

Consider Risk 1. It would be easy to skip the "cause" of this risk and simply say "missing class will result in a failing grade." That statement may be true for a class with a strict attendance policy. If so, the mitigation strategy may be to "not miss class at all costs." However, in a more lenient class, it is not the "missing class" that causes a student to fail; missing class results in a lack of understanding, which in turn causes the student to fail. This is an important distinction because it changes the mitigation strategy. Perhaps the student could schedule additional office hours with the professor and learn the material instead of accepting a failing grade.

Notice how precise the statement for Risk 2 is. By specifying the amount of time and money a mistake will cost, you can more accurately determine the severity of the risk. This risk might be a major issue on a project which was allotted a $200 budget; the risk would account for 25% of that budget – encountering that risk may be something the project

---

EXAMPLE 16.5, continued

---

teams must avoid at all costs. Alternatively, on a $10,000 budget, this risk is not worth giving any additional thought.

Risk 3 provides an example of why the "impact" statement is necessary. An ADC operating slower than expected will of course reduce sample frequency...but, so what? On some projects, that might not be an issue at all. By explaining that this risk would "result in signal aliases," a signal engineer knows to be on the lookout for this indicator. By specifying the Event, Cause, and Impact, the risk statement provides an engineer with multiple characteristics that can be monitored, making it easier to detect a negative event.

Mitigating Risks

There are four primary strategies to mitigate risks, they include the following:

- Avoid
- Transfer
- Contain
- Accept

Avoiding a risk means that an entirely different plan of action will be pursued. Avoiding Risk 1 in Example 16.5 may mean not missing class. Avoiding Risk 2 might be very difficult and probably is not the correct strategy. Risk 3 may be avoided by purchasing a higher quality ADC.

Transferring a risk means to shift ownership of the risk to a third party. The risk of failing class might, in part, be transferred by hiring someone to take notes for you; a PCB mistake may be transferred by outsourcing the PCB design to an expert. Risk 3 might be difficult to transfer.

Containing a risk means to take specific action to minimize the impact when the risk occurs (as opposed to taking action to avoid the risk from occurring). The risk of failing class might be contained by preemptively working with the professor to complete any assignments you might miss. Adding additional test and/or connection ports to the PCB may allow the PCB to be modified so that a connection mistake could be corrected. Filtering the signal at a particular frequency will eliminate possibility of aliases.

Accepting a risk means that no mitigating action will take place and that you will deal with the risk when it occurs. This strategy is inherently chosen for all unknown risks and many business risks. However, it is rarely an acceptable strategy for a known risk. The only exceptions are when the probability of a risk occurring is so low that efforts to avoid it are unnecessary or when the cost of mitigating a risk is prohibitive. Most likely, this is not applicable to Senior Design Risk Analysis.

16.7 Chapter Summary

In this chapter, you learned about the Critical Design process. During this final stage of the PLANNING phase, the Finalized System Design is produced and plans are made to efficiently conduct the EXECUTING phase. Additionally, a Rudimentary Risk Analysis is performed. All of this will be presented at the upcoming CDR. Here are the most important takeaways from this chapter:

* A Finalized System Design is detailed enough that a competent fabrication facility should be able to synthesize the design with no further input from the design team.
* The Fabrication Plan ensures an efficient workflow between team members. Mile markers should be included within the plan for conducting the low-level tests of the Verification Plan.
* The Verification plan must individually include all items of Requirements Document. The Specifications and most of the Requirements should be intermittently tested during the Fabrication process. The remaining Requirements and all of the Objectives should be tested on the System-Level Prototype.
* Identifying potential risks and predetermining a mitigation plan, significant time and resource may be saved during the EXECUTING phase. Risks should be articulate in the format of Event//Cause//Impact to be useful in a management plan.

Industry Point of View 16.1: Risk Analysis Considers Cause and Effect

Be accurate and precise when defining project risks. All too often a loose description of a risk results in an ineffective mitigation.

I was in charge of a multi-million-dollar budget to support our products worldwide. My vice president, Greg, came to me one day and asked me to allocate $200K to send a trainer to China for a year to train replacements for engineers who had recently left our company.

Losing those engineers was a significant risk to our project. However, our administration only focused on the problem, not the root cause. Greg saw the problem as a temporary shortage of five engineers but did not consider why those engineers resigned. In my opinion, due to a lack of experience in this market, we were not giving our employees a competitive salary.

As instructed by our administration, we trained five new engineers for that market. Within a year, each of them resigned. Our competition was offering nearly a 40% pay increase. Our competitors did not have the ability to train employees to the same level of expertise as we did. Their strategy was to hire experienced employees directly.

Industry Point of View 16.1, continued

Greg's solution to the problem was only a band-aide. He did not recognize that the underlying issue of problem was our pay scale. For the next year, we continued to train engineers only to have them immediately leave to work for our competitor. Eventually, a risk analysis was completed for this issue. Initially, we defined the project risk as "we have lost engineering staff and that has put our timelines at risk." Considering that statement, training new engineers is a reasonable mitigation strategy. However, there is flaw in this statement; there are only two parts, an event and an impact. This should have been an indicator that the risk was not fully defined. Eventually the risk analysis redefined the risk as "Competitors are offering high salaries, we are unable to retain quality employees, and the reduced workforce puts our project at risk." This three-part statement indicates that the risk is well defined and changes the mitigation. We convinced the administration to reallocate the training expenses to provide substantial bonuses for our engineers. The bonuses ensured that the engineers were well-compensated and employee satisfaction dramatically improved. I cannot recall losing another well-trained employee to our competition after that policy was enacted.

16.8 Case Studies

Let us look at a portion of our case study teams' Verification Plans.

Augmented Reality Sandbox

Table 16.2 displays a portion of the AR Sandbox team's Verification Plan for the Structure subsystem. Recall that the summary table provides a high-level description of the test plan and that each test needs a detailed test description. In this section, we will review a few of the AR Sandbox team's test descriptions.

Table 16.2 AR Sandbox Structure Subsystem Verification Plan

Performance Verification Summary Structure Subsystem				
Num	**Name**	**Test description**	**Pass Criteria**	**Planned Date**
1.1.1	Base must hold 300 lb	FEA simulation #1	No permanent deformation	11/5/2016
1.1.2	Sensor upright must hold upto 200 lb	FEA simulation #2	No permanent deformation	11/25/2016
1.1	Structure must withstand rough play	Weight will be added to prototype and observed for structural damage	Structure will hold weight for 24 h with no damage	2/11/2017

The test description for Requirement 1.1 read:

- *Requirement 1.1 – Prototype Weight Test:*
- *Setup/Environment: The Preliminary Prototype of the Structure will be used for this test. The test will be conducted in the Senior Design lab. The base will be filled with enough sand (200 lb) to fill the play area; additional weight will be added by hanging sandbags on the base and sensor arm.*
- *Equipment: A standard scale will be used to measure the weight.*
- *Procedure:*
  - *200 lb of sand will be added to the sandbox play area.*
  - *The structure will be observed after 30 min for damage.*
  - *A 20 lb sandbag will be mounted to the structure's upright in place of the Kinect sensor and projector.*
  - *The structure will be observed after 30 min for damage.*
  - *5–20 lb sandbags will be hung over the edge of the base, 1 one each side, and 3 along the front.*
  - *4–20 lb sandbags will be hung from the tip of the structure's upright.*
  - *The system will be observed after 30 min for damage.*
  - *5 additional 20 lb sandbags will be hung from the tip of the structure's upright.*

- *The system will be observed after 30 min for damage.*
- *The weight will be left on the structure for an additional 24 h.*
- *The system will be observed at the end of the test for damage.*
- *Data Collection:*
  - *Exact weight of each sandbag will be recorded.*
  - *Visual observations will be recorded.*
- *Repeatability: This test will be conducted one time.*
  - What went well: The team used simulation results to validate that their subsystem design would meet Specifications during the PLANNING phase (first semester) and then used the Preliminary Prototype to test the Requirement during the EXECUTION phase (second semester). Incrementally increasing the weight applied to the structure during Req1.1's test was also a good idea.
  - What could have been improved #1: The team should have kept the big picture in mind when writing these tests. Notice that the Requirement is that the structure withstand rough play. The KSDC's standard Specifications were that the structure should hold up to 300 lb and any uprights withstand 200 lb of force – *while in use*. The team "used up" 200 lb of their safety margin by including the sand necessary for play in the 300 lb that the base must hold.

    Moreover, although the exact design/built/test order is not explicit in the summary table, we can infer that the base was designed and tested by 11/05/2016 and *then* the sensor arm was designed and tested by 11/25/2016. This suggests that the weight of the sensor-upright was not included in the FEA simulation for Specification 1.1.1. This means that although the simulation proved that the base could with stand 300 lb, those 300 lb were composed of 300 lb of sand, 50 lb of the sensor-upright's material, and 40 lb of the sensor and projector. Now, consider Req1.1 "must withstand rough play." By considering 290 lb of project material in the 300 lb specification, the team has only ensured that children under 10 lb can safely play with the sandbox.

    This is an example where all Specifications can be met without meeting the related Requirement. What should have been tested in this case is 300 lb of additional weight after the system was fully functional. The specific weights required for the sand and sensor-upright may not have been known when the team conducted the test for Spec1.1.1. But, they could have estimated the additional system's weight and then over tested the Specification to provide a safety margin before the Requirements test was conducted.

Smart Flow Rate Valve

Table 16.3 shows a portion of the Smart Flow Rate Valve team's Verification Plan for the Control subsystem. A Simulink model was developed during the Preliminary Design process and used to verify a number of the control-parameters (i.e., Specifications). Then, during the Verification process, the physical system was used to verify the Requirements.

Table 16.3 Flow Rate Valve's Control Subsystem Verification Plan

Performance Verification Summary Control Subsystem				
Num	**Name**	**Test description**	**Pass Criteria**	**Planned Date**
2.1.1	Initial Rise time < 1 s	Simulink Simulation #2a	Simulation output TR.1 < 1 s	10.20.2017
2.1.2	Response to change during operation < 0.5 s	Simulink Simulation #2b	Simulation output TR.2 < 0.5 s	10.20.2017
2.1.3	Steady-state error $< 2\%$	Simulink Simulation #2c	Simulation output Ess $< 2\%$	10.20.2017
2.1	System must control flow rate accurately	Prototype test #3a. Measured parameters will be compared to simulation data	All values of physical system meet specificaitn and match simulation values within 10%	2.01.2018

The test description for Specifications 2.1.1–2.1.3 read:

- *Setup/Environment: MATLAB/Simulink*
- *Procedure: Model will be simulated, parameters will be compared to Specs*
- *Data Collection: T_{R1}, T_{R2}, and Ess will be measured on MATLAB graphs*
- *Repeatability: This test only needs to be conducted once.*
  - What went well: The team appropriately realized that the Specifications could be initially verified using the software model, but then would need to be re-verified with the physical product. The team also recognized that although the physical product's results must meet Specifications, it was also important that these results matched the simulation results. Of course, ultimately, the priority is for the physical product to meet Specifications. However, if the physical product does not match the model's perfor- mance, the *entire* model (and any related data) must be considered invalid.
  - What could have been improved: The test description for the Specifications is not complete, and it is too vague. The procedure should have stated the following:
    - The Simulink model will be simulated.
    - The modeled valve will start in the off position.
    - A worst-case scenario step-change from off to fully on will be applied.
    - The Rise Time (TR.1) and Steady-state Error (Ess) will be measured.
    - A step-decrease by 25% will be applied.
    - The Rise Time (TR.2) and Steady-state Error (Ess) will be measured.
    - Ten additional step-changes of varying magnitude and direction will be applied.
    - The Rise Time (TR.2) and Steady-state Error (Ess) will be measured for each.

    Furthermore, these changes to the procedure would require a change to the Pass Criteria as well. Passing Specification for variables TR.2 and Ess one time is not enough; these values must meet Specification for *each* step-change. The updated Pass Criteria should read "TR.2 < 0.5 s *for each input.*"

Robotic ESD Testing Apparatus

The previous two case studies exhibited properly formatted Performance Verification Summary tables for a particular subsystem. Although RESDTA also formatted their Verification Plan by subsystems, the author took the liberty to create a new table, Table 16.4, which only displays the summary information for Specification tests for purposes of this discussion. The test descriptions discussed in the previous case studies were not particularly technical in nature, but most of your tests will be technical tests. So, the RESDTA case study will discuss two well-written test descriptions, even though they do not happen to be part of the same subsystem.

Table 16.4 RESDTA Performance Verification Summary table

Performance Verification Summary				
Num	**Name**	**Test description**	**Pass Criteria**	**Planned Date**
4.3.2	Must deliver contact shocks with less than 1 lb of fource on DUT	Tip of gun will be applied to a load cell, force-limiter will be encacted	Subsystem must pass Spec on 10 consecuative trial.	3/15/ 2017
8.1.2	System must de-energize within 0.8 s of EPO	EPO will be tested using mock-loads and Oscopes	System must reach EPO standard "OFF" levels within EPO standard times on 100% of tests.	1/28/ 2017

The test description for Specification 4.3.2 reads:

- *Specification 4.3.2 – Tip Force must be limited to 1.0 lb on any DUT:*
- *Setup/Environment: A test-program will be written to extend the z-direction-motor by ¼ turn every 5 s. A software-interrupt will be included to monitor the system's force sensor and output a value for every ¼ turn of the motor. The tip will be applied to a digital load cell. The output of the load cell will be manually read and recorded throughout the test. The code will turn on an LED when 0.5 lb of force is achieved. This LED will indicate the moment in time when the system-code would have been forced to stop rotating the motor.*
- *Equipment:*
  - *Z-direction-motor and related controls*
  - *Digital load cell*
- *Procedure:*

  1. *The motor will be clamped to a table opposite the load cell at a distance of 0.2″ apart.*
  2. *The motor will be activated to move the ESD gun forward.*
  3. *The motor will pause every ¼ turn for 5 s to allow the operator to record data.*
  4. *For each ¼ turn, the code will display the force sensor's reading which will be manually entered into a table.*
  5. *For each entry of the table, the load cell value will also be recorded.*
  6. *There should be approximately ten recordings of 0 lb of force while the ESD gun is moving into contact with the load cell.*

7. *There should be at least four recordings of forces between 0 and 0.5 lb as pressure is applied to the load cell.*
8. *The code will activate an LED for any reading over 0.5 lb of force. The state of the LED during each recording will be documented,*
9. *Recordings will continue until there have been at least four recordings above 1.0 lb.*

- *Data Collection:*
  - *The load cell and force sensor data will be compared; to pass this test the data must match for each recording to within 0.1 lb.*
  - *The status of the LED will be monitored; to pass this test the LED must be ON for any recording above 0.5 lb.*
- *Repeatability: This test will be repeated ten times.*
  - What went well: The test description for Specification 4.3.2 provided the appropriate level of detail. That detail allowed the reviewers at the Critical Design Review to provide the team with three important suggestions:

    1. The test description explicitly states that the ESD gun (a $50,000 piece of equipment) was going to be used in the initial performance test. That is a huge risk and defeats the point of performing low-level testing before conducting system-level tests. The Stakeholders requested that step 2 of the procedure be rewritten as *"...to move the ESD gun-**mount** forward."*
    2. Although the test was to be repeated ten times, the test did not provide conclusive evidence of detecting the 0.5 lb threshold (in the opinion of the Sponsor). They argued that the LED could be turning on due to the distance the ESD gun-mount had moved rather than the sensor data. The Sponsor requested that the Repeatability Statement be rephrased as "this test will be repeated three time at 5 randomly selected distances."
    3. Concerns were raised whether 5 s was enough time for one person to accurately record two values and the status of the LED. This was a minor concern that could have easily been addressed in the middle of the test. However, you should remember that the test procedure is not written for *you*, the procedure is written for a qualified technician who does not know the specifications of your design. That means the operator may not have the necessary background to determine whether allowing additional time to record the data is an important test parameter or not. So, while this may sound like a trivial issue at CDR, small details like this can become important during the Verification process.
  - What could have been improved: There is not much to improve on this test description except what was already mentioned above. The point being made here is that a well-written test description will allow Stakeholders to understand precisely what you are planning to do and provide useful feedback. One small improvement that could have been incorporated with this description would have been to include a diagram of the test setup; although this description was straightforward enough that a diagram was probably not absolutely necessary, it might have been helpful.

The test description for Specification 8.1.2 read:

- *Specification 8.1.2 – System must de-energize w/in 0.8 s of an EPO (Emergency Power Off):*
- *Setup/Environment:* **This Specification will be tested using two resistors sized to simulate the power draw of the 5 V microelectronics and the 24 V motors. The E-Stop EPO switch will be connected as shown in** Fig. 16.7.

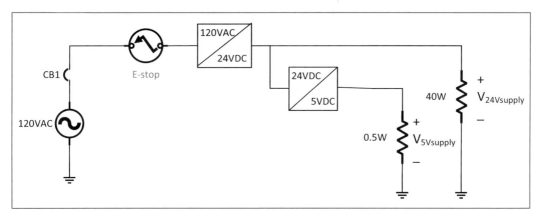

Figure 16.7 EPO test circuit

- *Equipment:*
  - *Oscilloscope,*
- *Procedure:*

  1. *The EPO switch will be toggled which should disconnect all power from the system.*
  2. *The output voltages of the microelectronics (5 V) and the motors (24 V) will be monitored.*
  3. *The time between hitting the switch and the DC supply voltages reaching acceptable de-energized levels will be measured.*
- *Data Collection:*
  - *The time required to de-energize a power bus after an EPO is defined as 0.8 s.*
  - *The "OFF" voltage of a 5 V bus is defined as 2.0 V.*
  - *The "OFF" voltage of a 24 V bus is defined as 7.5 V.*
- *Repeatability:* **This test will be repeated ten times.**
  - What went well: **The circuit diagram provided makes it easy for any Electrical Engineer to understand this test description. Doing so also reduced the amount of writing required in the test description.**
  - What could have been improved: **Ensuring that a test environment accurately represents the real system is important. Ideally, an emergency power switch would instantaneously de-energize a power system. However, energy storage elements (e.g., an inductor) prevent this ideality from ever occurring and are the reason why EPO standards include a timing specification. However, the team proposed to model the response of the 24 V power rail using a pure resistor (which does not exhibit a time delay). Therefore, the proposed model does not represent the real system and any data collected from this model would not be valid. The test description was modified to state that the 24 V load would be modeled with a pair of 24 V DC motor. Although the system design included many 24 V motors, at no time would more than two motors be running simultaneously. The modified model closely represented the real system.**

Chapter 17: **Critical Design Review**

The culmination of the Critical Design process is an event called the Critical Design Review (CDR). At this event, the design team presents their solution (in its *completed* form) of the Problem Statement to the other Stakeholders for approval. At this point in the Design Life-Cycle, the Stakeholders have risked a relatively small portion of the project resources, primarily during the Preliminary Prototyping process. The next processes, however, will require the bulk of the project resources and constitutes the largest financial risk.

> The purpose of the CDR is to verify that all Stakeholder's agree that the Critical Design process has produced a solution with a high-likelihood of success.

The purpose of the CDR is similar to the purpose of the PDR (0). At the PDR, the design team's objective was to convince the Stakeholders to accept a risk. The risk exposure was larger after PDR than it had been after PKM, but still relatively small compared to the risk of pursing the project after CDR.

The burden of proof at each review continually increases. Assuming a successful CDR, the project will be ready to exit the PLANNING phase and begin the EXECUTING phase, where the majority of project resources will be expended. Since the risk of accepting a team's CDR is more substantial than accepting a PDR, the design team will have an even higher burden of proof at CDR than they did at PDR. Your priority will be to convince the Decision Makers that you have a *complete* solution and that all aspects of the solution have been thoroughly considered.

As with the PDR, a well-presented solution will often allow Stakeholders to leverage their knowledge and experience to improve the design. But, hopefully, the improvements are related to minor details now because all the major issues have been designed during the Preliminary Design process, flushed out during the Preliminary Prototyping process, and finalized during the Critical Design process. Still, this is a good opportunity to seek further assistance from engineers who have more experience than you. This is the last chance to address any oversights in the design before they become extremely costly to fix.

There are three potential outcomes of a CDR. The ideal outcome is that the Stakeholders are convinced that your design has a high likelihood of success and agree to continue project support. Alternatively, if the Stakeholders are not satisfied with your design, they may ask you to repeat an aspect of the PLANNING phase. This could include expanding the scope of Preliminary Prototyping process to include additional subsystems and/or include addressing additional design concerns in the Critical Design process. If the Stakeholders are truly unhappy with the design, they may cancel the project outright. This is an extremely rare occurrence and one that the Design Life-Cycle was designed to avoid.

In this chapter you will learn:

- How to prepare and present a Critical Design Review.

© The Author(s), under exclusive license to Springer Nature Switzerland AG 2023
C.J. Mettler, *Engineering Design*, https://doi.org/10.1007/978-3-031-23309-8_17

17.1 Formatting a CDR

As we have already seen with the PKM and PDR, the exact formatting of a CDR is highly dependent upon the preferences of the primary Decision Makers who are contributing to the review. So, again, the exact format of what a CDR *should* be is not what is important here. What is important are the fundamental concerns of all CDRs and your ever-increasing ability to apply the organization's formatting preferences to *your* project.

As a reminder, these three topics are the focus of any PDR:

- Do you have a realistic solution to the Problem Statement?
- Does this solution directly address the Requirements Document?
- Have the critical aspects of the design been addressed in enough detail so as to reduce the risk of expending more effort and funds?

Compare how that focus is modified for a CDR:

- Do you have a *complete* solution to the Problem Statement?
- Does this solution directly address the *entire* Requirements Document?
- Has *all* functionality of the design been addressed in *complete* detail so as to reduce the risk of expending more effort and funds?
- *Has the integration of all aspects of the design been addressed to ensure the design will function cohesively?*
- *Have the budget and risks been analyzed in enough detail to provide confidence the design can be synthesized?*

You can see that the first three bullets are similar but require "completeness" to be accomplished correctly at CDR. There are two additional bullets for the CDR as well.

Figure 17.1 depicts the general flow of a typical CDR. The introduction to the CDR includes the Problem Statement, Necessary Background, and Design Concept. These topics should be essentially the same as the PDR. The Completed System-Level design is presented in enough detail to convince the Stakeholders that, if built, the conceptual design will successfully address the Project Motivation. Details of the Fabrication (how the system will be synthesized) and Verification (how the system will be tested) plans are presented. Both Budget and Risk Analyses are presented.

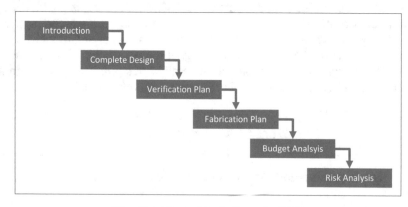

Fig. 17.1 Typical topic-flow of a PDR

The most important formatting issue to consider is:

> The complete design must be presented thoroughly, but concisely so that the Stakeholders will have complete confidence in the design.

17.2 A CDR Template

This section describes one method of presenting a CDR effectively. Keep in mind that the purpose of performing a CDR is to convince the reviewers that you have designed a complete solution that will successfully address the Project Motivation and fulfills the Requirements Document. The topics this method will cover are:

- Introduction (Problem Statement, Background, and Design Concept).
- Completed Design Solution
- Verification Plan
- Fabrication Plan
- Risk and Budget Analysis

Assuming that the primary Decision Makers are somewhat familiar with the project at this point is reasonable. However, you should not expect them to remember any of the *details* regarding the project. You can expect that they know who you are and can be quickly brought back up-to-speed. Therefore, the CDR does not start with an "Introduction of the *team*" as the Kickoff Meeting did. Instead, the CDR begins with an "Introduction of the *project*" similar to the PDR.

You may wish to state your team's name and welcome the audience to "your Critical Design Review" on the cover page of the slide deck, but this should require no more than a few seconds. The real introduction comes on the following slides.

Introduction

The introduction of the CDR includes the Problem Statement, Necessary Background, and Project Design Concept slides. These slides should be essentially the same as what was presented at the PDR (albeit improved if necessary).

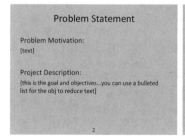
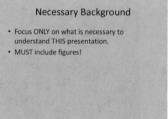
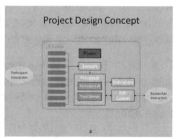

Completed Design Solution

The entire System-Level design is presented once the audience understand the high-level project details.

First, the solution to each subsystem is presented. Each team member should at least present their Critical Subsystem; however, all additional subsystems and/or interface-designs must also be discussed. Organize the order that the subsystems are presented to create a cohesive and understandable presentation.

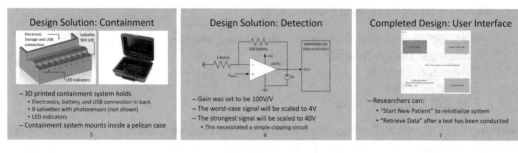

Many subsystems will have already been presented at the Preliminary Prototype Demo. If the subsystem was complete and already approved at PPD, you may simply copy/paste the same slide into the CDR presentation. Discussing the solution again is not necessary, simply introduce the subsystem and then state "this solution was complete at PPD." If the subsystem was not complete at PPD, introduce the subsystem and focus the discussion on the modifications made since PPD.

> The focus of the Design Solution discussion at CDR should be on any design aspects that were not covered at PPD, including system integration.

The final slide in the Complete Design Solution discussion is the System Integration slide. At PDR and PPD, the focus was on individual subsystems. At CDR, the focus should be on the *complete* system. One of the most important aspects of the System-level Design is that all subsystems integrate together.

This topic may require more than one slide. Include as many slides as necessary to present these aspects.

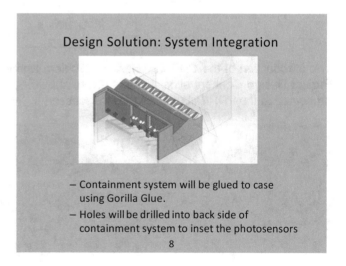

Verification Plan

Verification Plan details are provided next. The purpose of this discussion is to ensure that the team has testing plans that will provide confidence-inspiring results and that the plan is feasible within the allotted time and with the available equipment. The plan is usually presented into two parts: Low-level and System-level. Organize the Low-level discussion by discussing a set of Specifications and then the related Requirement. You may want to include the Low-level

Verification Plan : Detection

- Spec 1.1.1: Must detect removal of Salivette within 1.0s

Paper tubes will be created to simulate the containment system. Photosensors will be placed at the bottom of the tube with salivettes placed directly over the sensors.

Electronics will be built on a breadboard. OpAmp output will power an LED.

A camera will be used to simultaneously record removal of the salivette and LED status. The time between the salivette being completely removed from the tube and the LED turning on will be measured.

Test will be repeated in a variety of light conditins.

"Success" means that in low light conditions the LED turns on in less than 1s.

discussion for each subsystem immediately after the design solution for that subsystem. The Objectives (and any outstanding Requirements) should be discussed as part of the System-level tests. This discussion should be presented after the Complete System Design discussion. Refer to Sect. 16.5 for details on how to format a test description.

Fabrication Plan

The Fabrication Plan discussion starts with a high-level overview. Present this using a Gantt Chart with most of the tasks hidden so that you can present the larger categories of work.

The focus of this discussion should be on important milestones and team coordination.

The milestones presented here are those that are defined by the team; they are not limited to the milestones predefined by the Senior Design course structure.

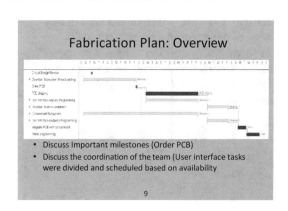

Fabrication Plan: Overview

- Discuss Important milestones (Order PCB)
- Discuss the coordination of the team (User interface tasks were divided and scheduled based on availability

9

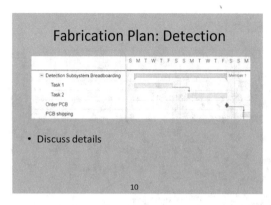

Team coordination is extremely important in order to complete the entire Fabrication Plan within the allotted time period. Discuss opportunities to split the workload to make sure the plan is enacted as effectively as possible.

After providing the high-level overview, break the workload categories discussed in the overview into more detailed tasks. Discuss how each subsystem or feature will be fabricated. Notice how the "Detection Subsystem Breadboarding" category is mentioned in both slides 9 and 10. The purpose of slide 9 is to explain how this task fits into the "big picture." The purpose of slide 10 is to provide the details.

It may not be possible to discuss every detailed task within the time allotted for a CDR. Focus on the first many weeks of the Fabrication process.

Risk and Budget Analysis

The last topics of the CDR are a Risk and Budget Analysis.

Stakeholders understand that all project inherently contain risk throughout the entire project. However, by identifying risks early, there is a better chance of avoiding the risk. Present approximately three of your most *concerning* risks on this slide. "Concerning" risks are ones with a significant impact to the project and which have the highest probability of occurring.

Project Risks

- Discontinuity of user interface subsystem development may cause issues interfacing the two pieces and result in additional programming bugs.

- Improper placement of the photosensors will modify the sensitivity. This could require recalibrating the signal conditioning system.

- The LED ports in the Containment system are very small, this could cause issues with the 3D printing process which would delay our Fab plan

11

Estimated Budget

- Show a table

9

Finally, include an updated budget. Every cost should be known at this point, so the budget should be relatively accurate. However, Stakeholders will appreciate a budget *estimate* that is still less than the *total* budget because there are always some unexpected costs.

Finally, conclude your CDR with a strong summary statement and open the discussion up to any Stakeholder's questions or concerns.

17.3 Chapter Summary

In this chapter, you learned about the Critical Design Review. The CDR is a milestone located at the end of the PLANNING phase. At CDR the complete design and EXECUTING phase plans are presented for a technical review. The purpose of the CDR is to identify that the risk of entering the EXECUTION phase has been reduced as far as possible. Here are the most important take-aways from the chapter:

- The CDR should include a top-down discussion of the design. This starts with the Problem Statement, discusses the design concept, and then provides the details of all the Subsystem Designs and the System-Level design.
- The Fabrications and Verification plans are also presented at CDR to verify that the Stakeholders agree that the product and results produced during the EXECUTION phase will ultimately be accepted at the end of the phase.

PART V: THE EXECUTING PHASE

"Excellence is never an accident. It is always the result of high intention, sincere effort, and intelligent execution; it represents the wise choice of many alternatives – choice, not chance, determines your [results]"
— Aristotle, Greek Philosopher

"Vision without execution is hallucination."
— Thomas Edison, American Inventor

"Engineers are not superhuman. They make mistakes in their assumptions, in their calculations, in their conclusions. Making mistakes is forgivable; that they catch them is imperative."
— Henry Petroski, Distinguished Professor Emeritus

On the evening of February 20, 1998, Tara Lipinski sat posed to execute a plan 7 years in the making [31]. After years of hard work and sacrifice, she found herself trailing her Olympic competition, but her plan was ambitious and she was prepared. Her program was considered one of the most difficult ever attempted at the Olympics; that night, Tara became the first female to ever attempt a Triple Loop/Triple Loop combination in competition [32]. Many consider her performance to be a "flawless execution" as she skated her way to Olympic gold; the youngest Winter Olympian gold medalist ever.

Tara's 1998 performance was not luck, it was not an accident. It was the result of setting a lofty goal, years of effort, and perfect execution. She began a dedicated skating regime at the age of 7. Her family moved across the country so that she could train with the Nation's top coaches at the age of 10. Those 3 minutes of perfection in 1998 were the result of dreams, sacrifice, and training – or in other words, planning – and flawless execution of the plan.

Tara had years, literally 1000s of hours, to prepare. Most engineers are not allotted that kind of time. Recall the Triple Point Constraint concept from Sect. 1.2, the time leg of the triangle is typically constrained. Recall one of the definitions of engineering design from Chap. 1 is producing a quality solution under constraints – one of which is *time*.

Field Marshal Bernard Law Montgomery and Lieutenant General George S. Patton were the two most effective officers in the European Theater of World War II [33]. Field Marshal Montgomery was known as a meticulous planner. His plans were flawless down to the minute detail, except for one small issue: often by the time his plans were ready to be executed, the war had passed him by and they were never enacted. General Patton was known as an aggressive battlefield commander, and – although somewhat unfairly – referred to as a "loose cannon." Patton's philosophy "a good plan enacted today, is better than a perfect plan enacted tomorrow" or in Montgomery's case "never enacted at all" made it appear to his contemporaries that he simply "did not plan." However, history has shown that Patton's planning was also effective – his style was just different. He set small, intermediate goals, and then aggressively pursued them. When up-to-date information deviated from expectations, he was able to quickly and effectively modify the short-term plan without losing sight of the long-term goal.

In today's management lexicon, we might refer to Montgomery's style of planning as "traditional" or "waterfall" and Patton's style as "agile" or "scrum." In the engineering environment, a combination of styles is often the most effective. The purpose of the PLANNING phase was to reduce project risk. As many details as possible were meticulously addressed (Montgomery's style). However, we must move forward and execute the plan in order to complete the design within our time-constraint.

During the EXECUTING phase, the plan will be carried out and the solution will finally be constructed. However, knowing that our plan is not perfect, careful attention must be paid to up-to-date information throughout this phase. This means that, while we aggressively pursue the end-goal, we must also seek new information and constantly incorporate that information into the plan (Patton's style).

In order to complete a successful Design Life-Cycle, we need a plan and we need to execute that plan. We do not necessarily have years to prepare the plan, but we also do not necessarily need to enact the plan flawlessly. What we do need to do is prepare a reasonable and well-thought-out plan and to enact that plan carefully. We also need to be able to adjust the plan to new information (i.e., testing results). We are not expected to be perfect like Tara Lipinski's gold medal performance, but we are expected to "catch" our own flaws throughout the process.

There are two primary processes in the EXECUTING phase: the Fabrication process and the Verification process. The adage "the sooner mistakes are found, the easier they are to correct" still applies. Therefore, it would be a poor decision to construct the entire solution and then test it. Rather, the Design Life-Cycle overlaps the Fabrication and Verification phases in order to continually reduce risk. This overlap was already addressed in the planning of these two processes. Now, as we enact the plan, critically think about the data collected during each Verification test and consider, "how does this new information affect the plan."

The EXECUTING phase can be broken down further into four processes; each will be discussed in this part of the textbook.

Processes of the EXCECUTING Phase	Presentation Structure
Each subsystem must be synthesized.	**Gantt Chart and Notebooks**
Each statement in the Requirements Document must verified.	**Subsystem Verification Documents**
All subsystems must be integrated into the final solution.	**System Prototype Demo**
Final solution must be verified	**Final Design Review**

Chapter 18: **Fabrication and Verification Process**

You were briefly introduced to the Fabrication and Verification processes in 0. At that point in the Design Life-Cycle, you were finishing the PLANNING phase by preparing for the Critical Design Review. At CDR, teams described the *plan* that will be executed during the EXECUTING phase. It is now time to enact that plan.

The first, and primary, processes of the EXECUTING phase are the Fabrication and Verification processes. One unique aspect of this phase of the Design Life-Cycle is that these processes that are intended to *overlap*. All the previous processes have been accomplished in a linear fashion – one process is finished and then the next one starts. Although *starting* the Fabrication process first is inherently necessary, it is not wise to *finish* fabricating the solution before initiating the Verification process. An overlapping timeline allows you to "test small and test often" in order to catch any issues as early as possible. (There are a few other processes in PMI's Design Life-Cycle which are also intended to overlap, but they are not covered in this textbook.)

During the Fabrication process, the System-Level prototype is produced by synthesizing the conceptual design into reality. The Verification process is used to validate each piece of the design before it is integrated into the next (larger) piece of the design. Eventually, the Verification process is designed to validate a project has fulfilled the Project Objectives and full addressed the Project Motivation.

This chapter will cover the management of both processes in a cohesive narrative.

In this chapter, you will learn:

- How to verify your Project Management tools were correctly developed.
- Tips for navigating the Fabrication process.
- Tips for presenting data collected during the Fabrication process.

18.1 Synergy of the Design Life-Cycle

Complete comprehension of the Project Management Design Life-Cycle requires a simultaneous understanding of the beginning, middle, and end of the cycle. In order to perform the initial tasks effectively, a Project Manager must consider the final tasks. In order to initially learn about the final tasks, a student must first experience the initial tasks. This dichotomy is due to the synergy of the entire cycle and creates a challenge when first learning about the Design Life-Cycle as a student.

Perhaps there have been times throughout your Senior Design experience thus far when you have felt that a certain task was unnecessary or that the level of rigor required on some document was overbearing. While understandable, these feelings stem from a temporarily incomplete understanding of the upcoming project deliverables. Hopefully, experiencing subsequent stages of the Design Life-Cycle has justified some of the initial project tasks in your own mind. If you ever have to develop another Requirements Document or perform another Project Kickoff Meeting, you will improve the quality of those deliverables by considering what you are about to learn in the Fabrication and/or Verification processes. Let us consider a few examples of where we have already seen this:

- The rigor with which the Requirements Document was developed was necessary to provide a detailed definition upon which the Alternatives Analysis could be performed. Clear and complete Specifications make the Alternatives Analysis easier. Whereas loose or missing Specifications leave too many options open, and that ambiguity makes drawing a final conclusion about Alternatives difficult to justify.
- A well-defined and organized Gantt Chart provides teams with a clear set of tasks which must be performed in a particular order. This clear understanding makes generating Action Item Reports, and keeping your team focused on the highest priority tasks, easier. Although Gantt Chart's initial development requires some effort, in the long run it saves time – you don't have to reconsider your task lists and priorities every week.
- Designs developed during the Preliminary Design process are typically improved during the PDR, often in unexpected and surprising ways. The Preliminary Design effort (i.e., "on-paper" and without expending resources) may not have been the *easiest* way to get started. But, the "on-paper" planning almost certainly prevented you from expending resources on a design that would eventually be modified.

Perhaps you can gain insight by looking back at how the Design Life-Cycle was applied on *your* project over the last many months. Can you identify any frustrations that were eventually justified by subsequent processes? Spoiler alert: We will do this more formally in the CLOSING phase with a process called the After-Action Retrospective (Sect. 23.2).

Recognizing the usefulness of the tools and processes you have already been exposed to may improve how effectively you use certain tools and processes during the EXECUTING phase. Remember, the most significant challenge of the EXECUTING phase is *integrating* all the remaining aspects of the project. The tools you have already been exposed to will assist you in addressing this challenge *if applied correctly.*

The end justifies the means.

Before completing any project, *you* must *prove* that your product successfully addresses the Project Motivation. Until you have personally experienced this process, underestimating what the statement "you must prove" requires is an easy mistake to make. Notice the statement is NOT "someone should provide evidence...," it says "*you* must *prove*..." That means that it will not be enough to demonstrate the system's functionality at a one-time event. Providing "proof" means rigorously testing the product, collecting a thorough dataset, drawing reasonable conclusions, and presenting all of that in such a way as to *convince* all the Stakeholders that you have been successful. This thorough process must be applied to every aspect of the project.

Many times, throughout this textbook it has been mentioned that "there is no *right* answer for..." The impact of all those statements will begin to become clearer at the end of the EXECUTING process. When there is a "right" answer, a student can simply *say* "see, here is my answer" and assume that the Professor will assess a grade according to how closely that answer matches the "right" or "known" answer. But, in Senior Design, the reviewer (i.e., the Professor) does not *know* what the right answer is. It is incumbent upon YOU to prove that your answer was a "right" answer. This high burden-of-proof comes as a shock to many students because of how *involved* that effort is. You likely have never experienced anything like it in academics before. But, do not worry, remember, this textbook was written to help you navigate exactly THIS issue (among other things, of course).

To do this, you will have to rigorously test your product and then analyze and interpret results for your reviewers. YOU have to show that analysis to demonstrate the answer is "right." This requires thorough testing and investigation. But, how do you *know* that you have thoroughly tested "every aspect" of a project?

The answer is: "*Go back to the beginning and use the Requirements Document to drive your Verification process.*" Recall that the Requirements Document is the technical definition of the Project Description, and the Project Description is a statement explaining how the Project Motivation will be addressed (i.e., how the problem will be solved). So, if the Requirements Document is shown to be completely verified, then you have conclusively proven that the Project Motivation was successfully addressed.

This is one of the many reasons, the most important reason, why the Requirements Document had to be so rigorously developed – it drives *everything* else we do throughout the project, but specifically:

> The Requirements Document defines the success of the project.

> *Your schedule reflects your priorities*

The ultimate priority of any project is to successfully address the Project Motivation *on-time* and *under-budget*. Secondary priorities are defined by the Requirements Document. The plan defines *what* must be done, and the schedule defines *when* it must be done. Project Managers must intentionally manage Project Schedules so that the plan can be accomplished within the given time period.

> Project Schedules must address all priorities and be closely followed to successfully complete a project.

Initially, the Project Schedule (i.e., the Gantt Chart) may have been cumbersome and time-consuming to develop. But, without the Gantt Chart tool, coordinating all the efforts of a complicated project is nearly impossible.

In the following sections, you will refine the tools you have already produced in light of the knowledge you have gained about the Design Life-Cycle thus far. When the tools are properly completed, they will be more effective. This effort, while at first may seem redundant, is meant to set you up for successful completion of the EXECUTING phase.

18.2 Modifying Requirement Documents

Requirements Documents define and drive all project development, tasks, and priorities. Therefore, verifying the document is accurate and complete is crucial at this stage.

Technically, the time to make changes to the Requirement Document is *long* overdue. In industry a change to the Requirement Document after the CDR *will* cost somebody real dollars. However, most Senior Design Requirement Documents were not developed *perfectly* all those months ago (i.e., before you knew what you were doing). There have been numerous opportunities to review and modify the document along the way, but until the document is considered in the context of a "definition of success" at the end of the project, detecting issues with the document can be challenging.

In the context of Senior Design, reviewing and, if necessary, modifying the Requirements Document before starting the Fabrication and Verification processes may save you a lot of effort and frustration. Consider:

- Is the actual value used in each Specification reasonable and necessary? If not, the value might need to be modified.
- Can a test be defined and data collected for each Specification? If not, the Specification might need to be modified or reworded.
- If all Specifications are met, will the related Requirement inherently be met also? If not, additional Specifications may be required.
- Can a unique test be defined and new data collected for each Requirement? If not, the Requirement might need to be modified or reworded.
- If all Requirements are met, will the related Objective inherently be met also? If not, additional Requirements may be required.

Modifying the Requirements Document

If an issue with the Requirements Document is detected upon the review described above, the document should be modified immediately!

Requirements Documents may be changed in order to correct a flaw in the document.

Notice that what is being discussed here is "flaws in the Requirements Document." This is not the same things as a flaw in the design. Changing the document because a Specification cannot be met by *your* design is not acceptable. There are other ways of handling this issue that will be discussed later. If handled correctly, a "failed Specification" rarely constitutes a "failed Senior Design project" and likely has little bearing on a semester grade. In Senior Design, the *process* is more important than the *result*, and changing the Requirements Document at this point is not the correct process.

Acceptable reasons to change the Requirements Document at this stage revolve around the writing of the document itself. Example 16.1 illustrates this point.

Example 18.1: Reasons to Modify a Req Doc

Consider the following two examples where Senior Design teams identified issues in their Requirements Document late in the Design Life-Cycle. Note: both teams ultimately produced very successful products.

Case #1: A lighting concern (2021)
> *A light source was to be included in a Little Library so that books could be previewed at night. Originally, a Specification of 500Lux was used. This value was chosen based on lighting standards in a public library. Upon further investigation, it was determined that 500Lux would be blindingly bright when applied in an extremely dark environment (as opposed to a well-lit building). Therefore, the team proposed to change the Specification to 100Lux, a comfortable reading level produced by reading lamps.*

Case #2: A power-draw concern (2019)
> *A device was designed to monitor and record exposure to UV light. One Specification was that the device had to operate for 14 h on a single battery charge. The power draw of the designed product was slightly higher than originally calculated, and there was doubt whether the device could achieve 14 h of operation. Therefore, the team proposed to change the Specification to 13 h of operation.*

Analysis:

In Case #1, the original Specification did not appropriately define a successful solution. The team chose 500Lux based on a reasonable, but ultimately unfitting, reference. By changing the Specification to 100Lux, the product was actually enhanced. Modifying the Requirements Document for this purpose is acceptable.

In Case #2, the Specification of 14 h appropriately reflected the Sponsor's desire. A reduction of this Specification to 13 h was not appropriate. Specs are not defined by the product's functionality, but rather by the Project Motivation. The proper way to handle this situation is to claim a failed spec and then attempt to justify why the Sponsor should accept the product with reduced functionality. This design only missed the 14 h mark by a few minutes and was still considered an excellent solution, even with the failed spec. Changing the document was not appropriate in this case but had little baring on the team's success or grade.

A written Change Order is required in order to make a modification to the Requirements Document at this stage. This Change Order must precisely describe why the change is necessary and propose a suitable modification.

All Stakeholders must approve the Change Order before the team is allowed to implement the modification.

In the context of Senior Design, the Change Order should start with the Instructor and Advisor, and if accepted should *then* include the Sponsor. Do not address a Change Order directly with a Sponsor before discussing the issues with the Instructor and Advisor.

18.3 Req Docs Drive Verification Plans

Requirement Documents also drive the Verification Plans. These plans should be completed at CDR, but students often find that their plans need modifications or updates after CDR. This is particularly important if you made any modifications to the Requirements Document. Let us review the Verification plans one more time.

- Start by reviewing the test descriptions for each Specification:
    - Is it clear what will be constructed for the test? If not, include a Circuit Diagram or 3D model of the device under test.
    - Is it clear what equipment will be required? If not, complete the list of necessary equipment.
    - Would another student (not currently on your team) be able to perform the test based solely on the test description? If not, include more details in your description.
    - Is it clear which data should be collected? If not, provide the necessary detail.
- Next, consider the test descriptions for each Requirement:
    - Does each test fulfill the four criteria of a successful Specification test as described above? If not, take the appropriate action as described for the Specifications.
    - Does each test independently verify a Requirement? (i.e., you should not repeat a Specification-level test). If not, redefine the test.
    - Does each test prove that a Requirement can be met by a *user*, or is the success of the test dependent upon who conducts the test? Reconsider the test in the context of a general user.
- Finally, consider the test descriptions for each Objective:
    - Is there any way that constructing the final system creates a doubt in the previously collected results for the Specifications and/or Requirements? If so, you may need to re-run a previous test to validate those results.
    - Is each Objective test conducted on the *final* system? If not, redefine the Objective test. Remember, if you modify the final system after successfully completing an Objective test, you must re-run that test on the modified system (in some cases it is acceptable to perform a shortened/modified test to prove that the previous test results are still valid).

18.4 Verification Plans Drive the Schedule

The Verification Plans drive project schedule. The System Prototyping process (including the System Verification) must be complete before the Final Design Review (FDR). The FDR is typically a milestone that is scheduled for the project team by high-level Stakeholders. A good rule of the thumb for a Senior Design project is to schedule at least 3 weeks for System Verification before FDR. Before starting the EXECUTING phase, review the Project Schedule. Consider:

- Are all features scheduled to be built and tested? If not, update the schedule.
- Is there any time in the schedule during which a team member is not fully tasked? If so, rearrange the schedule to make better use of the time.

- Is there any way to expedite the schedule? If so, rearrange the schedule.
- Is there any room in the schedule for when things go wrong? If not, attempt to create some extra room.

18.5 Schedules Drive Action Item Reports

The Action Item Reports in your Engineering Notebooks provide employers (Instructors) evidence that you have put forth the required effort for your project team. But, more importantly, they provide the team with a regularly scheduled sanity check. Each week every item scheduled on the Gantt Chart should be addressed by at least one team member's Action Item Report. During each weekly team-meeting, the team should quickly review the Gantt Chart and verify that *someone* has addressed every item from the previous week on their Action Item Report.

If an item was addressed, simply record where that item was documented in the Engineering Notebook. If the item was not addressed, discuss a plan to get caught up.

18.6 Tips for the Fabrication Process

Finally, the time has come for you to enact the plan! By now your Project Management tools should be set up correctly. You know what to do. This short section presents a few tips and best practices for efficiently managing your plan.

- Stick to the schedule! Any project activity should start with a brief review of the Project Schedule. Verify you have completed and documented any overdue tasks. Double check your teammate's progress as well. Although they are ultimately responsible for their tasks, any delay they experience will affect you eventually.
- Assume a task will take longer than scheduled! Things will go wrong, mistakes will happen, plan accordingly. Leave yourself a little wiggle room in the schedule. Get ahead and stay ahead, if possible. If you are behind by more than 2 or 3 days, find extra time to dedicate to the project.
- Order extra parts! Shipping delays are very costly at this point in terms of both time and money. If there is room in the Project Budget, order a few extra parts so they are on-hand if or when they are needed. If choices need to be made (e.g., due to limited budget), consider the cost, likelihood of failure or damage, and risk to the project in choosing where to keep backup parts.
- Work independently when possible! So far, you have had primary responsibility for particular subsystems. Soon, those subsystems will start to be integrated into larger systems. When this happens, it is not uncommon to see team members begin to work more closely with partners. Occasionally this duplication of effort is unavoidable, but the team will typically function more efficiently if one member claims responsibility for the larger (integrated) subsystem while the other member moves on to other tasks. This likely means more communication is needed. Closer integration means all team members need to be aware of what everyone is working on. Looming deadlines mean that making sure each team member is regularly producing a meaningful contribution becomes critical.

An important clarification about the term *Fabrication* should be made at this time. "Fabrication" may connote the idea that this process is strictly referencing the physical fabrication of a hardware system. However:

The Fabrication process is the task during which basic software and signal development are accomplished, along with all hardware construction.

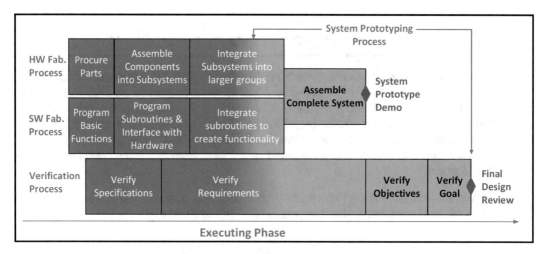

Figure 18.1 Coordination of Fabrication and Verification processes

Figure 18.1 displays a general timeline of how the Fabrication process, both the hardware and software versions, overlaps with the Verification process. This figure also introduces an additional process called System Prototyping process (light blue boxes). This process is unique because it is not a "stand-alone" process in and of itself. Rather, the System Prototyping process is really the culmination and merging of the Fabrication and Verification processes with the purpose of completing the final solution. The System Prototyping process will be explained in more detail in the next chapter.

Specifications should be verified during the early stages of the Fabrication process. The "Verify Specs" box is dark blue to indicate that these verifications should occur along with the earliest stages of the Fabrication process. The boxes entitled "Verify Objectives" and "Verify Goal" are light blue to indicate they should be conducted during the System Prototyping process.

Objectives and Goals *must* be validated using the full System Prototype.

The box entitled "Verify Reqs" is colored using a gradient from dark blue to light blue to indicate there is no clear dividing line as to when the Requirements should be verified. It is best to verify Requirements as early in the process as possible, but occasionally some Requirements may have to be validated using the System Prototype.

This figure also includes two important milestones (red diamonds). The first milestone occurs in the middle of the System Prototyping process. This is the date on which the system must be completely fabricated so that System-Level testing may being. The second milestone occurs at the culmination of the Verification process at an event called the Final Design Review. At this review, the final results of the Fabrication and Verification processes (with a focus on the System-Level results) are presented for final approval.

On the surface, the Fabrication and Verification processes are the easiest two processes to explain. The essence of the Fabrication process is "go build the thing approved at CDR." The Verification process boils down to "verify that the thing you built works." The challenge is coordinating all project tasks such that these two processes are accomplished *efficiently*.

18.7 Tips for the Verification Process

The primary purpose of the Verification process is to convince Stakeholders that the solution fulfills the Project Description and effectively addresses the Project Motivation. This requires *evidence.* Evidence, in this context, means data that was carefully collected from rigorous tests and effectively communicated. This section present tips for collecting and displaying data.

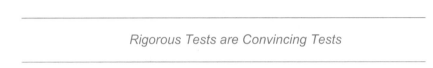

Rigorous Tests are Convincing Tests

Although this may seem counterintuitive, the purpose of the Verification process is NOT to pass all the tests and prove the design works. Passing all the tests are a byproduct of excellent engineering, but it is not the goal of this process. Rather,

The purpose of the Verification Process is to identify flaws in the design as early as possible.

Remembering this fact will change your mindset about testing. Develop tests which are meant to find the flaws as opposed to tests which are meant to be passed. Here are some best practices to follow:

- Consider whether the test might be passed (or failed) if a slight change was made to a particular condition. Would the test results be the same if:
  - The test was conducted in a different location?
  - The testing environment was different?
  - A different person conducted the test?
  Conduct the test in a worst-case environment before relying on the results.
- Do not conduct a test that has no chance of passing.
  - Conduct a controlled test first.
  - Then, conduct a test to push the limits second.
  Implementing this concept can be repeatedly scaled depending on the nature of the test. Perhaps this means that out of ten iterations of the test, the first two are conducted in a lab setting whereas the final eight iterations are conducted in the field.

Understandable Data is Convincing Data

Reviewers only have moments to assimilate the information you provide them during a presentation. If a reviewer does not understand where the data came from, or how it was collected, a shadow of doubt is immediately placed on the content you are attempting to deliver. At best, this doubt can be alleviated during the discussion portion of the review – but that wastes time. At worst, the reviewer may not believe your work and, therefore, will not accept the results.

There are four steps to presenting data in such a way that the reviewers easily understand the information and are convinced by your message. Figure 18.2 provides a visual for how you should present any data collected. If you anticipate the need to present these steps, and prepare for that eventuality while you conduct the test, you will save yourself time in the long run.

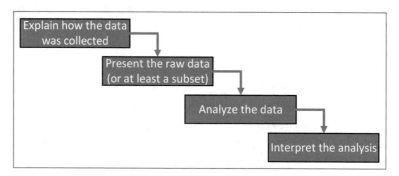

Figure 18.2 *Steps to present convincing datasets*

- STEP 1: Explain how the data was collected.
  The first step to presenting convincing data is to be clear how the data was collected. Here are some best practices to consider:

  1. Produce a communicative figure to explain the test environment and/or the Device Under Test (DUT).
     - If the DUT is a circuit, produce a Circuit Schematic *and annotate the measured signals using standard notation.* For example, if you are measuring the output voltage, label V_{OUT} with positive (+) and negative (−) polarity symbols.
     - If the DUT is a physical structure, produce a 3D drawing *and annotate dimensions and (if necessary) materials.*
  2. Take a picture of the test setup and surrounding environment while the test is being conducted.
     - Ensure that the picture clearly shows any necessary details. For example, do not allow a "rats nest" of wiring to obscure the important connections on a breadboard.
     - Include a reference, such as a ruler, to show distance.
  3. Record video of the test that demonstrates how the test was conducted.
     - Ensure the video is recording 1–2 full seconds before the event starts.
     - Verify the video is stable and in-focus.
  4. If the test includes a human subject, record who this person/group was. For example, if the test is related to the "user interface" of a device, the person conducting the experiment matters.
     - Obtain written permission to use the person's image if the subject was not part of the design team.
     - Record pertinent demographics (age, related experience, any physical attributes that matter to the test).

(Note: IRB protocols may apply when human subjects are used in some cases. Consult with your Instructor to determine whether your plan is subject to such protocols before conducting a test on human subjects.)

• STEP 2: Present the raw data.

After explaining how the data was collected, the data itself should be presented. When presenting data, use some sort of visual format (as opposed to a typed paragraph format). Typically, data is initially recorded in a table. That table *could* be presented during a review, but often formatting the data into a graph or chart perpetuates a more compelling argument. Two types of visuals are discussed below: graphs and charts.

1. If it is possible to interpolate data between samples, try formatting the data using a graph. For example, Fig. 18.3 shows a 2016 Senior Design team's data that displays Received Signal Strength (RSS) from a Bluetooth device with respect to distance away from the transmitter. Here are a few tips to consider when using graphs:

   • *Always* label the graph's axis (with units) and provide a title.

   • Data presented on the y-axis (RSS values) should be directly dependent upon the property presented on the x-axis (Distance). In other words, the x-axis is the property YOU control during the experiment and the y-axis is the value you measure.

   • When data presented on a graph is a subset of the complete dataset, a "scatter" plot should be used as opposed to a line-graph. Data points on a line-graph are *connected* suggesting that the information between data points is known. A scatterplot shows discrete data points; this indicates the data between points is not known.

   • A regression line can be used on a scatterplot to *estimate* a trend (or interpolation) between data points. This suggests that while the data between points is not known, it can be predicted.

   • Consider the resolution between data points. Recall: a poorly sampled dataset can create "aliases." For example, in Fig. 18.3, the regression line looks reasonable; however, consider the last data point. Omit the sample at 16 m and the regression line will drastically change. Perhaps the regression is correct, perhaps the trend is a decaying-exponential (as opposed to the linear trend shown) – there is not enough data between 8 m and 16 m to determine what is really happening. One more data point located at 12 m may be enough to resolve this ambiguity.

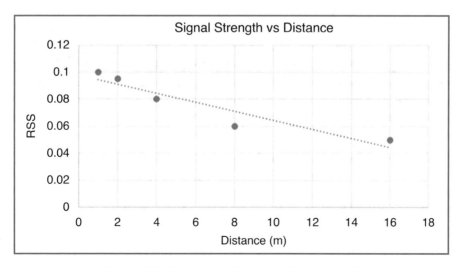

Figure 18.3 Example of data presented in a graph

2. When the data collected is not necessarily reliant upon an independent variable, try formatting your data using a chart. For example, Fig. 18.4 (left) shows a 2016 Senior Design team's data that displays whether a signal was received or not. In this case, there was no dependance (or correlation) to another variable as in Fig. 18.3: Example of Data Presented in a Graph. Fig. 18.4 (right) shows a data collection related to the usability of the solution's User Interface (UI). Here are a few tips to consider when using charts

- Pie charts are particularly useful when showing "part of a whole." In this example, the team was showing the number of successful attempts compared to the total number of trials.
- Pie charts should include a title and a legend. It is often helpful to annotate the chart itself too.
- Bar graphs are particularly useful when relating independent trials to a Specification. In this example, all three subjects failed to navigate the UI within the allotted time.
- Bar graphs should include a title, labels on the axis, and a legend.

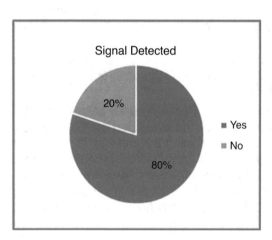

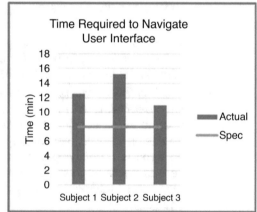

Figure 18.4 Examples of data presented in charts

Avoiding Common Pit Falls 18.1: Misleading Trendlines

Consider another graph created by the same Senior Design team that produced Fig. 18.3. This graph was produced much earlier in the Design Life-Cycle and the original dataset shows more variability in the data the refined dataset in Fig. 18.3. Regardless, the team still presented a linear regression line through the data.

Remember, a linear regression line can be produced for *any* dataset, that does NOT mean the trend is valid!

Avoiding Common Pit Falls 18.1, continued

For example, MS Excel was used to produce the regression line through the data shown below. It appears that RSS should equal about 0.08 at 6 m. Would you trust this prediction?

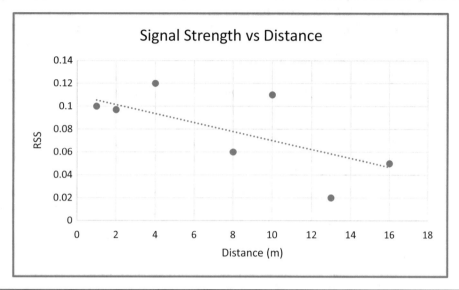

Analysis:

There is far too much "scatter" in this dataset to trust the regression. If establishing a trend in this data is important, then one option might be to make more measurements at each distance to characterize the noise inherent in this measurement.

- STEP 3: Analyze the data.

  Although presenting data in a visual manner, such as a graph or chart, communicates the information more effectively, you should not assume the reviewer automatically understands the information. As the presenter, you must articulately interpret the data for the audience. Here are some tips to consider when analyzing data:

  1. Summarize the data for the audience. Here are some example analysis statements regarding the figures above (assume that the presenters have already properly described the test itself) that would have been appropriate during the respective teams' presentations.
     - In Fig. 18.3: "Here we can see that as the distance from the transmitter increased, the RSS decreased. During our initial data collection, we doubled the distance between each sample, from 1 m to 2 m to 4 m, and so on. Although the trend looks reasonable, the jump between 8 and 16 m may have been too large."
     - In Fig. 18.4 (left): "A test was conducted to analyze whether a signal can be reliably transmitted. The test was conducted 5 times. This pie chart shows that the signal was received 80% of the time."
     - Figure 18.4 (right): "*The time required for each of the three test subjects to complete the navigation was recorded. Their time*s were compared to Specification 2.3.1 (shown in orange). As you can see, each subject failed to complete the navigation within the allotted time."

Notice that none of the previous three examples actually explain what the data *means*. What will be done about the jump between 8 and 16 m? Is receiving the signal 80% of the time acceptable? Is it "good" or "bad" that the subjects completed the navigation *over* the spec-line? Explaining what the analysis means is the last step of the process – we will get to that shortly.

2. Engineers *quantify* results! This includes presenting *averages*, and most students do that part well. However, quantifying results often constitutes more than that. Here are some additional values you may want to consider:

    • Sample size. Baseball commentators are notorious for presenting "silly" averages. "David Ortiz is hitting ground balls up the third base line at a rate of .333 (33%) when facing Joe Nathan…at home…in the rain…on a Tuesday." Of the 176 ground balls Ortiz hit in 2013 and 2014, he hit a total of three ground balls (2%) "up the third base line," one of which just happened to be against Joe Nathan. The sample size here makes a big difference in the reliability of these averages: 3/176 vs 1/3.

    • Standard Deviation. The reliability or consistency of a dataset is another important value to provide reviewers. This is typically done using "standard deviation." The data points presented in Fig. 18.3 were the average of 5 trials at each location. The raw data (all shades of orange) is plotted along with the average values (blue) in Fig. 18.5.

        – Here we can see that the data is relatively consistent across all 5 trials at 1 m, 2 m, 4 m, and 8 m but drastically less consistent at 16 m. This suggests that the average value of the first four distances probably can be trusted, but the average at 16 m is not a reliable indication of what will happen at that distance.

        – Table 18.1 displays the raw values for each trial, along with the average and standard deviation. The standard deviation for the data at 16 m is highlighted as an anomaly. While the table makes it clear that there is an issue at 16 m, the effect of that issue is difficult to visualize without the graph.

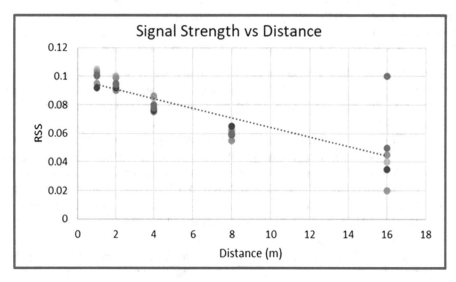

Figure 18.5 Graph with variable data

Table 18.1 RSS values measured at different distances

Distance (m)	Trial 1	Trial 2	Trial 3	Trial 4	Trial 5	Ave	St. dev
1	0.095	0.105	0.092	0.103	0.101	0.099	0.004915
2	0.090	0.100	0.092	0.099	0.094	0.095	0.003899
4	0.075	0.085	0.076	0.086	0.078	0.080	0.004604
8	0.062	0.059	0.065	0.055	0.059	0.060	0.003347
16	0.020	0.040	0.035	0.045	0.100	0.048	0.027313

- Error bars. Occasionally, displaying a table such as Table 18.1 is sufficient to illustrate your analysis. Often, displaying a figure such as Fig. 18.5 is a more effective way of illustrating the same point. However, in the case of Fig. 18.5, including the *values* of the standard deviation may be beneficial. You *could* present both the graph and the table, but that requires additional space and time.
    - A better method of presenting all this information cohesively is to plot error bars on the graph. Figure 18.6 displays the same *average* data as Fig. 18.5, but rather than the raw data points, it includes "error bars." Error bars show the one standard deviation (1sigma) of the data points at each measurement. The error bars at the first four distances (1 m, 2 m, 4 m, and 8 m) are significantly smaller than those at 16 m. This suggests that the data collected at 16 m is less reliable than the data collected at the other distances.
    - Sometimes, the raw data and error bars are not mutually exclusive information. Including both on a figure is often the best option. The raw data was omitted from Fig. 18.6 only to highlight the error bars for the purpose of *this* discussion about error bars.

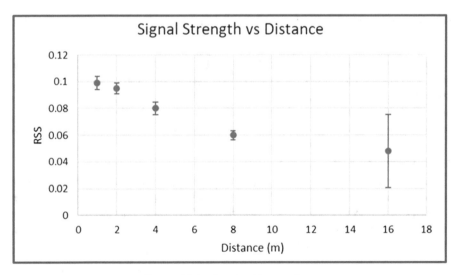

Figure 18.6 Graph with error bars

- Percent Error. Engineers quantify results. Avoid qualifying terms such as "better than," "too slow," or "sufficient size." Instead, use quantifiable statements such as "20% higher than," "3 s slower," or "5 gigabytes more than Specified." Consider quantifying your results against theoretical/expected values. Additionally, remember, that ultimately, *all* results should be quantified against your Project Requirements.

Avoiding Common Pit Falls 18.2: Presenting the "Correct" Answer

Unexpected results such as the spread in the data at 16 m in Fig. 18.5 can cause students frustration and stress, but it is important to put these results into the appropriate context so that the level of stress remains reasonable.

A Senior Design *review* is not a "test," there are not necessarily "right" or "wrong" answers. The assessment on a review is about your ability to design an experiment, collect data, and analyze that data. Meaning that the WAY you present is much more important than WHAT you present.

This type of assessment is *much* more indicative of how you will be assessed in your careers. Consider the impact of presenting Fig. 18.3 as opposed to Fig. 18.5.

SCENARIO #1: Fig. 18.3 suggests that the system is working and that RSS values will be an accurate indicator of distance. The project team likely breezes through FDR. But, when asked to demonstrate the working system, the distance prediction is only accurate some of the time – meaning the product is a failure. There is no time to redesign, modify, or account for the flaws. Worse, not only does the product not work, but the failure comes as a surprise to the Stakeholders. There is now a contentious relationship between the team and the Sponsors.

SCENARIO #2: Fig. 18.5 conveys that the students are meticulously verifying their results and are working on the problem. When confronted with troubling results at a review, Stakeholders will leverage their experience to support the team. Solutions can be found. There is now a supportive relationship between the team and Sponsors. In the best of cases, there is time to eliminate the issue. In the worst case, the issue persists and the product ultimately fails to meet Requirement, but the relationship with the Sponsor has likely been salvaged.

If your early results are not perfect, continue to work the Design Process. Collect, analyze, and present the data using the best engineering methods. Demonstrate that you are working the problem. Seek help if necessary. Most importantly, do not panic or allow yourself to become overly stressed – your grades will work out in the end.

• STEP 4: Interpret results.

The final step in presenting data is to *interpret* the analysis, sometimes this is called "drawing conclusions." Assuming that the "data speaks for itself" is a common mistake, particularly when an individual has been fixated on the testing for so long; they arrive at the review mistakenly thinking that what is obvious to the individual should be obvious to the group – it rarely is. Here are some tips for interpreting analyses.

1. Explain what must be done based on the analysis. Remember, one way to define *Engineering Design* is *"the process of making informed decisions based on engineering data."* So, you collected data, displayed, and analyzed it, now what?

   • Consider the data displayed in Fig. 18.5, recall that the analysis was "data collected at the first 4 distances were reliable and produced a trustworthy trend, the data at 16 m had a large variability and reduced the trustworthiness of the trend at distances over 8 m." So, what does that mean? To conclude the discussion about this verification, an explanation of "what is next" is required.

    – Noting that the one high outlier at 16 m is dramatically pulling the average high, perhaps more data should be collected to verify whether this is truly an outlier or whether it is a valid data point.

    – Noting that the data appears to be trustworthy through 8 m and untrustworthy at 16 m, perhaps more data should be collected between 8 m and 16 m to determine where the device's performance limitations are located.

    – Perhaps the team must request that the performance parameters of the device should be reduced to 8 m.

- There are many courses of action which *could* reasonably be enacted based on this data. The question is *which course of action do YOU believe is the best based on the Project Motivation and Requirements.*

2. Assume that every analysis requires a "conclusion" or "summary." Explain what the data means in a larger context. "Larger context" could refer to the overall system design, impact on the schedule or budget, or the usability of the product for the end user.

- Let us assume your analysis is that "the data shows the results to be under spec by 2%." Does that mean you failed spec by 2% or passed by 2%?

- If you meant to say that the results *failed* spec by 2%, is that enough to warrant a redesign? Perhaps this is so close to the Specifications, that the sponsor is willing to accept this result. Perhaps not.

- If you meant to say that the results *passed* spec by 2%, is that a sufficient margin of error based on the sample size you collected? Technically, you *did* pass your Specification, but perhaps the results are not sufficiently trustworthy – the sponsor may desire more data.

18.8 Chapter Summary

In this chapter you learned a number of tips that will support you during the Fabrication and Verification processes. Here are the most important takeaways from this chapter:

- The Design Life-Cycle works amazingly well when considered as a whole. Each process developed in the early stages support the processes at the end of the cycle.
- Before embarking on the EXECUTING phase, it is useful for students to consider the "big picture" of the Design Life-Cycle, and reconsider whether all the previously designed tools were developed in a manner that will enhance the EXECUTING phase processes.
- The Design Life-Cycle uses a top-down approach during the STARTING through PLANNING phases, but transitions to a bottom-up approach during the EXECUTING through CLOSING phases. It is important to build and test small, then build bigger and test bigger, until the completed system is used to validate the Project Goal.
- Careful consideration should be used when presenting data. Remember to consider the four-step process of discussing data for each new piece of data presented. Format the data using informative figures to allow reviewers to personally analyze the data during a review, but remember to lead them to the correct analysis yourself.

18.9 Case Studies

Let us look at data collected by our case study teams during the Verification process. You may want to review the test descriptions provided in 0 for each of these tests in order to fully understand the data discussed here.

Augmented Reality Sandbox

The AR Sandbox team started their Structure subsystem verification using FEA analysis, Fig. 18.7-left, and then by applying weight to the physical prototype, Fig. 18.7-right. The worst-case results are shown in Table 18.2.

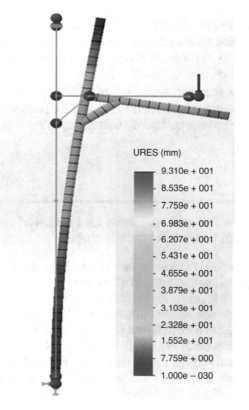

URES (mm)

9.310e + 001
8.535e + 001
7.759e + 001
6.983e + 001
6.207e + 001
5.431e + 001
4.655e + 001
3.879e + 001
3.103e + 001
2.328e + 001
1.552e + 001
7.759e + 000
1.000e − 030

Figure 18.7 Displacement analysis of structure's upright

Table 18.2 Worst-case displacement

	FEA (inches)	Physical (inches)	Difference (inches)
Worst case deflection	3.44	4.01	0.57

- **What went well:** The data showed that the upright structure was designed well and that it could hold the necessary weight.
- **What could have been improved #1:** The team should have stayed consistent in their choice of units. The FEA graphic presents results in millimeters and the table's results are in inches. Unit consistency is important to maintain throughout the entire Verification process, but at the very least within a single test description.
- **What could have been improved #2:** The difference in Table 18.2 is presented in inches. Discrepancies between theoretical and empirical (i.e., measured) data are typically presented in terms of percent error. A difference of 0.52 inches may not seem like much at first, but a reviewer may question a 16.5% error.
- **What could have been improved #3:** The FEA analysis showed that 3.44 inches (9.31e1 mm = 93.1 mm = 3.44 inches) of displacement would not cause plastic (i.e., permanent) deformation, nor structural damage. However, the physical structure displaced (i.e., bent forward) 4.01 inches. The team *assumed* that an additional 0.57 inches would not cause damage, but they never proved that assumption was valid. One way the team could have accomplished this was to further stress the FEA analysis by adding additional weight to the model until a 4.01 inch displacement was achieved. Then, the results could have been analyzed for failures. Assuming there were no failures at 4.01 inches of displacement, even more weight could have been added until a failure or plastic deformation occurred. Then, the team could definitively claim that the 4.01 inches of displacement would not cause failure *and* put a safety margin on that claim.

Smart Flow Rate Valve

The Requirements Document for this project defined Requirement 2.1 as:

System must control flow rate accurately.

To verify this, the prototype was placed in a test apparatus where the ball valve could be used to control the flow rate of a pressurized water tank. The system's flow meter was used to record the actual flow. The results were graphed along with the model's simulated results.

The first condition tested simulated the sprayer being turned on for the first time. The worst-case scenario would occur when the sprayer was commanded to output the maximum flow rate immediately. So, the first data collection started both the model and the prototype with a flow rate of 0 GPM and demanded a step-change to 150 GPM. The control system was specified to respond within 1 s. Figure 18.8 is a graph of the "Measured" data from the prototype compared to the "Predicted" data from the model.

- **What went well:** Most importantly, the prototype worked! From the graph, we can see that the Measured and Predicted steady-state values (values after 0.75 s or so) were perfectly set at 150 GPM. This means the Steady-State Error (Ess) was zero and met Specification. The Rise Time (T_{R1}) is the time required to reach steady state. It appears that the graph reaches the final condition (i.e., 150 GPM) in less than 0.7 s. The Specification was that the output must reach steady state in less than 1.0 s, and so T_{R1} also met Specification. Moreover, the real (i.e., measured) output matched the simulation (i.e., predicted) graph perfectly, except for some real-world rippling around the curve of the graph.

- What could have been improved: This graph should have included the input signal. The definition of Rise Time is the "time between the input signal's command and the output response reaching steady state." This graph leaves us to assume that the input occurred at 0 s, and therefore the system met Specifications. However, if 0 s on the graph is defined by when the output response started to change, and that change did not start for many moments after the command was given, then we have incorrectly interpreted the Rise Time results. By including the input signal, we would not have to assume where to start the rise time measurement.

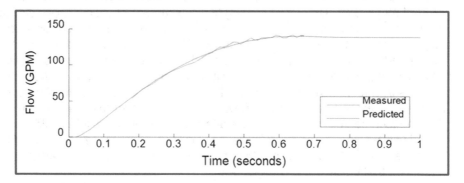

Figure 18.8 Initial flow rate response

The verification test for Requirement 2.1 continued by modifying the input command to produce a variety of step-changes. This was to simulate changes in flow rate demand while the farmer was moving through the field. The system was required to response within 0.5 s under normal operational conditions. Figure 18.9 presents the results for a sequence of four step-changes. The blue (Desired Flow) line represents the input command. The yellow (System Response) line represents the output flow rate we care about for this discussion.

- What went well: Fortunately, the team appropriately included both the input and output data on this graph. We can see that there is a lag time between the input command and the start of the output response after all. This suggests that the data in Figure 18.8 was, in fact, measured incorrectly (although it turned out that the results still met specification once measured correctly).
- What could have been improved: Although it is difficult to tell from Fig. 18.9, the system did not initially meet Specifications for T_{R2} (i.e., operational rise time < 0.5 s). A table should have been provided of Rise Time and Steady-State Error results for each step-change analyzed so that the reader could easily interpret the results. Fortunately, the team recognized that T_{R2} required longer than the 0.5 s allotted and worked to modify the control parameters. They were eventually able to meet this Specification.

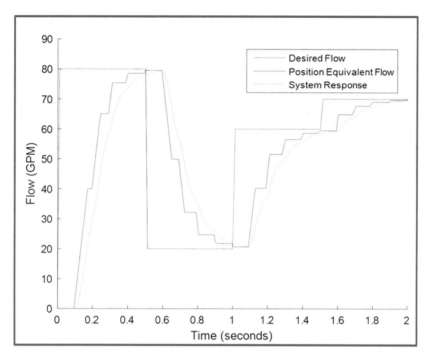

Figure 18.9 Operational flow rate response

Robotic ESD Testing Apparatus

After ensuring that the force-limiting subsystem was accurately measuring the applied force at the tip of the ESD gun with the Specification 4.3.2 test discussed in 0, the RESDTA team moved on to verifying Requirement 4.3. The Requirement test consisted of lowering the ESD gun-mount onto a load cell representing the DUT to verify that less than 1.0 lb of force was applied to the DUT. The load cell was placed at different locations and different heights to simulate a variety of DUTs.

The team had initially established a software threshold of 0.5 lb of force in order to allow a margin of safety below the 1.0 lb Specification. The Specification test discussed in Sect. 16.8 used this 0.5 lb threshold. However, the team had recently learned that 0.5 lb of force was not enough to deliver the ESD shock. Therefore, they had increased the force threshold to 0.8 lb.

Figure 18.10 displays the results of the initial 50 consecutive tests.

- What went well: Showing the results of each test on a graph like Fig. 18.10 with a threshold line indicating the Specification limit makes this dataset very easy to analyze. A reviewer can clearly interpret that all 50 trials produced less than 1.0 lb of force and therefore the system passed this test.
- What could have been improved: Technically, this test met its Pass Criteria. However, one of the data points (#42) *almost* failed the Specification. Considering that the Sponsor's real ESD testing procedure could require up to 500 test points for each DUT, having 1 sample out of 50 trials come so close to missing the Specification casts

some doubt on the validity of the data. There were not enough trials conducted to produce a high statistical confidence level in this dataset. Had the data not included this outliner, the Sponsor may have accepted the initial test results, but with that outliner, the Sponsor was understandably concerned. At the Sponsor's request, the team ran 1000 more iterations of this test and had 2 failures! The team reduced the software threshold to 0.75 lb of force and ran the test another 10,000 times without failure. Only then did the Sponsor accept the results.

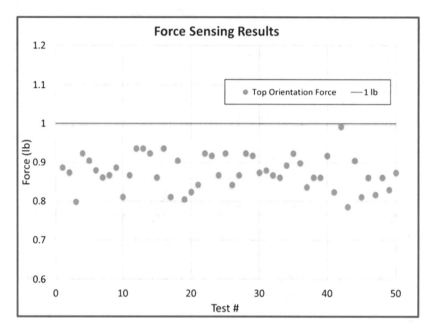

Figure 18.10   Force sensing results

RESDTA was also required to include a safety feature wherein the entire system could be powered off in an emergency. This is called an Emergency Power Off (EPO). EPO standards specify the allotted time a system has to de-energize after an EPO is triggered. These standards also define the maximum voltage allowed during a "de-energized" state. For example, systems like RESDTA are required to reduce the 24 V bus to 8 V within 0.8 s and a 5 V bus to 2.4 V within 0.4 s. These standards were tested during RESDTA's Specification 8.1.2 testing process. The test description for this Specification required the team to run ten trials of this test. The graphical results for the first trial are presented in Figure 18.11 and the entire dataset is summarized in

- What went well: The team recognized that, for this test, drawing conclusions from tabular data would be more effective than from the raw data graphs. When presenting this data, they first used an oscilloscope screen shot, Fig. 18.11, to clarify exactly how the data was measured. Then, they summarized the entire dataset in a table, Table 18.3, where average values could easily be calculated and presented.
- What could have been improved #1: The team used the "One Second" bracket to put the results into perspective. However, while this was an excellent idea, the Specification was limited to 0.8 s, not 1.0 s. It would have been better to space the brackets to the Specification limit.

- **What could have been improved #2:** The timing Specification in question here is the time between the EPO being triggered and the respective voltages dropping below their voltage thresholds as defined by the EPO standards. However, the EPO signal is not displayed. We are left to assume that the EPO event occurred immediately as the 5 V supply starts dropping (i.e., where the start of the "One Second" bracket is located). However, since the 24 V signal did not respond for some time after that *assumed* start time, it may be reasonable to assume that the 5 V signal also had a delay, meaning that the actual EPO event occurred before the 5 V signal started to drop – this of course would change all the timing data collected. As it turns out, the EPO time *did* occur where the "One Second" bracket was labeled, so the data is valid. However, the team was required to reperform the test to prove that assumption.

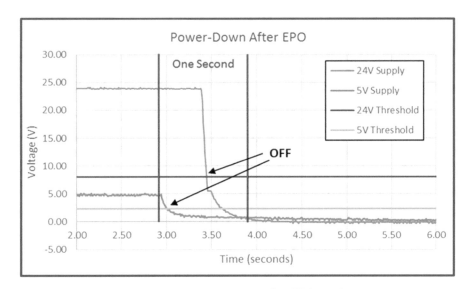

Figure 18.11   Power down after EPO results

Table 18.3   Power down after EPO data

Trial	5 V discharge (ms)	24 V discharge (ms)
1	100	390
2	85	500
3	90	490
4	75	520
5	90	490
6	90	500
7	80	500
8	90	510
9	85	485
10	95	510
Average	**88**	**489.5**

Chapter 19: **System Prototyping Process**

The last process of the EXECUTING phase is called System Prototyping (or System-level Prototyping). The purpose of the System Prototyping process is to complete the fabrication of the whole system and the Verification Process. This makes the System Prototyping process unique in the fact that it is not truly a stand-alone process. The previous two processes (Fabrication and Verification) were two independent processes that were conducted in tandem. Rather, the System Prototyping process is the culmination and combination of the Fabrication and Verification processes.

In this chapter, you will learn:

- The unique structure of the System Prototyping process.
- How to perform a System Prototyping Demo.
- How to complete the System Prototyping process.

C.J. Mettler, *Engineering Design*, https://doi.org/10.1007/978-3-031-23309-8_19

19.1 System Prototyping Process

The point at which the System Prototyping process begins is more difficult to describe than any of the previous processes. Fortunately, the *exact* starting point is not of any particular consequence. A project will gradually transition from the independent Fabrication and Verification processes into the System Prototyping process. To understand this, let us start by reviewing the structure of the *Preliminary* Prototyping process before discussing the unique structure of the *System* Prototyping process. Table 19.1 describes the important aspects of the Preliminary Prototyping process. Notice that the starting point of the Preliminary Prototyping process is the PDR and the culmination is the PPD; this is quite different from the System Prototyping process. Table 19.2 presents the same information for the System Prototyping process. The starting point of the System Prototyping process begins when the Fabrication process transitions from subsystem synthesis to system synthesis. Conceptually, this is easy to define; in reality, the transition is gradual because a real workflow rarely has clear dividing line between subsystem and system synthesis. You do not need to specifically define exactly when this transition happens, but you do need to shift your mindset from subsystem- to system-level issues a few weeks prior to the Final Design Review (FDR).

Table 19.1 Important aspects of the Preliminary Prototyping process

Entry point:	Preliminary Design Review (milestone)
Project status:	Nearly complete subsystems design
Purpose:	Implement aspects of the project to gain information by which the subsystem design can be completed
Culmination:	Preliminary Prototype Demo (milestone)

Table 19.2 Important aspects of the System Prototyping process

Entry point:	Point at which Fabrication begins on the System-Level
Project status:	Most of the subsystem have been built. Specs and most Reqs have been verified at the subsystem-level
Purpose:	Complete system fabrication and verification
Culmination:	Final Design Review (milestone)

The start of the System Prototyping process

Notice that the Preliminary Prototyping process begins *and* ends with an important milestone (the PDR and PPD, respectively). The System Prototyping process *ends* with a milestone (FDR), but this process does not *begin* with one.

The exact moment at which the System Prototyping process begins is ambiguous. The transition from the Verification and Fabrication processes into the System Prototyping process is gradual. Let us review a very generalized explanation of how these processes relate in order to understand this transition. When we verify our system, we:

1. Use Critical Components to construct "small pieces" of subsystems → Verify Spec
2. Use previously constructed "small pieces" to construct subsystems → Verify Req.
3. Use previously constructed subsystems to construct final system → Verify Obj.

In this three-step description, step 3 would be the System Prototyping process. However, the dividing line between step 2 and step 3 is rarely cut-and-dry. In reality, the EXECUTING phase follows a less-linear approach, such as:

1. A few Critical Components are developed into "small pieces" of subsystems. → A few Specifications are verified.
2. The "small pieces" are combined into subsystems along with additional components. → Remaining Specifications are verified along with a few Requirements
3. Initial subsystems are combined together at the system level and while additional subsystems are fabricated. → Additional Requirements are verified
4. Eventually the complete system emerges. → Final Requirements and all Objectives are verified

Therefore, in practice, the System Prototyping process intermittently starts and stops as project teams progressively work on steps 2 and 3. Figure 19.1 shows how the Fabrication, Verification, and System Prototyping process are correlated. Here we see that the "Integrate Subsystems into Larger Groups" and the "Verify Requirements" task-boxes are colored using a gradient to indicate that the System Prototyping process starts during these tasks, but the exact starting point is ambiguous. Fortunately, there is no need to precisely define the starting point of this process.

To avoid causing the reader confusion, let us review the evolution of this figure. This concept was first introduced in Fig. 16.2 to collaborate the Fabrication and Verification processes. The concept evolved in Fig. 18.1 when we explained that the term Fabrication referred to both hardware and software. We also briefly mentioned the System Prototyping process at that time. Now, in Fig. 19.1 we further develop the idea of that the start of the System Prototyping process is a gradual transition.

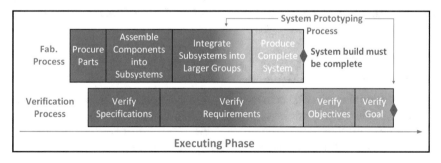

Figure 19.1 Coordination of Fabrication and Verification processes

System Prototype Demonstration

Since the starting point of the System Prototyping process is ambiguous, it is helpful to schedule a "System Complete" milestone (indicated in Fig. 19.1 by the red milestone). This is the point in time at which the system must be completely fabricated in order to allot enough time for system-level testing (e.g., Objective and Goal testing). A good standard for Senior Design is to leave at least 3 weeks prior to the FDR to complete the system-level tests. However, if you have left multiple Requirements to verify during the System Prototyping phase, you may need to allot additional time than the standard.

This milestone is occasionally accompanied with an event called the System Prototype Demo (SPD). In industry, the SPD is typically not a stand-alone event, but rather included as part of the FDR. In an academic environment, splitting the SPD and FDR often improves project results. There are a number of reasons for this difference.

- Splitting the SPD and FDR allows students to receive additional feedback on their progress before presenting at the FDR (typically a significant portion of a Senior Design course grade).
- Including the SPD as an intermediate deliverable ensures that students will have enough time to complete the system-level testing before the FDR.
- Completing an FDR that *includes* a prototype demonstration within an hour (a typical length for a student's FDR) would be difficult; there are simply too many topics to discuss. Industry FDRs can easily require many hours to multiple days, allowing engineers significantly more time to present all the necessary topics.

The System Prototype Demonstration will be discussed in more detail in Sect. 19.2.

System Prototyping Process's Culmination

The System Prototyping process, the EXECUTING phase, and therefore final the technical aspects of the Design Life-Cycle all culminate with the Final Design Review. The FDR is similar to the PDR and CDR, but focus is obviously different. The focus of the first two reviews is "the plan" (i.e., future-tense) whereas the focus of the FDR is "the results" (i.e., past-tense). This event will be discussed in Chap. 20.

19.2 System Prototype Demonstration

The purpose of performing an SPD is to demonstrate to Stakeholders that you are now able to initiate the System-Level Verification tests. Ideally, the *entire* system has been fabricated. This means that *every* feature (as defined by the Requirements Document) has been implemented. In reality, fabricating every feature by this point in the Design Life-Cycle would be an ambitious requirement for most academic projects. So, most Senior Design programs will still allow for a little wiggle room at this event. The minimum requirement at the SPD is that you show you are able to *begin* System-Level testing. This means that perhaps some packaging, cosmetic, or secondary features have not been integrated with the final system yet. However, if this is the case, you must also convince reviewers that, if you add additional features after you have started System-Level testing, that there is no possibly way the additional features will affect the test results. You may be required to reperform some of the System-Level testing you if you add features after start.

This section will define what is necessary to present at SPD and what type of features can be addressed in the final weeks of the Design Life-Cycle. This section will also discuss how to present one version of an SPD.

How "complete" must your system be at SPD?

To start this discussion, remember that ideally *every* item on the Requirements Document should be synthesized with the complete system by the time of SPD. This high standard is often not achievable by many Senior Design teams. Therefore, it is understandable if you have a few minor outstanding details to fabricate after SPD. The vast array of unique Senior Design projects makes it difficult to provide an exact description of the acceptable minimum level of completeness at SPD. What is presented here are general guidelines that can be commonly applied across a wide range of projects. You should verify with your Instructor that what you plan to present at SPD will be considered "complete enough."

- When compared to the Requirements Document:
  – At least 80% of the items on the Requirements Document should be implemented on the final system.
  – Any remaining items on the Requirements Document (no more than 20%) must still be demonstrated using subsystem prototypes.
  – All the Objectives should be significantly addressed.
- When compared to the Project Description:
  – At least one *full* process should be implemented, meaning the system is able to accept an input, process that input, and produce an output.
  – The most important features that define the project must be implemented.
- When compared to the final deliverable:
  – Any electronics should be implemented on a PCB.
  – The system should be housed in a package that is representative of the final packaging.

Here are some examples of omissions that are still commonly accepted at SPD:

- A flaw was detected on the PCB. The flaw was temporarily corrected on the original PCB using jumper wires to bypass the flaw. A second version of the PCB is scheduled to be produced to fix this flaw.
- The electronics and the housing have been fabricated, but not integrated due to the need for continual access to test points. For example, once the housing has been sealed, collecting additional measurements may not be possible.
- All important features of the system have been implemented, but some noncritical or cosmetic aspects are not yet complete. For example, the mounting structure required for final installation may be incomplete, assuming the installation structure was not a critical aspect of the project. (In this case, noncritical means "does not impede verification of anything else.")
- Multiple (identical) units are to be produced, but only one fully working prototype is completed for SPD. The remaining units can feasibly be completed and tested by the FDR.

Example 19.1: Acceptably Complete Systems at SPD

This example presents two descriptions of complete SPD systems.

1. A system was designed for a corn-harvesting combine in 2018 that would take a photograph at the start of every pass across a field without input from the driver. It would then tag GPS coordinates to each image. A user interface was developed to allow the user to scroll through Google Maps and display the appropriate image when the curser hovered over the appropriate points.
   - The system demonstrated at SPD included a camera which interfaced with the combine's steering software. The camera was triggered at the start of a pass and would produce an image. Multiple images could be uploaded to a database. The first image uploaded was tagged with GPS data and could be reviewed in Google Maps.
   - The system presented at SPD was not able to GPS-tag and review multiple images. A clear explanation was provided on how the system would be scaled and tested over the coming weeks.
2. In 2017 a system was designed to alert a user to a processing error based on a complicated set of conditions. A proprietary algorithm was developed to react to a set of seven sensors which monitored these conditions. The algorithm could not function without receiving all seven sensor's data simultaneously. The alert included an audible and visual alarm, along with a message that was uploaded to a webserver.
   - The system demonstrated at SPD included all seven sensors and the algorithm that generated the alert. The outputs were limited to the audible and visual alarms.
   - The team also demonstrated that the alert message could be generated by a subsystem-level prototype of the algorithm. Once generated, the message could be sent to the web server and displayed. The team provided an explanation of how the final subsystem would be integrated.

Analysis:

Neither of these teams produced working system by SPD. However, in both cases, the teams had synthesized all the required features and integrated most of them (including all the critical features). More importantly, both teams addressed the shortcomings of their System Prototype and presented an articulate plan for completing the _____without affecting the System-Level Verification testing.

Topics in a typical System Prototype Demo

The purpose of the System Prototype Demo event is to convince Stakeholders that you are able to begin System-Level testing; it is not necessarily a formal review. Therefore, this event is slightly less technical than the previous presentations (e.g., CDR). The focus of the SPD is typically a live demonstration of the product, although occasionally a live-demo is not feasible for certain types of projects. If a live-demo is not possible, consult with your Instructor to set expectations for your unique situation.

This section will provide guidelines for one method of SPD, assuming you are able to produce a live-demo. The topics this method will cover are as follows:

- Introduction (including the Prob. Statement, Background, and Design Concept)
- Demonstration content
- Outstanding issues
- Remaining Schedule

Introduction

Any presentation's introduction should specifically outline key information the audience requires to understand the body of the presentation. The content of the previous reviews has been primarily technical – an explanation of the subsystems you designed. The content of the SPD demonstrates that the *system* is ready to be tested. Therefore, the introduction should focus on the high-level system rather than the low-level details. The Problem Statement and Necessary Background have always been "high-level" and make a great starting point for the SPD's introduction. Conversely, thus far, the Project Design Concept (i.e., Block Diagram) slide has been used to discuss the *details* of the Requirements Document. During the SPD, the Project Design Concept should still be discussed, but limited to the high-level System Objectives.

> Focus on explaining "what the system does" as opposed to "what is required by the project's Requirement Document."

Explain "what the system does" by focusing on the how the Objectives were achieved. Provide enough technical detail so that reviewers understand the project, but only at a System level. Use the Block Diagram as a reference to assist the audience in visualizing the solution. Likely, the discussion of each Objective will include multiple blocks in the diagram. Organize your thoughts into a very articulate explanation that follows some sort of logical path through the discussion (typically this is from input-to-output, but that is not necessarily true for all projects).

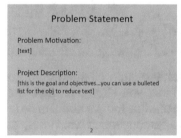 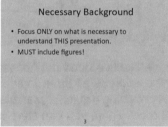 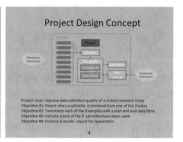

Demonstration Content

The demonstration discussion should be organized by Objectives as well. For each Objective, explain what the Objective was and how it was achieved. Provide enough detail for the audience to understand the demonstration. Remember, this is not a technical presentation;

therefore, it is not necessary to provide as many low-level details as was required in previous reviews.

Make good use of figures, schematics, and diagrams to support your discussion. Notice on *this* example slide, the first bullet-point discusses the physical placement of the Photosensors. Rather than use a figure to present this, the presenter used the physical product as the necessary visualization tool. The work probably should have been documented, but presenting the physical hardware makes for a more conclusive presentation. Conversely, using the circuit schematic on this slide was a more effective visualization tool than presenting the electronics for these topics.

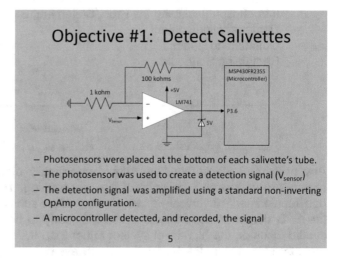

Objective #1: Detect Salivettes

– Photosensors were placed at the bottom of each salivette's tube.
– The photosensor was used to create a detection signal (V_{sensor})
– The detection signal was amplified using a standard non-inverting OpAmp configuration.
– A microcontroller detected, and recorded, the signal

5

After you have provided the audience with the necessary technical explanation, you may begin the demonstration itself. There are three general steps to presenting a demonstration, they are (in order) as follows:

- Explain what the audience should expect to see.
- Perform the demonstration.
- Interpret the demonstration results.

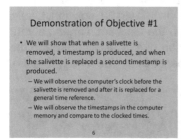

Demonstration of Objective #1

- We will show that when a salivette is removed, a timestamp is produced, and when the salivette is replaced a second timestamp is produced.
  – We will observe the computer's clock before the salivette is removed and after it is replaced for a general time reference.
  – We will observe the timestamps in the computer memory and compare to the clocked times.

6

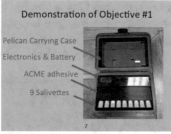

Demonstration of Objective #1

Pelican Carrying Case
Electronics & Battery
ACME adhesive
9 Salivettes

7

Interpretation of Demo #1

- The timestamps were consistently accurate to within 100 ms.

- Projects Specifications required the timestamp to be accurate to within 1 s.

- The demonstration proves the product will successfully address Objective 1.

8

The demonstration of a particular Objective does not necessarily require all three of these slides, but it does require that you address all three topics. First, you should prepare the audience for the demonstration by providing a quick explanation of what you are about to do or present. This typically requires a slide. Then, you can start the demonstration which may or may not require additional slides. If you do not have content to place on either of the following slides, you might consider including a slide with a picture as a place holder.

The content of the demonstration itself is extremely project-specific. You might perform a physical demonstration in the typical sense, but there may be better ways to showcase your work. Perhaps you might embed a video of the demonstration in the slide deck. You might simply provide a summary slide with data for *a few* features (only use this option if you truly cannot demonstrate the product directly).

After the demonstrations, be sure to interpret what the audience just observed. Explain what the "flashing light" meant, interpret any numerical outputs, and explain how "what we just saw" means you successfully completed the Objective. You may include some summary data on a slide, if appropriate. Remember, however, that the purpose here is not data analysis, it is demonstrating that you can produce the data which you will analyze in the coming weeks.

You have a lot of freedom to structure this part of the SPD so that the presentation showcases your work best. However, stay consistent throughout the presentation by presenting each Objectives with essentially the same structure and content.

Outstanding Concerns

Outstanding Concerns

1. The photosensors need to be mounted into the final housing more securely.
2. The output report must be properly formatted.
3. Final (version 2) PCB must be populated and installed.

9

After all the Objectives have been demonstrated, document any outstanding concerns. In particular, this list should include any fabrication steps that are necessary to complete before delivering the final product to your sponsor. It is acceptable if *minor* issues remain at this point, so long as they can be addressed in the coming weeks while you begin the initial System-Level tests. However, any outstanding issues *must* be revealed (documented) here.

Remaining Schedule

The final content slide is used to display an updated Project Schedule for the remained of the System Prototyping process. This schedule must address any outstanding concerns presented on the previous slide, as well as the remaining Verification process tests.

Notice, this is a very tight schedule, typically only 3 weeks long. The allotted 3 weeks *includes* the week you deliver the SPD. Progress during the week of SPD is usually necessary in order to complete the System Prototyping process on-time.

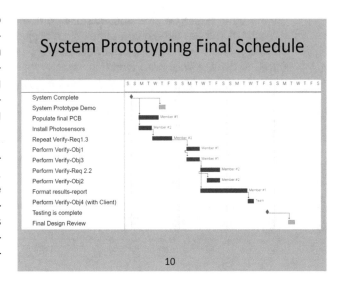

19.3 Completing the System Prototyping Process

Unlike the Preliminary Prototyping process, the System Prototyping process does not *end* with the demonstration. The System Prototyping process more-or-less *begins* with the SPD and ends with the Final Design Review (FDR). The FDR is the culmination of the System Prototyping process and in-turn, concludes the EXECUTING phase.

The remaining effort within the System Prototyping process is dedicated to completing the System-Level tests. There is not much difference here than what you have already been doing during the Verification and Fabrication processes. However, the focus is different.

During the initial processes of the EXECUTING phase, the focus was verifying the technical characteristics (Specs and Reqs) of the solution. During the System Prototyping process, the focus shifts toward System-Level concerns. Here are a few System-Level items to consider:

- The primary focus of the System Prototyping process is to complete the System-Level fabrication (if necessary) and verification.
- Testing at the System-Level is about "defining functional parameters" rather than proving a certain benchmark was met. This means that:
  - Your testing should push the limits of the product (without destroying it) so that you not only show that you met an Objective, but can define exactly how the product will perform.
  - Occasionally, a technical benchmark (Spec) cannot be met by a particular design. Ask yourself whether the product can still be considered "successful" (i.e., meets Objectives) and address the Project Motivation, albeit with reduced functionality.
  - What will the end-user's experience be when implementing your product? Consider the user's point-of-view, rather than the designer's or engineer's.
- Carefully plan and conduct your tests. You cannot afford a mistake that damages the system now! Just in case:
  - Record everything that you do so that you have evidence of your progress if something does go wrong.
  - Perform the tests that exhibit the highest risk either first or last. If performed first, you might have enough time to recover if something goes wrong. If performed last, at least all the other testing will be successfully completed and documented.
- Finish strong! You are almost complete with a project that was likely the most complicated engineering effort you have ever undertaken. Put the polish on the product and get prepared to display your hard work!

19.4 Chapter Summary

In this chapter, you learned about the System Prototyping process. This process is the culmination of the Fabrication and Verification processes. The process starts when you begin to integrate subsystems into the final system. The process ends when you have completely validated the Project Objectives and Goal. Here are the most important takeaways from this chapter:

- The process does not have a clear starting point. Rather the process intermittently starts as the Fabrication process begins to focus more on the system than on any individual subsystem.
- Part of the process includes a System Prototype Demonstration. The purpose of this event is to demonstrate that the system is ready to enter System-Level tests.
- It is acceptable that the system has not been fully fabricated at the SPD, so long as the remaining features are minor and that there is a clear plan to complete the integration of those features before FDR. Any feature necessary to at least begin System-Level testing must be implemented at this stage.
- The System Prototyping process concludes with the Final Design Review.

19.5 Case Studies

Each of our case study teams produce a fully working system prototype. Let us review each team's final results.

Augmented Reality Sandbox

The AR Sandbox prototype, Fig. 19.2, was a fully functional and interactive system. Children were able to witness a topographic map update in live-time as they moved the sand around within the sandbox. A specific hand gesture initiated a simulated rainfall and the resulting hydrology-runoff could be observed. Another hand gesture simulated the tilling and planting of corn in desired areas. The average temperatures during a simulated growing season could be defined. The predictive crop-yield algorithm calculated the health of the simulated crop in 1-inch segments across the planted area and the resulting corn-model for each square updated with an image representative of the health of the crop in that specific square. The total yield across the field was calculated and displayed for the previous five simulated growing seasons. Children had the option to simply play with the sand and topographical map or challenge themselves to grow the healthiest crop of corn possible.

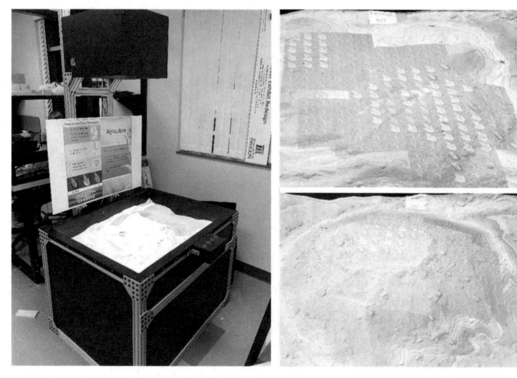

Figure 19.2 AR Sandbox System-Level Prototype. Left: Complete structure with operating AR environment. Upper-Right: Crop planted on flat ground. Lower-Right: Corn planted on top of large plateau

In Fig. 19.2, we can see the system operating in the left-hand image. In the upper right, we see a field was created on relatively flat ground. The area in the upper left was provided sufficient rain for a moderately healthy crop to grow as indicated by the yellow leaves on the corn models. In the center area, the same amount of rain was provided to the crop over a larger area. There is a natural depression in the center of the field where was able to pool up and support a healthy crop (green leaves on the model). The remainder of the field lost water due to the depression and the crop was moderately healthy or unhealthy. The remainder of the field did not receive enough water to support a crop at all. In the lower right, we see a field was created high on a plateau. Nearly all the water provided ran off the plateau to the right and created a large lake on the parameter of the image. Overall, this crop was unhealthy.

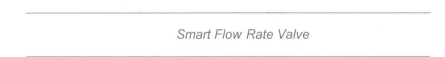

Smart Flow Rate Valve

The Smart Flow Rate Valve team produced a fully functional proof of concept for Raven. The system successfully passed every Verification test. The system had the ability to communicate with a user interface to determine the desired flow rate for a particular area of a field and then to produce and maintain that exact flow rate for as long as desired. The system responded correctly to both desired changes to the flow rate as well as any unexpected disturbances.

More importantly, the product was fully contained in a single package that significantly reduced manufacturing and installation costs. A prototype of the housing was 3D-printed as part of the proof of concept. The students also developed plans for upgrading the prototype housing to use an injection-molding solution if and when the Sponsor decides to mass-produce the product.

A cut-a-way version of the housing was also printed for demonstration purposes. This demonstration housing is shown in Fig. 19.3.

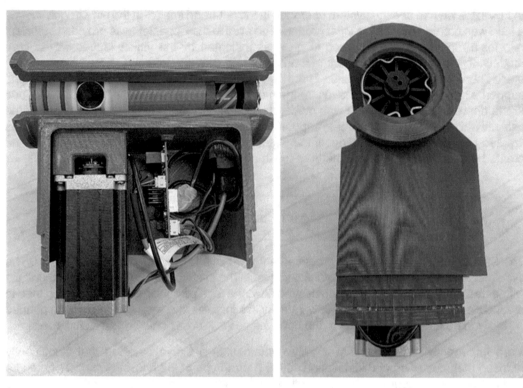

Figure 19.3 Smart Flow Rate Valve System-Level Prototype

Robotic ESD Testing Apparatus

The RESDTA prototype, Fig. 19.4, successfully addressed every item of a 12-page Requirements Document that included nine Objectives! The prototype successfully completed multiple automated ESD tests on real hard drives before delivery to the customer. Although the project was considered "expensive" by Senior Design standards, the $24,000 budget was estimated to have a payback period of less than 3 months.

The apparatus worked similarly to a 3D printer in that it articulated in the X and Y directions around the surface area of the DUT. Then dropped the ESD gun in the Z direction onto the DUT's surface. The user interface communicated with the ESD gun's commercial software to trigger a shock. The user interface also allowed the user to locate the specific contacts points for the ESD shocks and then store them in a DUT-specific file to reduce the setup time for similar DUTs.

One difficult Specification to meet stated that the DUT must be manually placed with 0.005″ tolerance. The team simply included a permanently mounting-jig on RESDTA's table. The corner of the DUT could be placed flush against the jig and orient the DUT perfectly every time. This simple solution solved another potential issue. Some of the ESD test points are located on the side of the DUT and the ones near the surface of the table can be difficult to reach with the ESD gun. At the very least, to shock these points, the head of the ESD gun was expected to rotate 90°. However, since the DUT was so easy to place, the team convinced the Sponsors to relax the Project Constraint which states that "once a test begins, no further operator interaction would be required." The final operating procedure required the operator to turn the DUT on its side partway through the test. This simple modification to the Requirements reduced the project budget by at least $5000, meaning the project was able to be completed on-budget. More importantly, there is no way the Senior Design team could have designed the extra complexity of a rotating head given the time-constraints of Senior Design. The team successfully used the Triple Point-Constraint theory in their argument to convince the Sponsor to relax the Constraints.

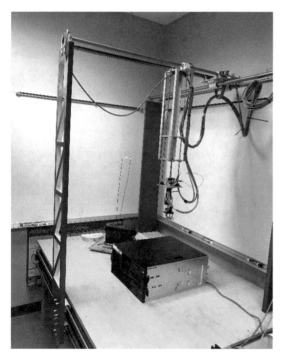

Figure 19.4 RESDTA System-Level Prototype

Chapter 20: **Final Design Review**

The culmination of the System Prototyping process and, ultimately, the EXECUTING phase is an event called the Final Design Review (FDR). At this event the design team presents the verification of their final solution. This presentation is different from all previous presentations because at this point in the Design Life-Cycle, you are no longer requesting permission to advance to a "riskier" process. Rather, you are attempting to convince the Stakeholders that you have, in fact, successfully completed the project.

> The purpose of the FDR is to verify that all Stakeholders agree that the solution has successfully addressed all items of the Requirement Document.

The burden of proof has, yet again, increased at this event as compared to all previous reviews. It is the responsibility of the project team to convince the Stakeholders that the project has been successful. This does not mean that you simply present your test results. Proving that you have met your Requirements Document means that you *use* your test results to formulate a strong and convincing argument that the project was successful. This means that you have a *complete* solution and that *all* aspects of the solution have been thoroughly considered.

As with the PDR and CDR, a well-presented solution will often allow Stakeholders to leverage their knowledge and experience to improve the results. But, hopefully, the improvements discussed at FDR are related to minor details because all the major issues have been well-designed during the PLANNING phase and executed correctly during the EXECUTING phase. Still, this is a good opportunity to seek advice from the Stakeholders to assist you in completing the final Design Life-Cycle phase (i.e., CLOSING phase).

As was the case in previous reviews, there are a number of potential outcomes of an FDR. The project may be considered a success, an acceptable attempt, or a failure. Ideally, the project is considered a complete success and the team moves forward into the CLOSING phase. However, it is relatively common to consider a project an "acceptable attempt." As we learned in Sect. 1.2, *most* projects are not successfully completed on-time and under-budget. But, there is a significant difference between an "acceptable attempt" and an "outright failure." If a project is considered an acceptable attempt, the project will most likely advance to the CLOSING phase. If the project is considered a failure, the project might return to a previous step in the Design Life-Cycle in order to improve the results or the project might be canceled.

Section 20.1 will explain the difference between a successful project, an acceptable attempt, and a failed project; this section will also describe each scenario's potential impacts to help you put *your* project results into context before FDR. The remaining material in this chapter will assist you in preparing your FDR.

In this chapter you will learn:

- What constitutes a successful project compared to an acceptable attempt and/or outright failure.
- How to format and present a Final Design Review.

© The Author(s), under exclusive license to Springer Nature Switzerland AG 2023
C.J. Mettler, *Engineering Design*, https://doi.org/10.1007/978-3-031-23309-8_20

20.1 Successes, Attempts, and Failures

Before preparing for the Final Design Review, you should understand the potential outcomes of the FDR. Figure 20.1 depicts a range of these outcomes with outright failure on the left and complete success on the right. *Understanding* this range is *vital* in an engineering profession because, as we already learned, most projects are not 100% successful (refer to Figs. 1.2, 1.3, and 1.4 back in Sect. 1.2).

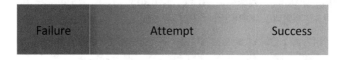

Figure 20.1 *Range of potential project outcomes*

First and foremost, the term "failed project" should be put into the proper context.

A failed project does not mean that YOU (or your team) failed!

In academics, students occasionally fail a test and have to admit "I failed that test." That is a drastically different statement than "this project failed." An academic test is intended to be a direct assessment of your ability, a project assessment is not. All successful engineers experience failed projects. Do not let this deter you from trying again. Even the most successful, high-quality engineers rigorously and meticulous apply the Design Life-Cycle skills and sometimes fail. What defines the engineer as "successful" and "high-quality" includes, among other characteristics:

- Effort and dedication
- Appropriate application of fundamental engineering skills
- Ability and willingness to learn from mistakes and improve

If you ever start to doubt yourself, consider for a moment all the projects Thomas Edison, Albert Einstein, and Alan Turning *failed* at before becoming famous household names. They continued to fail projects even *after* they became famous. Yet, they are still considered some of the elite engineering and technical minds of the twentieth century. Similarly, a failed Senior Design project does not directly correlate to *you* being a poor student or poor engineer!

The Range of Project Outcomes

A "failed" project is simply defined as a project that did not meet the Project Requirements, on-time, and under-budget. By applying the Design Life-Cycle and proper Project Management techniques we have done our best to reduce the likelihood of this happening but have not eliminated the possibility.

Obviously, on the other end of the spectrum are projects defined as "successful." These projects have fulfilled the Requirements Document, on-time, and under-budget. These are rare and to be thoroughly enjoyed.

Most projects can be classified as *Acceptable Attempts* (i.e., "Attempts" for short). The range of this category is very broad and extremely subject to personal interpretation. Understanding this range will allow you to produce better and more articulate arguments at the FDR. Obviously, the goal was to produce an unquestionably successful project. In reality, most projects (particularly the first time they are conducted) have some sort of shortcoming. The way you "sell" your project will dictate whether reviewers interpret the project closer to the success or the failure line.

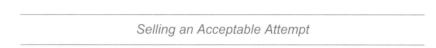

Selling an Acceptable Attempt

At FDR you will want to convince reviewers that your project was an Acceptable Attempt at meeting the Project Goal. It is incumbent upon you to provide a convincing argument for this case which reviewers understand and agree with. Of course, that is relatively easy if your product literally passed 100% of all Verification Tests. But, even then the argument is still necessary. Most projects cannot be defined as 100% successful in all aspects under consideration, but many of these projects will still please the Stakeholders. Providing a positive interpretation of the results is sometimes called *selling* your project.

To be clear:

Selling is not synonymous with *misleading*, either by falsehood or intentional omission.

Selling, in this context, is the art of convincing someone to interpret a situation in a positive light. This is not usually required of engineering students in traditional courses. For example, in a junior-level Linear Controls course, a student may be asked on a test to determine the necessary parameters of a certain type of controller to achieve a steady-state error of 2%. Of course, the professor created the problem herself and knows the precise answer(s). If you understand and apply the appropriate skill set, your answer will match the professor's and, therefore, be considered "correct." Any deviation from this known answer would be considered "incorrect."

Project assessment is not nearly so cut-and-dry. Theoretical calculations used on a test typically do not include complex nonidealities, environmental noise, or component variability. Perhaps in reality, your calculations suggested that your design should achieve a 2.0% error, but the product demonstrated a 2.1% error. Is this a success or failure? Technically, it cannot be considered a success – you missed a Specification. But, does that mean the project was failure? That all depends on how well you justify your work.

- Perhaps the Specification could be allowed some additional tolerance because the Requirements and Objectives were still met by your solution.
- Perhaps you *could* have met 2.0% error but at a significant cost increase which would have resulted in the project being over budget. Is it better to have an "unsuccessful" project because of slightly missing a Specification or due to a drastic cost overage?

- Alternatively, perhaps you *could* still meet the Specification by replacing certain components with higher tolerance versions, but this will require a time overrun. It may be better to conclude the project with the failed Specification and address the known issue on the "next revision."

The point is that there are so many additional realities that a project must consider other than just the technical values. It is necessary to constrain an academic test in order to fairly assess an individual student, but a design engineer does not have that luxury. At FDR you will need to put all the project results into context and draw reasonable conclusions. Do your best to convince Stakeholders to view your results as positively as possible.

Get support before presenting

Of course, *surprising* a Stakeholder with a less-than-ideal result at an FDR is never a good strategy. Communicate any issues long before the FDR and do your "selling" then. If you have an external Project Sponsor, discuss the issues with your Instructor prior to contacting the Sponsor. Procure everyone's buy-in (i.e., agreement) *before* the FDR so that there are no surprises. Document these agreements in your Project Notebook.

20.2 Formatting an FDR

As we have seen a number of times, the exact formatting of a review is highly dependent upon the preferences of the primary Decision Makers who are contributing to the review. So, again, the exact format of what an FDR *should* be is not what is important here. What is important are the fundamental concerns of all FDRs and your ever-increasing ability to apply the organization's formatting preferences to *your* project.

Figure 20.2 depicts the general flow of a typical FDR. The introduction to the FDR yet again includes the Problem Statement, Necessary Background, and Design Concept. The design and verification of each subsystem is presented in enough detail to convince the Stakeholders that the appropriate Specifications and Requirements were successfully addressed. Then, the integrated system is presented along with the high-level verification (i.e., the Objectives and any outstanding Requirements). Next, the Project Status and Final Budget are reviewed.

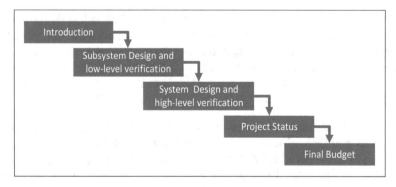

Figure 20.2 Typical topic-flow of an FDR

The most important issue to consider in structuring and organizing your presentation is:

> Presenting the *complete* design thoroughly, but concisely, so that the Stakeholders will believe the Verification results.

The order which the design topics of the FDR are presented is typically the same as the CDR. You will present the first subsystem's design and the related verification. Then, you will present the design and verification of the second subsystem and eventually the integration of verification of the system. Of course, at CDR the verification discussion was about the *plan* and at FDR the discussion is about the *results*. While your CDR slides will provide a reasonable starting point for developing your FDR slides, consider any confusions or questions the reviewers had about your CDR presentation and improve your presentation to avoid repeating the same issues. Sometimes even a slight modification to how you present this material can alleviate those questions and concerns.

20.3 An FDR Template

This section describes one method of presenting an FDR effectively. Keep in mind that the purpose of performing an FDR is to convince the reviewers that you have designed a complete solution that has successfully addressed the Project Motivation and fulfilled the Requirements Document. The topics this method will cover are:

- Introduction (including Problem Statement, Background, and Design Concept).
- Subsystem design and Low-level Verification
- System Design and High-level Verification
- Project Status
- Conclusion (including Final Budget, Bill of Materials, and Summary)

Introduction

The introduction of the FDR includes the Problem Statement, Necessary Background, and Project Design Concept slides. These slides should be essentially the same as what was presented at the CDR (albeit with any improvement or updates necessary).

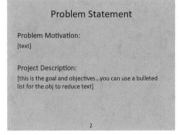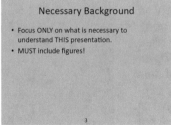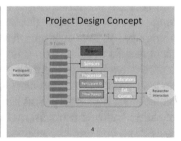

Subsystem Design and Verification

After the introduction, present the Finalized Designs and Verification Results for each subsystem. Start with a slide showing the Final Design of the subsystem, and then present a set of slides documenting the Verification Results specifically related to this subsystem. Repeat the pattern for all important subsystems. You may not have time to cover every detail of every subsystem within the allotted timeframe of the presentation, and you need to prioritize the upcoming System-Level discussion over the Subsystem-Level discussions. So, you will need to be judicious about what you present.

At this point, there is not a distinction between "Critical" and "non-Critical" subsystems. If you have already thoroughly presented the details and results of a given subsystem, you could perhaps limit the FDR discussion on this topic and prioritize other important aspects of the design. Do not go so far as to eliminate the entire discussion of what was once a Critical Subsystem in order to present a trivial topic. Rather, prioritize the material to provide representative evidence of the overall quality of your work. Review Sect. 18.7 for best practices on presenting data (i.e., Present, Analyze, and Interpret data).

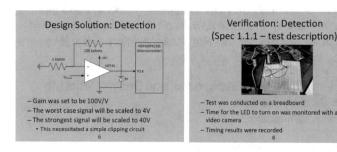

There likely will not be enough time for you to present all the data for every Verification test you performed on a subsystem. In the interest of time, choose a representative subset of verification topics to present for each. Perhaps one method to do this is to select:

- One Specification test
- One Requirements test
- Any/all test that failed to pass

If you do limit the amount of Verification Results presented for a subsystem, you should still document a summary of the subsystem's complete results in a table on a final slide before advancing to the next subsystem. A detailed explanation of this table is not necessary. You may simply state "here is a summary table of results, as you can see every item not previously discussed, passed."

Rely on effective engineering figures to present this material. Effective figures will reduce the amount of time you will require to present each topic.

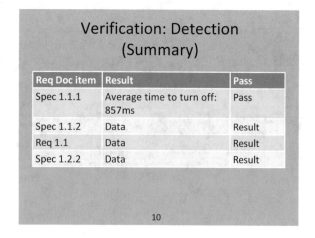

System Design and Verification

After presenting the subsystem material, you will present the System-Level Design and Verification

> The System-Level Design and Verification should be the primary focus of the FDR.

This means you will have to prioritize and scale the subsystem material to leave sufficient time for this discussion. A general guideline is to spend more than a third of the allotted time on the System-Level topics.

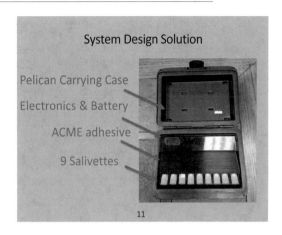

During the System-Level Design discussion, present the completed system's hardware, electronics, and high-level processing. When organizing your presentation, you might consider presenting the product as if you were doing so to the end-user for the first time. Consider, what would the end-user need to know first, and then build upon that by adding more detailed information. Focus on the integration of the subsystems which you have already presented. Consider displaying the physical system as a supplement to the presentation slides. Take caution here: displaying the physical system does not replace the need to completely document the Design and Verification in the slide deck.

After presenting the System Design, the System-Level Verification should be discussed. At least every Objective should be addressed. If time allows, you could also discuss any of the Requirements which you tested during the System Prototyping process.

If your design did not perfectly address every Req Doc item, *sell* your solution. Convince the Stakeholders that this project resulted in an Acceptable Attempt. Do not hide any flaws or shortcomings, but do explain how the system might still be useful to the end-user, if at all possible.

Project Status

Project Status

- Containment system and electronics are complete
- During System Prototyping process:
  - The battery was determined to be undersized by 5%. A replacement battery with 20% larger capacity is on order, should arrive this week
  - The sponsor has requested minor modifications to the output formatting. This will be addressed within the week

The Project Status is summarized after the System Design and Verification is presented. Ideally, you are able to state "the project is complete" on this slide. More likely, there are a few loose ends that need to be wrapped up.

Specifically document what aspects of the project are complete and which tasks still need to be finished. Be sure to document *when* the tasks will be addressed. If there are more than two simple tasks that need to be addressed, present a schedule with a Gantt Chart.

Conclusion

Typically, the FDR concludes with three additional slides. Present the finalized Project Budget, the Bill of Materials (BoM – pronounced *bomb*), and the Summary.

The Project Budget should clearly display how much of the allotted budget was spent throughout the entire project. Additionally, many sponsors will want to know the cost of reproducing a single iteration of the final solution (i.e., unit cost). Occasionally, sponsors may also want to know the itemized cost of a scaled production, *for example*, "what is the cost of one product if 50 products are simultaneously produced." Normally, the budget does not have to be presented in much detail at this time; a summary is sufficient. In fact, if the Budget Table does not fit on a single slide, the table can be reduced to a summary-version of the full table. Of course, if there is an issue with the budget (e.g., you expended more than the allotted amount) you may need to spend additional time on this slide and present the full set of details.

The BoM is a list of any and all items that would be required to reproduce a single iteration of the solution. For each item, specify the quantity and the purchasing information (i.e., provide a link to order the item). This slide also does not have to be presented in any detail. Simply introduce the slide and move on.

The last slide of the FDR is a Summary (i.e., Conclusion) slide. When you are developing your Summary slide, review exactly what you stated on the Problem Statement slide. Phrase your Summary to directly correlate with the Problem Statement. Address whether the solution met the Project Description and what the greater impacts of the project in the context of the Project Motivation.

PART VI: THE CLOSING PHASE

"It is time for us all to stand and cheer for the doer, the achiever – the one who recognizes the challenges and does something about it."
– Vince Lombardi, Head Football Coach & Civil Rights Forerunner

"The only real mistake is one from which we learn nothing."
– Henry Ford, American Industrialist

"You cannot connect the dots looking forward.
You can only connect them looking backwards."
– Steve Jobs, Business Magnate

Steve Jobs cofounded Apple, Inc. in 1976 [34] with Steve Wozniak. Their first product, the Apple I, spearheaded what would become the most valuable company in the world by 2018 [35]. But, Jobs did not have "it all figured out" in 1976.

The Apple I was originally shipped without a keyboard or monitor, no pre-loaded software, and no casing (the wiring and circuitry were exposed). Their first Sponsor, Byte Shop, almost did not accept delivery and had Jobs not "stared down [the owner of Byte Shop] until he agreed to take delivery and pay," Apple might have died right then and there [36].

Without a doubt, the Apple I eventually became a resounding success story, launching Jobs and Woz out of their garage and into international recognition. Fortunately, Jobs recognized that even though the Apple I was a success, there were issues to be addressed. When the Apple II was released in 1977, it was sold with a keyboard and color graphics. More importantly, by 1979, the groundbreaking machine was sold with software, VisiCalc – the predecessor to the modern-day spreadsheet.

In the early 1980s, Jobs continued to innovate. At that time, he led the development of a new product, the Apple Lisa. The Lisa was powerful (it was one of the first computers to boast a Graphical User Interface (GUI)) but failed to catch on commercially. Declining revenues and massive research expenditures had put the company in trouble, and by 1982 Jobs was removed from the project leadership. The Lisa was released in 1983, without Jobs at the helm, but the product never lived up to revenue expectations. In the meantime, Jobs had been reassigned to the development of the Apple Macintosh where he again demonstrated the ability to learn from previous experience. The Macintosh GUI was improved compared to Lisa's, and its significantly lower price-point made it a commercial juggernaut. But most notably, the "attractiveness" of the casing was, for the first time, a primary design consideration – an attribute that Apple has remained committed to today.

Although the Macintosh was another success, it did little to improve Apple's market share. In the mid-1980s, IBM *dominated* the technology world, at one point accounting for 6.4% of the S.&.P.500, making it 2.35 times larger than the next largest company, Exxon [37]. No company has ever had a larger lead on its competition before or since.

At the same time, Apple was still fighting to survive. Not surprisingly, Steve Jobs, the cofounder of the company and leader of its two biggest products, *was not* who the board members wanted leading the company. The problem was that Jobs had earned a reputation for being difficult to work with. "He sweated the details, often at the expense of his team's feelings" [34]. His projects were notorious for scope-creep and often missed important deadlines.

After losing a "boardroom battle" with then-CEO John Sculley, Jobs resigned from the company he founded on September 16, 1985 [38], and started a new company, NeXT. Possibly the most important contribution NeXT ever made to the world was to provide Jobs with experience as a CEO. In that position, he refined his management style, and on September 16, 1997 (12 years to the day), Jobs returned to Apple as a competent and capable leader.

With Jobs now at the helm, Apple developed into a company with an extremely high employee satisfaction. Job's philosophy that "it doesn't make sense to hire smart people and tell them what to do; hire smart people so they tell you what can be done," infused employees with a sense of mission and encouraged them to "think different[ly]."

Neither Apple, nor Jobs himself, demonstrated a story of *continual* success; there have been many ups and downs throughout their history. One attribute that sets Apple (and Jobs) apart from the competition is the impressive ability to learn from past experience and constantly innovate better solutions.

The cliché phrase, "learn from your mistakes" is important, but not enough. Jobs continually learned from his failures *and successes*. The Apple I was successful, but the Apple II was better. The Lisa was ultimately a failure, but the same mistakes were not repeated on the Macintosh. Early packaging solutions were functional, but Jobs recognized the value of *attractive* products, a feature which has set Apple apart for its competition. The young Steve Jobs of the 1980s alienated teams to his detriment; the experienced Steve Jobs of the 1990s fostered a challenging environment engineers fought to be a part of. He never gave up "sweating the small stuff," but he learned to do so in a way that inspired greatness, rather generate contempt. Steve Jobs continually recognized "the great challenges" and did something about them.

The CLOSING phase of a project is an opportunity to "look backwards and connect the dots." Whether the product was a smashing success or a dismal failure, there are always lessons to be learned from the experience. These lessons are often impossible to recognize while you are in the midst of the experience. The CLOSING phase is a time to review the technical product *and* the non-technical soft skills applied to the project. It is a time to consider what was done well, what could have been done better, and what opportunities there are for "the next adventure."

During the CLOSING phase, the technical work is completed and documented, information is archived, and the team is formally released to pursue other endeavors [39]. Just as importantly, processes are also taken to improve the future success of the organization and the individual. There are many processes in the CLOSING phase, including Confirmation of Project Completion, Completion of Paperwork, Deliverable Hand-off, Release of Resources, conduct an After-Action Retrospective, Documentation Archival, and...most importantly...Celebrate. Many of these tasks are not applicable to a Senior Design project. This textbook will focus on those tasks that are applicable to you [40].

Processes of the CLOSING Phase	Presentation Structure
Confirmation of Project Completion	**Design Fair**
Completion of Paperwork	**Final Project Report**
Deliverable Hand-off	**Final Sponsor Meeting**
Post-Mortem	**Lessons Learned Checklist**

Chapter 21: **Closing Phase Overview**

This chapter will provide a summary of the CLOSING phase processes and explain which are typically addressed in Senior Design.

Although it is tempting to overlook much of the CLOSING phase, this phase of a project is equally impactful to an organization as any of the previous phases were – particularly, when you keep in mind that a project was defined as a "temporary endeavor." Most organizations have business goals that far exceed the timeframe of any individual project. The purpose of the CLOSING phase is to ensure the impact of a project far-outlives the project team. The CLOSING phase benefits the organization in a number of ways [41]. Some of the process in this phase are meant to protect the organization from legal risk by ensuring all of its obligations have been met. Other processes attempt to capitalize on the anticipated benefits of a project – in particular, they make sure the organization is properly compensated for the engineering work provided. Yet other processes are meant to protect the organization from losses. With the legal and financial issues already addressed, the "loss" this final category of processes is concerned with are typically *information, quality, or skill* losses. Essentially, the lessons learned are reviewed and archived for the future benefit of the organization and/or individual engineer rather than to benefit the project itself. This chapter will provide a brief overview of a few of these processes that directly relate to Senior Design. This chapter will not attempt to cover every process within the CLOSING phase.

In this chapter, you will learn about:

- Confirmation project completion
- Completion of Paperwork
- Formally transferring all project deliverables
- Post-Mortems
- Archiving information
- Releasing of resources

Each of these topics are significant processes in their own right, but due to the compressed nature (and other logistics of Senior Design), not all the process are directly applicable to you. The following sections will provide a brief background on each process and insight as to how these processes relate to Senior Design. The following chapters will discuss the details of the few remaining processes that are necessary to complete the course.

C.J. Mettler, *Engineering Design*, https://doi.org/10.1007/978-3-031-23309-8_21

21.1 Confirmation of Project Completion

The CLOSING phase should start by confirming that the project is, in fact, complete. This process should start with a review of the Project Charter and any Change Requests that were approved. This may seem trivial on your Senior Design project, but on a much larger, more complex project, it is not inconceivable that important items were inadvertently omitted. By conducting a formal review, you ensure that nothing was overlooked.

This effort is often undertaken in stages. There should be a thorough internal review before conducting a review with superiors, Sponsors, and/or the public. This might mean that there are two, three, or many more reviews before the team is confident that *all* Stakeholders agree that the project is complete.

Typically, a live demonstration is conducted at the finale of this review. In Senior Design, that event is called the Design Fair or Design Expo. This event is an important public occasion at which the design team demonstrates to a large audience that the project was completed successfully.

21.2 Completion of Paperwork

The Completing Paperwork process can be an involved and complex process. It may include a legal review of contracts and/or a financial review of invoices and payments. These reviews will involve departments internal to your organization (such as human resources or marketing) and external organizations (such as vendors and subcontractors). Obviously, this portion of the process is to ensure that "if the work was not completed, the payment is not exchanged" [40]. This probably does not relate to most Senior Projects.

Another important aspect of this process that *does* relate to Senior Design is the need to compile the final documentation. Typically, a Final Project Report (FPR) is produced that documents how each item of the Project Charter was addressed. This includes a detailed design description and explanation of the verification results. The FPR may also contain information related to budgets, future work, and other important information which an organization does not wish to lose.

21.3 Transfer of Deliverables

All project deliverables must be transferred to the appropriate owners. Obviously, this includes the product itself but often includes other deliverables as well. The support-team will require user manuals, the sales-team will require technical specifications, and the installation-team will require construction instructions, just to name a few.

This process applies to a *subset* of Senior Design projects. Probably the most common nontechnical deliverable is a user manual. You should review the "Additional Project Concerns and Deliverables" section of the Charter and verify that you have produced and delivered all items requested by your Sponsor.

A Final Sponsor Meeting should be conducted at some point near the end of the project. Often Senior Design teams are not able to formally deliver the product before the Design Fair, but waiting until after the Design Fair presents problems as well; namely, if the Sponsor has any final concerns, you will have very little time to react to them. It would be a good strategy to meet with your Sponsors during the Verification Process and again during the System Prototyping process in preparation for the final meeting with the Sponsor.

Before the final meeting with the Sponsor, ensure the following:

- The sponsor is well-aware of the status of the project – no surprises at the end.
- You are prepared to deliver all aspects of the project to the Sponsor – this means that you have coordinated with the Sponsor and have agreed upon what will be delivered.

Sponsors often wish to see a demonstration of the product as well as the details of the Verification Results at the end of the project. This event can be time-consuming and it may not be possible to limit the discussion arbitrarily to 1 hour like most course-required events. You may need to plan a full (or part) of a day to complete delivery of the project. Occasionally, students may travel to the Sponsor's facility for this event. Ensure you have arranged travel if necessary (this may require some coordination between the university and the Sponsor). Presenting some of the data and/or demonstration during pre-meetings may reduce the duration of the final meeting.

21.4 After-Action Retrospective Meetings

A CLOSING phase event is often held at the end of the project for the purpose of analyzing the soft-skills of a project after the project is finished. This particular event has many different names including Project Debrief, Post-Mortem Meeting, or After-Action Retrospective Meeting. Regardless of what your organization calls this event, an effective Project Manager dedicates time at the end of each project for the purpose of improving the organization's Design Life-Cycle strategies.

Occasionally these meetings are considered a "waste-of-time" and it can be difficult to get Stakeholders to support the effort - after all, this event does not improve product functionality, and always requires additional resources and time. But only people who lose sight of the big picture adhere to that viewpoint. The completion of a project should *never* be your end goal. Ultimately, you should be striving for a successful career, and the organization should be developing a sustainable business. These goals are not achievable if you (or your organization) do not learn and improve from every experience.

> The After-Action Retrospective Meeting is one method of ensuring continual improvement for the individual, the team, and the organization.

This effort is incredibly important to Senior Design students. You may not have fully comprehend the *entire purpose and impact* of every task which you were asked to undertake this year, particularly early in the Design Life-Cycle. Hopefully, you gained valuable experience and insight as you progressed through the curriculum. Taking a moment at the end of the project to "look backwards and connect the dots" may uncover insights you did not even realize you gained. This effort should solidify the lessons you did learn so that you can more effectively contribute to the next project and the one after that.

21.5 Archiving Information

You probably were surprised at the volume of documentation you created for this project, and in the broad scheme of things, your Senior Design project and team were relatively small compared to a complex industry project! Image the documentation that would necessarily be developed for a multimillion-dollar, multi-year engineering construction project? That information *must* be archived in such a way that the information is not lost and can easily be reviewed in the future. Losing documentation can cause companies to lose law suites and patents. That loss will certainly cost time and efficiency in the long run.

While this process is critically important to companies in industry, it is probably less of a concern to a Senior Design project. Few Senior Design projects are ever revisited, and it is difficult for a university to archive the volume of documentation created by all the teams over all the years. Just in case, consult with your Sponsor and Advisor to verify whether there is anything special about your specific project that requires archiving.

Industry Point of View 21.1: Formally Close All Projects

I used to work for a company that did not appreciate the value of following good Project Management practices. I was part of a ten-person team writing proprietary user interface software. It was a massive job with a projected schedule of 3 years. After 2.5 years, upper management halted work and directed us to research and evaluate alternative methods against what our strategy was. We spent the next 4 months doing the research and concluded that our original plan was the correct one. Management agreed and instructed us to continue software development. Looking back, I now realize, these 4 months would not have been wasted had we performed an appropriate Alternatives Analysis early in the project.

About 1 month before we completed the work, our project was canceled because the company's priorities had evolved. Our design team was dissolved and we each went on to work on other projects for the same company. Six months later, management reversed its decision; the company really did need our project completed. I was asked to reassemble the team and complete the software. But, I knew this was essentially impossible. Due to our failure to adhere to good practices, we had not archived our work, we had not developed any lessons learned documentation. When the project was canceled, we all just dropped everything and went on to the next assignment – I mean, the project terminated, we assumed we would never pursue that idea again, why waste time getting started on the next opportunity?

I estimated it would require at least 1 month just to gather the team considering they were currently assigned to other priorities. We would need another 3 months for everyone to back trace the work that had been done just to figure out what was left to complete. Then, we would have to complete the unfinished work that was originally scheduled to require 6 months. In the end, the company decided the project would require too much effort to resurrect and that the benefits were not worth the effort. They killed the project a second time.

A few days of proper CLOSING phase efforts when the project was first terminated would have saved months of effort later and possibly allowed the company to peruse the project benefits when they were most needed.

21.6 Releasing of Resources

Projects are defined as "temporary endeavors" and eventually all good things must come to an end. A project team is designed to specifically support a particular project, meaning that project teams do not outlive projects. One of the last tasks a Project Manager must conduct is to formally release the project resources to be free to pursue other opportunities. For a large company, this may simply constitute notifying Decision Makers that the project team-members are now available to be reassigned to other projects. For a small consulting firm, this may constitute a major restructuring of the entire company.

One of the most important considerations a young engineer might have during this process is to preemptively procure a reference from team leaders or managers. Requesting a reference immediately after a successful project completion will most likely result in the most positive and comprehensive review you could possibly expect from that manager. Whereas waiting until you actually need the reference might mean that the success is not quite as fresh in the manager's mind.

This process is critically applicable to Senior Design. It may just be the most momentous Release of Resources process you ever take part in. Of course, in the university, we do not call it "Release of Resources," we call it "Graduation." To prepare for that event, course instructors will assess your work and assign a grade as in most courses. Additionally, Senior Design students often request Letters of Recommendation from Instructors and Advisors. A more unique reference can sometimes be procured from your Project Sponsor. If you worked closely with this person and produced an Acceptable Attempted project, a positive reference from your Sponsor will impress many potential employers down the road. Consider how valuable that reference might be if your goal during an interview process is to "stand out." Many student-applicants provide employers with Letters of Reference from university faculty, but not many can provide a positive reference from an engineering client (i.e., your Sponsor).

21.7 Chapter Summary

This chapter provided a review of six categories of CLOSING phase process groups. Within these groups, there are four processes which are applicable to all Senior Design projects. These include the following:

- Design Fair
- Final Project Report
- Deliverables Meeting
- After-Action Retrospective Meetings

Additionally, you may want to consider whether any of the following issues may apply to your unique project:

- Any outstanding invoices
- Any archival needs
- Any references you may want to procure

Chapter 22: **CLOSING Processes (Completion)**

Processes of the CLOSING phase can be split into two broad categories, each of which address an important project-related concern. One category of process ensures that all project-related obligations have been completed. The other category ensures that the knowledge and skills developed during the project are not lost after the project team disassembles.

This chapter will focus on ensuring that all project related obligations have been completed. At first glance, it may seem that these processes are redundant after the Final Design Review. However, that is not true! The purpose of the FDR was to demonstrate that *what had been built was built successfully.* The purpose of these CLOSING phase processes are to demonstrate *that everything needing to be addressed was in fact addressed.*

These two purposes support each other but are distinctly different for two important reasons. First, the FDR demonstrates what now exists, whereas the CLOSING phase addresses what potentially does not yet exist. This becomes more important the larger the project becomes; there is a reasonably high probability that something was "missed" on a project whose Requirements Document was a few hundred pages long, which has been muddled with changed orders, and whose Project Manager role was covered by 18 different individuals. Fortunately, these are not typically concerns for a Senior Design project and this set of processes will be relatively straightforward for you.

The other reason that the CLOSING phase is still necessary despite having preformed an FDR is that the FDR only addressed the technical aspects of the project. However, the CLOSING phase processes review the technical and nontechnical aspects. Nontechnical aspects may include user manuals, financial statements, and/or closing subcontractor agreements.

Each of the processes within this category may have *many* different tasks depending on the complexity of the project and project team. This chapter will limit the discussion to one task in each of three processes directly applicable to completing your project.

In this chapter, you will learn:

- How to produce a Final Project Report.
- How to prepare for and present at a Design Fair.
- How to prepare for and conduct a Deliverables Meeting.

© The Author(s), under exclusive license to Springer Nature Switzerland AG 2023
C.J. Mettler, *Engineering Design*, https://doi.org/10.1007/978-3-031-23309-8_22

22.1 Final Project Reports

The Final Project Report (FPR) document is related to the technical aspects of the project. This includes the Problem Statement and Project Requirements, the details of the Technical Solution, and the Verification methods and results. You have already presented all of this information in previous presentations. This section will provide some high-level insights into writing these sections, but you already have developed most of the skills required throughout the last many months. Apply all the best practices you have thus far developed to produce a readable, informative, and comprehensive report.

This section discusses the categories of topics required in the FPR.

Preface Pages

The FPR typically starts with a few preface pages. These will include a Cover Page, Signature Page, and Technical Summary, followed by a Table of Contents. You may also wish to include a List of Figures.

The Cover Page is typically standardized by your organization so that all formal reports have the same official appearance. You should not change the Cover Page text, but you may wish to add a picture of your product or project team for embellishment.

When you sign the Signature Page, you are claiming that "this is your best work" and that you base your professional reputation upon it. This would be where a Professional Engineer (P.E.) would "stamp" the design and quite literally stake their career on the veracity of the work documented within. Often, an Advisor is also asked to sign this page certifying that the students' work is acceptable to be released to the Sponsor – this is particularly important if the Sponsor is external to the University.

The Technical Summary provides a high-level abstract of the project for an executive. Often, the Decision Makers will not read the entire report, they will only read the Technical Summary. So, the entry should include the Problem Statement, Design Concept, and an overview of the Technical Solution and Results. Be sure to conclude the Summary with a strong conclusion paragraph clearly stating whether the project was a success or not. In some cases, it is appropriate to include one or two pieces of data (probably graphs or charts) that clearly depict a critical result.

Introduction

The Introduction material includes the Project Motivation, Project Description, and Project Background. Most of this should be nearly-identical to same sections you wrote for the Project Charter. Of course, you should review verb tenses throughout these sections. The Project Charter described "what needs to be" (future tense) and the FPR documents "what was" or "what is" (past or present tense).

You may also wish to consider whether any of the Background material should be updated with new or additional information. For example, you may have focused on a certain technology as a potential candidate when preparing the Charter, but then during the PLANNING phase chose to implement a different technology. There is no need to keep the material on the original ideas, but you should include some background on what was actually implemented.

Project Requirements

The Project Requirements material should also be nearly identical to the Project Charter, albeit any necessary verb changes. As you recall, this section explains the Objectives to a technical audience and documents the Project Requirements and Constraints. If any item was formally changed or modified since the Charter was submitted, you should include the *updated* version of the item in this section along with a footnote. The footnote should document when the modification was made and where the Change Order is located. The Change Orders should be included in an FPR-appendix. Each Change Order should be given its own appendix that is labeled with the Requirements Document item number. An example of a Change Order appendix might be:

"**Appendix C:** Spec 1.2.3 Change Order"

You may choose to order these in numerical order based on the Requirements Document item number or possibly chronologically in the sequence which the Change Orders occurred.

Technical Design Solution

The Technical Solution material should start with an introduction of the Block Diagram and the upcoming subsections (recall that is a list and requires a formal introduction). A high-level overview may be provided in the first subsection but occasionally may be unnecessary depending on what is presented in the section's introduction.

The subsequent subsections should each present the details of one aspect of the design. Most likely, these aspects should each be related to a particular subsystem just like everything else we have done this year. However, there are some projects that are better described by categorizing the solution into a differently – for example, sometimes it is better that each subsequent subsection relates to a feature, rather than a subsystem. If you feel that your solution would be better described using a different method than by subsection, you should talk with your Instructor before committing to the change.

The final subsection should present the System Integration. In this subsection, you should present the integration of all project aspects discussed in previous subsections.

This material should be presented in meticulous detail so that the reader could reproduce your design precisely as you produced it. To provide that level of clarity, you start each subsection with a short introduction and a presentation of a figure. The majority of the discussion should then describe the figure in detail. Most topics will require multiple figures. Consider whether one viewpoint of an image is sufficient. Perhaps you should provide front and back views of the same images. Consider providing a full image in the introduction of a subsection and include "zoomed in" or detailed version of the image during the supporting text.

One important difference between the technical content of the FPR and previous documents is that now this material should include both the "paper" design *and* details on the synthesis or implementation of the design. Typically, this means that the design is initially discussed in technical (or theoretical) detail. Circuit Schematics or 3D drawings are used to support this first paragraph. Then, in a subsequent paragraph of the same subsection, a description of how the subsystem was constructed is provided. Photographs are often used to support the second paragraph.

The *design* of every subsystem of a project should be discussed in explicit to start each subsection. This may require many paragraphs and figures to provide sufficient detail. The amount of fabrication or implementation detail that you are required to provide varies greatly from project-to-project and even from subsystem-to-subsystem. For example, you do not need to describe how each component was soldered to a PCB, but you may need to describe how each element of a structure was assembled. Provide enough information so that an experienced engineering could easily reproduce your design.

Performance Verification Results

All of the Performance Verification Results must be compiled and documented in the next FPR section. Whichever way you chose to organize the Technical Design Solution section (e.g., by subsystem) should be used to organize the Performance Verification Results. The introduction of this material should include an explanation about the organization of the upcoming subsections. We will assume you have organized your work by subsection for the remainder of this chapter.

The first subsection should present the Verification Results specific to the first subsystem. Provide the summary table for that subsystem followed by the narrative describing those specific tests. Remember to update the verb tense used at CDR; at CDR you were proposing the *plan*, now you are discussing *what you did* and the related results. Each subsequent subsection should discuss the "next" subsystem, until you ultimately can discuss the System-Level testing.

The introduction to each subsection should start by presenting the appropriate summary table and a brief overview of the success of that subsystem. Each subsection is further divided by Verification Test. For each Verification Test, describe the Setup, Equipment, Procedure, and Data Collected. Make sure to present the data in the same format that you learned Sect. 18.7 and that should have been implemented at the FDR.

Each test discussion should include at least one figure of the subsystem under test (e.g., a Dimensioned Drawing) and a figure representing the data (e.g., a graph). Additionally, most tests should also include at least one photograph captured during the test. Many tests may require additional figures to sufficiently document the work.

Remember, one important Technical Writing technique is to heavily rely on figures. The more figures you include, the less verbiage you usually need to write.

Example 22.1: Amount of Implementation Details in an FPR

The most important aspects to include within each subsection of the Technical Design Solution material in an FPR are the conceptual design descriptions and final implementation results.

Hopefully, you can determine how to present the conceptual design descriptions at this point since you have done so many times already. The new challenge is to sufficiently describe the implementation results. The amount of information required for a given subsystem varies dramatically. This example will provide you with some ideas to think about when describing your subsystem.

1. Complicated Circuit Designs:
   Hopefully when you developed your circuit designs, you did so in stages relative to the Design Life-Cycle process. You probably sketched an ideation, produced a simulation, built the circuit on a breadboard, and eventually fabricated the circuit on a PCB. It is not necessary to provide the details of each one of these steps in an FPR.
   Keep in mind that the purpose of the FPR is to provide the reader with enough information to reproduce your final design. So, the finalized circuit schematic should be used to discuss the design concept. If significant Verification Data came from the simulation, the simulation should be mentioned. Certainly, the final results will have been tested on a PCB, so that fabrication step should be mentioned.
   Describing HOW the simulation was built or how the PCB was "laid-out" is not necessary. The simulation software and PCB manufacturer may be pertinent information. Providing the PCB layout artwork or Pin Diagrams may be useful.
2. Physical Mounting Structure:
   The physical mounting structure of a design might initially be developed in a drawing software package, loads on different aspects of the structure calculated hand, certain aspects of the design manufactured by a machine shop, and the system assembled by the design team.
   In the body of the FPR, the 3D model should be provided along with a summary of the calculations and the details of the calculations included in an appendix. Then, you might simply state that "the system was manufactured by a machine shop" (provide contact information in the BoM). So far, this discussion might be fairly short. However, now you will need to describe how the system was assembled. This may require a lengthy discussion and many detailed photographs of the assembly processes.

Analysis:

The circuit subsystem may require a significant technical description with a number of equations, but the fabrication process may only require a few short sentences. Conversely, there might not be a lot to discuss about the design of the physical mounting structure or fabrication, but the implementation may require a lengthy description. It should be noted that these are just arbitrary examples. When writing your own FPR, you should include enough detail for another engineer to recreate your design.

Conclusion and Summary Material

In the Conclusion material, you should summarize the Technical Solution. To do this, reread the Motivation and Description sections. Then, answer the questions:

"Did your project successfully fulfil the Project Description" and "What was the project's impact upon the Project Motivation?"

Also, discuss any Future Work that could or should be conducted. If there were aspects of the project which were not addressed due to time constraints, explain what those were and what should be done about them. If the project was not successfully completed, discuss what might have been done differently. Even if the project *was* successful, there is usually something you can contribute here. Remember the Steve Job's example: there are important lessons that should be learned even on a successful project. Consider what more could be done to further impact the Project Motivation or additional features that could be added to the project in version 2.0; how would you recommend the Sponsor make this project even better?

A final Bill of Materials (BoM) should be included. At minimum the BoM should provide a complete list of every expenditure the project team made during the course of the project. This list constitutes every item necessary to complete the *project* and explains exactly how the project budget was consumed. Often, sponsors will also want to know the final cost of reproducing a single *product* or perhaps the cost of producing some quantity of products. If so, producing multiple (scaled versions) of the BoM may be necessary. Occasionally, items may be donated to a design team. These items should be included on the BoM. Rather than include a cost on the project-related BoM, state "donated" since that item did not expend any project resources. However, on any necessary product-related BoM, be sure to estimate the cost or value of that donation – your Sponsor will need to account for that cost if they ever manufacture another iteration of your design.

Finally, you may wish to acknowledge important contributors to the project. At the very least, an external sponsor should be thanked and appreciated. If you received notable support from an individual that was not officially part of the project team, you should also acknowledge this person and their contribution.

22.2 The Design Fair

Another important aspect of the CLOSING phase is exhibiting your final solution to demonstrate that you have, in fact, completed the project. Design Fairs are one type of event at which projects are demonstrated publicly. Typically, there is a lot of energy and excitement surrounding a Design Fair, and many students find the event one of the most enjoyable Senior Design experiences.

The venue of a Design Fair is dramatically different than all previous presentations you have produced thus far. The venue is designed to encourage meaningful interactions with a wide range of people. Typically, there will be many booths set up in a large auditorium where the presenters remain stationary and the audience mills between booths that strike their personal interests. This facilitates an engaging discussion between presenters and interested audiences. You may end up speaking with dozens of people of various backgrounds over the course of the event.

Your audience at a Design Fair will range from the mildly curious to the technically astute, so you will have to prepare a flexible presentation which can be spontaneously adapted to each person you may speak with. Rather than prepare multiple presentations, one for each audience-type, presenters should prepare a *layered* presentation that starts very generalized and progressively provides more detail. During the discussion, the presenter must gauge the interest and experience of each individual and determine how much of the presentation to provide. This is an advanced presentation technique that some engineers spend a career developing. This section will provide some suggestions and best practices to help you navigate your first Design Fair.

Audience Description

As you already know, it is important to consider your audience's demographics before preparing any presentation. This audience, by far, is the most complex audience you will have to prepare for because you do not know precisely who will visit your booth at any given moment. As a starting point for this discussion, let's categorize the audience into four groups: the Mildly Interested, the Nontechnical, the Inexperienced, and the Experts. This subsection will characterize each group so that you understand your audience when preparing your presentation.

The Mildly Interested audience is generally interested in new ideas and products. This person may be a Sponsor or family member of an unrelated project who is wandering the exhibit to kill time. They will be pleasant to talk with but will have very little experience or knowledge about the technology related to your project. They will be interested only in a very high-level description of your Project Motivation and System Ideation.

The Nontechnical audience is more interested in your project solution than the previous audience group. This group will have a specific reason why they stopped at your booth, meaning they have some tangential knowledge about your project or topic area, perhaps even an informal education related to the technology you used. They may have a deep understanding of the application or business aspects of your project. They will understand your Project Description and try to understand the high-level design concepts, at least in nontechnical terminology. You should present a summary of your results and conclusions as well.

An Inexperience audience member is someone that does not have much experience in your project specifically but does have the technical background to essentially understand most of your project. These are the other engineers and technicians in the attendance. To be clear, these are NOT inexperience people, they are simply inexperience in *your* specific project. These people will essentially understand the details of the Problem Statement and high-level design concepts. They will want an explanation of your System-Level solution and potentially some of the details of your subsystems. They will have the background necessary to completely understand your data and results assuming you present them well. They may even challenge your analysis and conclusions which can develop into meaningful technical discussions.

The Experts category consists of people with a formal education or experience directly related to the technology used to solve your project. The people in this category include professors and engineers experienced in fields related to your project. They may not be directly experienced with *your* project but easily will understand the technology, methods, and data you present. They are also likely experienced project reviewers. They will desire an in-depth explanation of how your solution resolved the Problem Statement, perhaps even down to the subsystem or component level. They are likely to have suggestions on how to improve your design further; you should take note of these and perhaps include them in the Future Work section of the FPR.

If your Sponsor plans to be in attendance, you should be cognizant of which category they fall into and perfect that portion of the presentation. Sponsors typically fall into the Inexperienced or Expert categories and never fall into the Mildly Interested group.

Developing and Presenting a layered Presentation

Knowing that your audience will consist of people with varying degrees of experience and interest, you should develop a presentation that can be spontaneously scaled to the specific person you are speaking with at any given time. Consider the following steps to develop your presentation, but be flexible when applying these guidelines to *your* project.

Start your presentation with information that anyone from any audience group will understand. Include the following:

- The Motivation Statement and perhaps a sentence or two of background to ensure that all audiences understand the Motivation.
- Then, in one or two sentences describe how you addressed the Motivation. This is not quite your Project Description. Rather it is a summary of the System-Level Ideation.

Now, you need to gauge the person's reaction. Are they engaged and seem to be tracking you? They probably are and you should continue on with your presentation. However, if you sense they are not tracking you, they are probably part of the Mildly Interested audience. In that case, you may want to pause and interact with the person at this level. Perhaps, ask them a question such as:

- Would you like to know how we accomplished this?
- Do you have any questions about our solution that I could answer?

Most people will still be tracking you at this point and it will be unnecessary to interrupt your presentation. If so, discuss your:

- Project Description, including the Project Goal and Objectives.
- System-Level technical solution (use the block diagram).

You may lose the Nontechnical audience at this point, so keep the technical discussion relatively brief. Be prepared to wrap-up your presentation and have a conversational discussion with these people, perhaps offer a product demonstration. If the person is still engaged, you are speaking with at least an Inexperienced audience member and should continue on. If so,

- Provide your System-level results and analysis.
- Demonstrate the product's functionality.
- Ask whether the person has any questions.

You will likely be able to assess whether the person is an Inexperienced or Expert audience based on the questions they ask. Be prepared to handle both general and detailed questions from either audience type.

Most importantly, remember that regardless of who you are speaking with, have fun at this event. Each person you visit with specifically took time to visit *your* booth; that means they are an engaged and friendly audience. Conduct yourself with a corresponding level of high energy and excitement. By the way, your project IS cool and you should be proud of the accomplishments you made – let that show.

Using Visuals

Rely heavily on visuals when presenting any portion of your presentation. At the very least, you should have at your disposal a poster created specifically for this event, along with the physical product you produced. You may also want to consider displaying the following:

- Flyers of results and analysis that do not fit on the poster
- Break-a-way, partially completed, or unpackaged portions of your physical product that provide additional insight to your solution
- Additional monitors exhibiting software interfaces
- Videos of the product in use

Project Posters

The Project Poster is an important visual required at most Design Fairs. This document qualifies as an official publication. Often, universities will display the Project Posters in a public area for at least a year after the Design Team graduates. Therefore, the Project Poster should be a stand-alone document which fully describes the project without further explanation. Additionally, it should be a colorful and eye-catching display.

The Project Poster should include the following:

- Problem Statement (including Objectives)
- Important Specifications and/or Requirements
- Block Diagram
- Technical Description of the Solution
- Important Results
- Conclusions and/or Future Work

It would be difficult to provide further guidelines on how this document should be formatted because the content of Project Posters is extremely dependent upon the project's content and success. You should review this material with your Advisor before producing the poster. You may even wish to create a mock-up of the poster for your Advisor's approval before completing any of the details.

Dress Codes

A Design Fair is a professional event. You should present yourself this way. This means that you should dress professionally. Doing so does not *necessarily* mean you need to dress in a business suit, although that is appropriate in some cases. Shorts, open shoes, and hats are probably not appropriate. Cohesive teams often chose to match appearance to some extent. At a professional trade show, employees typically wear logoed, three-button, polo-style shirts with casual dress pants. A common substitute student-teams sometimes make is to simply all wear a shirt of the same color. This shows team unity and cohesiveness. It is a nice touch, but not required.

22.3 Deliverables Meeting

Arguably, the *most* important process of the CLOSING phase is called Transfer of Deliverables. While in practice, this may be a complex and involved process, in Senior Design this process can usually be completed with a Deliverables Meeting with the Sponsor.

By the end of the EXECUTING phase, or early in the CLOSING phase, the Deliverables Meeting should be scheduled as a milestone on the Project Schedule. This meeting should occur around the same times as the Design Fair, but keep in mind that you will be required to display the physical product at the Design Fair – so do not deliver the product too early.

Conducting at least one pre-meeting before the official event is often a good idea. Sometime early in the CLOSING phase, coordinate with your Sponsor to define what will be expected at the Deliverables Meeting and how the final transfers will take place.

Here are some ideas to consider before conducting the official (final) transfer:

- Will a technical review (similar to the FDR) with the Sponsor be required?
- Will an in-person demonstration (similar to the SPD) with the Sponsor be necessary?
- Will the Sponsor require a training session in order to use the product in the future?
- What specifically will be transferred at the Deliverables Meeting?
- How and where will the Deliverables Meeting be conducted?
- Is it necessary to complete the final transfer in-person, or can the product be shipped to the sponsor after the Design Fair has been conducted?
- Will physical hardcopies of any reports be required, or would the Sponsor prefer digital softcopies?
- How will any digital documents (including reports, licenses, code, schematics, drawing, etc.) be transferred? This could include any combination of emails, dropboxes, USB drives, etc.

When you conduct the Deliverables Meeting, consider the event a formal one. Dress accordingly and perform the meeting with the utmost professionalism. The only impression more important than the "first impression" is the final one. Rehearse any demonstration or presentation. Create a checklist of the deliverables and double check that list before the meeting. Be sure to thank the Sponsor for their support!

22.4 Case Studies

Each of our case study teams performed well at their respective Design Fairs and delivered successful products to their Sponsors. Here is a summary of the project results.

Augmented Reality Sandbox

The AR Sandbox was presented at the 2017 Design Fair and was awarded first place in the consumer-product category. Shortly after the Design Fair, the AR Sandbox was installed at the Kirby Science Discovery Center in Sioux Falls, South Dakota, an event covered by local news media. Over the next 5 years, the exhibit was one of the most popular in the museum. Over time the fine-grain white sand required for a clear topographic image created enough dust to degrade the quality of the projector lens and eventually damaged the projector itself. This eventuality was anticipated and the projector mount had been designed to easily contain a variety of replacement projectors. However, rather than purchase a new projector for this exhibit, the KSDC administration elected to exchange the exhibit for a new product, a strategy they regularly employ to keep the exhibits fresh and exciting for returning visitors. The AR Sandbox was a success by all accounts and the team performed exemplary. James went on to develop image processing software for NASA. Moe earned a Master of

Science degree in image processing and is a Project Manager today. Trever earned his PMI certification. Logan and Tucker started a small business together and regularly employ Project Management skills on a daily basis.

Smart Flow Rate Valve

The Smart Flow Rate Valve was delivered to Raven in the spring of 2018 after many weeks of rigorous system testing on a specialized test apparatus. The design was fully functional and met Raven's expectations for the new product. After observing Tyler's performance as he successfully led this complex project, Raven offered Tyler an engineering position which he holds to this  day. At the time of writing this textbook, the prototype the student team delivered to Raven remains on Tyler's work desk.

Both Gannon and Christen have received their PMI certifications and work as engineering team leaders for large manufacturing companies. They regularly employ many of the Project Management skills they developed during their Senior Design experience in their career positions.

Thein now owns his own business. Although the business is not engineering-related, he is in the process of procuring his PMI certification as well.

Robotic ESD Testing Apparatus

The 2016/17 RESDTA design team produced a prototype that exactly matched the agreed upon Project Requirements by one of the largest Requirements Documents developed for a Senior Design project. The students presented in the nonconsumer category of the 2017 Design Fair and were awarded first place by a landslide, receiving nearly twice as many votes as the second-place team. The product was delivered to their Sponsor shortly thereafter.

As is typical with most first-integration projects, over the course of the project the Sponsor identified a few shortcomings in the original Requirements Document which needed to be addressed before utilizing RESDTA. During the summer of 2017, a student intern was assigned by the Sponsor to address these issues. The Sponsor has remarked many times since the conclusion of the project that had the 2016/17 design team not produced such complete and articulate documentation, the student intern would have had no chance at successfully modifying RESDTA. Therefore, due to the excellent documentation efforts by the design team, the intern was able to the necessary modifications and have RESDTA fully operational by September 2017.

The time and effort saved by the automated system proved to be valuable for the Sponsor. The payback period was initially estimated to be 3 months. However, that was based on a projected workflow that never quite occurred. Since the number of drives that were actually tested proved to be less than the original estimate used in the payback analysis, the project costs were recovered after 9 months. The company typically expects payback periods between 1 and 2 years, so RESDTA's payback still exceeded expectations.

Kaitlin has pursued her interest in Project Management since RESDTA's completion. She is currently employed as a Project Manager and develops substations for large renewable energy plant. Her teams are composed of Electrical and Civil/Structural Engineers. She plans to gain her PMI certification in order to advance her career and allow her to become a Senior Project Manager in the near future.

Chapter 23: **CLOSING Processes (Retention)**

As mentioned in the introduction of Chap. 22, there are two broad categories of processes in the CLOSING phase. The first category (covered in Chap. 22) ensures that every aspect of the project is complete.

The other category of processes in the CLOSING phase ensures that the information, skills, and competencies developed during the project are retained for posterity. Projects are never conducted for the sole purpose of performing a project. Projects are conducted to generate profits, develop lasting solutions, improve situations, or otherwise impact the Project Motivation in a lasting and meaningful way. This means that the project's impact should persist long after the project team disbands. The most important process in this category is called the After-Action Retrospective meeting.

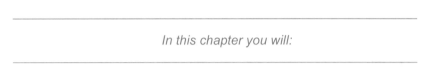

In this chapter you will:

- Learn about After-Action Retrospective meetings.
- Conduct your own After-Action Retrospective of your Senior Design experiences.

C.J. Mettler, *Engineering Design*, https://doi.org/10.1007/978-3-031-23309-8_23

23.1 After-Action Retrospective Meetings

The After-Action Retrospective meeting process is a set of meetings in which the Stakeholders reflect on and assess the project experience. Often, this is conducted at multiple levels with different groups of individuals. At the very least, the project team should conduct an internal After-Action Retrospective. Additionally, the project team may wish to conduct After-Action Retrospective meetings with their Advisor, Sponsor, Subcontractors, Legal teams, Financial or Human Resource departments, or anyone else who may have insight to the project performance.

Keep in mind that the After-Action Retrospective meeting is conducted to improve the Organization and/or the Design Team, not necessarily the product itself. Hopefully, you have already met the Project Goal and confirmed that the project is complete in the earlier CLOSING phase processes. Now, in this last process, the focus should be on improving the efficiency of the Project Management Implementation, Design Life-Cycle, and/or the Design Team. Certainly, if there are technical improvements that could be made to version 2.0 of your project, these could be discussed as well – but not at the expense of discussing the larger issues.

Maintaining a cordial and productive atmosphere during the meeting is important. The purpose is not to point fingers at anyone for any perceived shortcomings but rather collectively identify what was done well and what could be improved. Asking questions in such a way as to facilitate this positive atmosphere while also gaining actionable information requires experience and skill. Here are some examples of questions [42]:

- Which part of the project are you most proud of?
- What was the most frustrating part of the process?
- Did the leadership team and other Stakeholders offer enough support?
- Did the project team effectively use the support the leadership team provided?
- What would you do differently if you could do it again?
- How well did the project team follow the initial project plan?

Providing a structured agenda may elicit more specific (and therefore actionable) responses. To do this, create a timeline of the Design Life-Cycle, including the STARTING, INITIATING, PLANNING, and EXECUTING phases (and maybe the CLOSING phase if needed). Identify approximately two process milestones within each phase. For each process or milestone, ask 1–3 questions focusing on what was successful and what could be improved. An example of this method will be provided in the next section.

Recruiting a mediator for this event is an important strategy. This person's role is to encourage a positive atmosphere if tensions start to get high, to keep the conversation moving and focused on actionable criticism as opposed to complaints, and to act as a notetaker. This person does not typically contribute to the discussion itself. Your Advisor may function as the mediator for your Senior Design project's After-Action Retrospective, but in practice the Senior Engineer (i.e., the Advisory role in industry) should be *part* of the discussion panel. One potential idea for Senior Design teams is to include the Advisor as part of the discussion panel and recruit a fellow classmate to act as the mediator, and then return the favor for that person's team. Most faculty Advisors can effectively handle both roles simultaneously, so this additional complexity is not absolutely necessary.

23.2 A Senior Design After-Action Retrospective

This section will help you organize and conduct an After-Action Retrospective meeting specifically for Senior Design. Break the Design Life-Cycle into phases and select 1–3 processes or milestones within each phase to review. Figure 23.1 provides a starting point. You may wish to add a few more milestones.

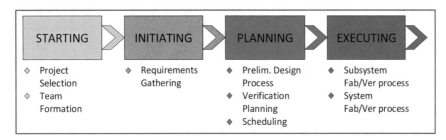

Figure 23.1   Key milestones within each design phase

Next, pose 1–3 questions for each milestone. Maintain a balance between "what can be done better" and "what was done well" style questions. Table 23.1 provides a few examples of questions you might ask for each phase.

You may also wish to pose a few general questions, such as:

- What is the most important aspect that you learned about Project Management during this experience? Which aspects do you think you will use most in the future?
- How has your communication improved? What is the most important communication skill you developed this year?
- What part of Project Management did you enjoy the most?
- Which was the most frustrating? How might you personally modify this aspect while still achieving project success?

Preparing for the After-Action Retrospective Meeting

The meeting coordinator or moderator is responsible for generating the list of questions and setting the agenda. The team members should receive the questions-list a few days prior to the meeting in order to prepare for the discussion.

Each team member should individually review the questions and topics which will be covered at the After-Action Retrospective meeting in order to contribute effectively to the discussion. Prepare detailed (but concise) answers to a number of the questions, perhaps one-per-category, more if you have time.

Table 23.1 Suggested after-action retrospective questions

STARTING PHASE

♦ **Project Selection**

    1. Were you able to directly provide a significant technical contribution to this project?

    2. Is there a benefit to pursing this type of project again in the future?

♦ **Team Formation**

    1. What caused this team to function well?

    2. What type of team where you (Pseudo, Potential, Real, Performance)?

    3. What is one action you could have taken to advance the level of your team by 1 step?

INITIATING PHASE

♦ **Requirements Gathering**

    1. Were there any project characteristics that could have been defined more thoroughly?

    2. Which characteristics were difficult to verify? Could those characteristics have been defined differently?

    3. Are there any questions you wished you had asked the sponsor during this phase that you didn't think to ask?

PLANNING PHASE

♦ **Preliminary Design Process**

    1. Were the Critical Subsystems (and/or Components) defined and assigned well?

    2. Which contribution or idea are you most proud of?

    3. What was the most frustrating part of this phase and what could you have done differently to reduce the frustration?

♦ **Verification Planning**

    1. After having performed the Verification Process, would you have written any of your test descriptions differently?

♦ **Scheduling**

    1. What aspects of the project should have been provided additional time in the schedule?

    2. Could any aspect of the project been completed faster, would having done so been beneficial?

    3. Did any of your teammates go "above and beyond" to ensure the team maintained the scheduled?

EXECUTING PHASE

♦ **Subsystem Fabrication and/or Verification Process**

    1. How did the work completed during the INITIATING and/or PLANNING phase support/reduce the effort required here?

    2. Was there any aspect of the project which should have been planned better so that this process could have been completed faster?

    3. What was the most frustrating part of this process? What could your leadership (Instructors/Advisors) done better to help you?

♦ **System Fabrication and/or Verification Process**

    1. How did the team function well while integrating subsystems?

    2. Was there any aspect of the project which should have been planned better so that this process could have been completed faster?

    3. What did you enjoy most about the EXECUTING phase?

An agenda should be released shortly before the meeting as well. The agenda should allot time to each phase of the Design Life-Cycle but prioritize the topics within each phase. The agenda for a 50-minute Senior Design meeting may look like this:

• Welcome & Kickoff:	2 minutes
• STARTING phase discussion:	10 minutes
• INITIATING phase discussion:	10 minutes
• PLANNING phase discussion:	10 minutes
• EXECUTING phase discussion:	10 minutes
• Closing comments:	8 minutes

Conducting the After-Action Retrospective Meeting

Maintaining focus and staying on-schedule is challenging during an After-Action Retrospective meeting. The agenda, and a good moderator, will help keep the discussion on track. An effective moderator will attempt to keep the conversation flowing because reviewing the *entire* Design Life-Cycle is important. This means, at times, the moderator may have to interrupt a particular discussion point and force the meeting to move on in order to stay on-schedule.

If you happen to be cut off on a particular point, make a note as a reminder. If you feel strongly that your point should have been completed, you are encouraged to follow-up individually with the moderator, or possibly send an email to the group.

Allowing time for each team member to fully address each question on the agenda is probably not possible during a time-constrained meeting. Each individual moderator will have their own personal style to get the most out of the meeting. One method the Moderator might use to conduct the meeting is as follows:

1. Pose the first question and select one team member to answer.
2. Allow about 30 seconds for the selected team member to provide their response.
3. Ask the panel if anyone agrees with the response and allow 1–2 minutes for comments.
4. Ask the panel if anyone disagrees with response and allow 1–2 minutes for comments.
5. Move to the next question.
6. Initially cover about 2–3 questions per Design Life-Cycle phase over the course of about 40 minutes.
7. During the time allotted to closing comments, ask the group "would anyone like to address a question that we have not had time to cover?"
8. If anyone has a response, allow 2–3 minutes for that discussion.
9. Wrap-up the meeting on a positive and constructive note.

The moderator should take notes and follow up with anyone after the meeting if necessary. The notes should be disseminated to the entire team, and anyone wishing to provide additional information should be allowed to provide a written response to any or all of the topics which were not completely covered during the meeting. Occasionally, a follow-up meeting is conducted if time runs out and the participants feel there were important topics which were not covered.

Chapter 24: **The Final Process – Celebration**

As Vince Lombardi proclaimed in his speech "Leadership," *it is time for us all to stand and cheer the doer, the achiever – the one who recognizes the great challenges and does something about it.*

The final process of the CLOSING phase is to CELEBRATE your achievements!!! You have accomplished something no one else has, you have produced something unique, you have left your mark. That is something worth celebrating. Specifically dedicate some time immediately after the Design Fair or your Deliverables Meeting to enjoy your success with your teammates. Take time to feel good.

Let us take a minute to consider where you are today compared to where you were a year ago. Obviously, you have made a technical achievement by producing an engineering solution to a problem of your choosing. YOU used your knowledge, skills, intuition, and creativity to develop something new, something no one else has ever built. Perhaps your project was wildly successful, perhaps you found the 2001st way to *not* build a lightbulb. Either way YOU have added something to the engineering body of knowledge and *that* is an achievement to be proud of!

However, that is not your only achievement this year. You have developed the "capstone" skills necessary to be a well-rounded, functional engineer. Consider how you spoke about your project at the beginning of this experience compared to how you do so now. You have become more articulate and more precise. Consider how you thought about your project early on. Did you realize how many layers you would eventually uncover? You have learned to think and analyze engineering problems on a much deeper, more complex basis. These developments are also worth cheering.

C.J. Mettler, *Engineering Design*, https://doi.org/10.1007/978-3-031-23309-8_24

A note from the Author

Engineering is so much more than calculating the numbers your professors predicted you *should* calculate or solving *rote* problems someone developed in an arbitrarily constrained environment. In fact, the theory and calculations you were taught in your typical engineering courses, while foundationally important, are the *EASY* part of engineering. In this profession, Engineers must regularly apply that theory but are also required to navigate legal, contractual, financial, and interpersonal challenges – many of which you were exposed to this year.

I had one overarching purpose for writing this textbook, and that was to bridge the gap between academics and industry – in essence, to expose you to the real challenges engineers face *beyond* the technical issues. My goal was to provide you a more-complete skill set than your technical classes alone could have provided. In doing so, I hope you have begun to develop the tools necessary to stop solving someone else's problems and are now able to begin pursuing solutions to new problems which are of importance to you personally. This pursuit will be so much harder than your course work, but also that much more rewarding.

Engineering is a great profession, not because our skill and intelligence makes our jobs easy, but because our skill and intelligence allows us to address the most difficult of problems. Engineers regularly have dreamed of a better world and then gone out and made that dream a reality. But, changing reality is no easy task and doing so will require a mastery of both the soft skills you have recently learned and the technical competencies you previously developed in those other courses.

From the Roman Aqueducts to the da Vinci Robot, engineers throughout history have recognized daunting challenges *and did something about them*. History only remembers the successes, but these engineering feats required years of effort, alternatives failed, and progress, at times, was slow and frustrating. Fortunately, individuals persevered because the challenges were *worth* pursuing, and in so doing, they changed the world.

The great challenges have not all be solved. More than ever, engineers are needed to recognize the new challenges facing society – and do something about them. Climate change, biotechnology, microfluidics, clean energy, and feeding the world's ever-growing population present just a small subset of the challenges today's society faces. So how will you leave your mark? Which challenges will you pursue?

I hope that a few years from now, when you look back at your college education, that this course will stand out as different, challenging, but mostly, applicable.

It is said that "Engineers turn dreams into reality [43]," I, for one, hope you dream BIG.

Sincerely,

Cory Mettler, M.S.E.E.

References

1. "Project Management Institute," [Online]. Available: https://www.pmi.org/about/learn-about-pmi/what-is-project-management. [Accessed 4 May 2021].
2. L. LaPrad, "Triple Constraint Theory in Project Management," [Online]. Available: https://www.teamgantt.com/blog/triple-constraint-project-management. [Accessed 10 May 2021].
3. A. Baratta, "The triple constraint, a triple illusion," in *PMI Global Congress*, Seattle, Wa, 2006.
4. R. W. Lacher and R. Bodamer, "The new reality of agile project management," in *Agile, SCRUM*, 13 October 2009.
5. Project Management Institute, "Project Management Body of Knowledge (PMBOK)," 6th ed., Newton Square, PA, PMI, 2017.
6. FORTUNE, [Online]. Available: https://fortune.com/company/ibm/fortune500/. [Accessed 4 May 2021].
7. M. Wideman, "Max's Project Management Wisdom," [Online]. Available: http://www.maxwideman.com/papers/managing/lifecycle.htm. [Accessed 4 May 2021].
8. Defense Acquisition Guidebook, [Online]. Available: https://www.dau.edu/guidebooks/Shared%20Documents/Chapter%203%20Systems%20Engineering.pdf. [Accessed 4 May 2021].
9. Raven Industries, [Online]. Available: https://ravenind.com/.
10. D. D. R. Williams, "NASA Space Science Data Coordinated Archive," National Aeronautics and Space Administration, 12 Dec 2016. [Online]. Available: https://nssdc.gsfc.nasa.gov/planetary/lunar/ap13acc.html. [Accessed 24 May 2021].
11. Brett & K. McKay, "Art of Manliness, Lessons in Manliness from Gene Kranz," 20 Jul 2009. [Online]. Available: https://www.artofmanliness.com/articles/lessons-in-manliness-from-gene-kranz/. [Accessed 24 May 2021].
12. Wikiquote, "Eugene F. Kranz," [Online]. Available: https://en.wikiquote.org/wiki/Eugene_F._Kranz. [Accessed 24 May 2021].
13. V. Briand, "Apollo 13: five crisis management lessons from a successful failure," 21 Apr 2020. [Online]. Available: https://medium.com/@virginiebriand/apollo-13-five-crisis-management-lessons-from-a-successful-failure-1202da0cc744#:~:text=The%20Apollo%2013%20mission%20is,to%20get%20the%20crew%20home. [Accessed 24 May 2021].
14. J. Katzenback and D. K. Smith, "The discipline of teams," *Harvard Buisness,* vol. 71, pp. 111–120, 1993.
15. H. F. Hoffman, The Engineering Capstone Course, fundamentals for students and instructors, Switzerland: Spring International Publishing, 2014.
16. J. Donnell, S. Jeter, C. MacDougall and J. Snedeker, Writing Sytle and Standars in Undergraduate Reports, 3rd ed., College Publishing, 2016.
17. M. Dugard, Farther Than Any Man: The Rise and Fall of Captain James Cook, Allen & Unwin, 2003.
18. St John's College, University of Cambridge, [Online]. Available: https://www.joh.cam.ac.uk/library/library_exhibitions/schoolresources/exploration/southerncontinent. [Accessed 09 05 2021].

C.J. Mettler, *Engineering Design*, https://doi.org/10.1007/978-3-031-23309-8

19. Britannica, [Online]. Available: https://www.britannica.com/place/Northwest-Passage-trade-route. [Accessed 09 May 2021].
20. A. Lansing, Endurance: Shackleton's Incredible Voyage, Hodder & Stoughton, 1959.
21. C. Alexander, The Endurance: Shackleton's Legendary Antartic expedition, Bloomsbury, 1998.
22. Project Managment Institute, "Project Managment Body of Knowledge (PMBOK)," 6th ed., Newton Square, PA, PMI, 2017, pp. 698–726.
23. Merriam-Webster, "merriam-webster.com," [Online]. Available: https://www.merriam-webster.com/dictionary/charter. [Accessed 08 June 2022].
24. Project Manager, [Online]. Available: https://www.projectmanager.com/blog/project-charter. [Accessed 08 June 2022].
25. "Patent Drafting Catalysis," [Online]. Available: https://patentdraftingcatalyst.com/antecedent-basis-patent-claim/. [Accessed 09 June 2022].
26. "IPWire – The Patent Expert's Resource," [Online]. Available: http://ipwire.com/stories/antecedent-basis-lesson/. [Accessed 10 June 2022].
27. "Online Paralegal Degree Center," [Online]. Available: https://www.online-paralegal-degree.org/lists/5-examples-of-leading-questions/. [Accessed 09 June 2022].
28. "Bio of Conrad Anker," The North Face, [Online]. Available: https://www.thenorthface.com/about-us/athletes/conrad-anker.html. [Accessed 13 Aug 2021].
29. P. Chu, Embedded SoPC Design with NIOS II Processor and VHDL Examples, Hoboken, New Jersey: Wiley, 2011.
30. Project Management Institute, "Project Management Book of Knowledge (PMBOK)," 6th ed., Newton Square, PA, PMI, 2017, pp. 87, 405.
31. B. Whitney, "Tara Lipinski – American Figure Skater," Britannica, Encyclopedia, 06 June 2022. [Online]. Available: https://www.britannica.com/biography/Tara-Lipinski.
32. "Tara Lipinski's history making performance," Impossible Moments in Olympic History, [Online]. Available: https://olympics.com/en/original-series/episode/tara-lipinski-s-history-making-performance-impossible-moments.
33. T. Brighton, Patton, Montgomery, Rommel – Masters of War, Crown Publishing Group, November 2010.
34. Business Insider, "This is why Steve Jobs got fired from Apple – and how he came back to save the company," 31 Jul 2017. [Online]. Available: https://www.businessinsider.com/steve-jobs-apple-fired-returned-2017-7. [Accessed 23 Jun 2022].
35. Investopedia, "How Did Apple Get So Big?," 10 Sept 2021. [Online]. Available: https://www.investopedia.com/articles/personal-finance/042815/story-behind-apples-success.asp. [Accessed 23 Jun 2022].
36. Macworld, "History of Apple: the story of Steve Jobs and the company he founded," 25 Apr 2017. [Online]. Available: https://www.macworld.com/article/671584/history-of-apple-the-story-of-steve-jobs-and-the-company-he-founded.html. [Accessed 23 Jun 2022].
37. The New York Times, "Apple Won't Always Rule, Just Look at IBM," 25 Apr 2015. [Online]. Available: https://www.nytimes.com/2015/04/26/your-money/now-its-apples-world-once-it-was-ibms.html. [Accessed 23 Jun 2022].
38. WIRED, "Sept. 16, 1985: Jobs Quits Apple Sept. 16, 1997: Jobs Rejoins Apple," 15 Sept 2000. [Online]. Available: https://www.wired.com/2008/09/sept-16-1985-jobs-quits-applesept-16-1997-jobs-rejoins-apple/. [Accessed 23 Jun 2022].
39. Project Management Institute, "Project Management Book of Knowledge (PMBOK)," 6th ed., Newton Square, PA, PMI, 2017, pp. 121–123 & 634.

40. Project Manager, "5 Steps to Project Closure," 15 Feb 2022. [Online]. Available: https://www.projectmanager.com/blog/project-closure. [Accessed 24 Jun 2022].
41. E. E. P. Aziz, "Project Closing – the small process group wiht big impact," in *PMI Global Congress 2015*, London, England, 10 Oct 2015.
42. Redbooth, [Online]. Available: https://redbooth.com/templates/post-mortem-analysis. [Accessed 27 Jun 2022].
43. H. Miyazaki, *Movie director and producer.*

Index